NUMERICAL ANALYSIS FOR APPLIED MATHEMATICS, SCIENCE, AND ENGINEERING

NUMERICAL ANALYSIS FOR APPLIED MATHEMATICS, SCIENCE, AND ENGINEERING

Donald Greenspan
University of Texas at Arlington

and

Vincenzo Casulli
University of Trento, Italy

Addison-Wesley Publishing Company, Inc.
The Advanced Book Program
Redwood City, California • Menlo Park, California
Reading, Massachusetts • New York • Amsterdam
Don Mills, Ontario • Sydney • Bonn • Madrid
Singapore • Tokyo • San Juan • Wokingham, United Kingdom

Publisher: *Allan M. Wylde*
Production Administrator: *Karen L. Garrison*
Editorial Coordinator: *Pearline Randall*
Electronic Production Consultant: *Mona Zeftel*
Promotions Manager: *Celina Gonzales*

Library of Congress Cataloging-in-Publication Data

Greenspan, Donald.
  Numerical analysis for applied mathematics, science, and engineering / Donald
Greenspan and Vincenzo Casulli.
    p. cm.
  Bibliography: p.
  Includes index.
  ISBN 0-201-09286-7
  1. Numerical analysis. I. Casulli, Vincenzo. II. Title.
QA297.G725 1988
519.4--dc19   87-37391
                                                     CIP

Cover: "Metered Man" courtesy Bettmann Archive.

This book was typeset in Microsoft Word on a Macintosh. Camera-ready copy
printed on an Apple LaserWriter.

ABCDEFGHIJ-AL-898

This book is designed for a first course in numerical analysis. It differs considerably from other such texts in its choice of topics. Our concern has been with the needs of science, engineering, and applied mathematics students, who, in increasing numbers, form a majority of the students in numerical analysis courses. Thus, we have presented a broad spectrum of topics of applied interest, which, in fact, brin g the reader to various frontiers of the subject. We have been able do this by purging the traditional curriculum of many topics which are not likely to be encountered in either science or technology. These topics include, for example, number systems, computer operations, the secant method, Weddle's rule, Richardson extrapolation, and the methods of Graeffe and Milne. Related decisions were guided by a desire to include nonlinear equations, in addition to linear ones.

The need to solve a wide spectrum of nonlinear problems has been increasing since the end of the nineteenth century. For example, accurate simulation of dynamical processes related to planetary motion, rocket propulsion, turbulent flow, chemical oscillators, diffusion processes, and elastic stress demand the ability to solve nonlinear equations. Accurate fitting of data derived by use of laser technology, atomic clocks, electron microscopes and/or radio telescopes requires greater sophistication than that available from a linear least square fit. Quantitative methodology applicable to large classes of nonlinear problems became available only with the development of modern digital computers, and the result has been, and probably will continue to be, an explosion of knowledge.

In the final chapter, we have included a study of the Navier-Stokes equations, a fully nonlinear system of fluid dynamical equations which, interestingly enough, can be derived from both the macro, or hydrodynamic, approach and a micro, or molecular, approach. These equations are among the most challenging in contemporary numerical analysis and are fundamental to diverse studies relating to such areas as weather prediction, aerodynamics, petroleum recovery, cardiovascular circulation, heat convection, ocean currents, and the not-so-ordinary flow of ordinary water in pipes. Interest in the Navier-Stokes equations is so broad that we felt their inclusion to be appropriate.

One major goal in the writing of this book was to develop methodology for which the numerical solution of a given problem has the same qualitative behavior as the analytical solution, for, thereby, the numerical solution preserves the physics of a given mathematical model. Another major goal was to develop numerical analysis in such a fashion that the reader would be able to apply the methods thoughtfully and within a reasonably short time to problems which he or she finds both interesting and significant. For this reason, methodology, theory, and intuition have been interwoven throughout.

The book is suitable for a junior, senior, or first year graduate course. Only a familiarity with computer programming and ordinary differential equations is assumed throughout. We have set stars before various sections to denote material of relative difficulty. Chapters 1-5, exclusive of the starred sections, provide ample material for a one semester undergraduate course, assuming computer implementation by the student. In their entirety, these chapters can be used for a one semester graduate course. Chapters 6-9 are designed for the second half of a full year course. These chapters deal with partial differential equations, and, in each, the first few sections develop the necessary mathematical background. Note that a star has been affixed to the title of Chapter 9 to indicate the advanced nature of the entire chapter. Observe also that the exercises have been divided into two sets, basic ones and supplementary ones. Among the supplementary ones are several unexpected surprises.

In our own teaching, our philosophy has been that computer implementation by the student is essential. As a consequence, the time required for study of various topics has been greater than that required in a purely theoretical lecture course. Nevertheless, from an applied point of view, a numerical algorithm that does not run on a computer is useless, and the student should verify an algorithm's viability by direct computer implementation. Theory is important in numerical analysis, but theory, alone, will gail to meet the needs of the majority of our students.

Finally, we with to thank those undergraduate students at the University of Trento and those undergraduate and graduate students at the University of Texas at Arlington who contributed in so many ways to the final structure of the book. Unfortunately, the list of names is too long for inclusion.

D. Greenspan
V. Casulli
1988

CONTENTS

5. Approximate Solution of Boundary Value Problems for Ordinary Differential Equations

6. Elliptic Equations

7. Parabolic Equations

8. Hyperbolic Equations

*9. THE NAVIER-STOKES EQUATIONS

1

Algebraic and Transcendental Systems

1.1 INTRODUCTION

Science is study of Nature. We study Nature not only because we are curious, but because we would like to control its very powerful forces. Understanding the ways in which Nature works might enable us to grow more food, to prevent normal cells from becoming cancerous, and to develop relatively inexpensive sources of energy. In cases where control may not be possible, we would like to be able to predict what will happen. Thus, being able to predict when and where an earthquake will strike might save many lives, even though, at present, we have no expectation of being able to prevent a quake itself.

The discovery of knowledge by scientific means is carried out in the following way. First, there are experimental scientists who, as meticulously as possible, reach conclusions from experiments and observations. Since no one is perfect, not even a scientist, all experimental conclusions have some degree of error. Hopefully, the error will be small. Then, there are the theoretical scientists, who create models from which conclusions are reached, often using mathematical methods. Experimental scientists are constantly checking these models by planning and carrying out new experiments. Theoreticians are constantly refining their models by incorporating new experimental results. The two groups work in a constant check-and-balance refinement process to create knowledge. And only after extensive experimental verification and widespread professional agreement is a scientific conclusion accepted as valid.

Scientific experimentation and observation require mathematical methodology for handling and analyzing data sets. Scientific modeling requires mathematical methodology for solving equations and systems of equations. In this book we will develop basic, constructive techniques for both these areas of endeavor. And, though the development will be relatively elementary, we will, throughout, use the exceptional arithmetic power available through the application of modern, high-speed digital computers. The resulting methods are called numerical methods.

It is natural, then, since the arithmetical operations performed by modern computers are those of classical algebra, that we begin with the study of algebraic and transcendental systems of equations.

1

1.2 MATRICES AND LINEAR SYSTEMS

For $n \geq 2$, the general linear algebraic system of n equations in the n unknowns x_1, x_2, x_3, ..., x_n can be written in the form

$$a_{11}x_1 + a_{12}x_2 + a_{13}x_3 + ... + a_{1n}x_n = b_1$$
$$a_{21}x_1 + a_{22}x_2 + a_{23}x_3 + ... + a_{2n}x_n = b_2$$
$$a_{31}x_1 + a_{32}x_2 + a_{33}x_3 + ... + a_{3n}x_n = b_3$$

(1.1)

$$\qquad . \qquad . \qquad . \qquad . \qquad .$$
$$\qquad . \qquad . \qquad . \qquad . \qquad .$$
$$\qquad . \qquad . \qquad . \qquad . \qquad .$$

$$a_{n1}x_1 + a_{n2}x_2 + a_{n3}x_3 + ... + a_{nn}x_n = b_n .$$

If matrix **A** and vectors **x** and **b** are defined by

(1.2)
$$\mathbf{A} = \begin{bmatrix} a_{11} & a_{12} & a_{13} & \cdots & a_{1n} \\ a_{21} & a_{22} & a_{23} & \cdots & a_{2n} \\ a_{31} & a_{32} & a_{33} & \cdots & a_{3n} \\ . & . & . & & . \\ . & . & . & & . \\ a_{n1} & a_{n2} & a_{n3} & \cdots & a_{nn} \end{bmatrix}, \qquad \mathbf{x} = \begin{bmatrix} x_1 \\ x_2 \\ x_3 \\ . \\ . \\ x_n \end{bmatrix}, \qquad \mathbf{b} = \begin{bmatrix} b_1 \\ b_2 \\ b_3 \\ . \\ . \\ b_n \end{bmatrix},$$

then system (1.1) can be written compactly as

(1.3) $\mathbf{A}\,\mathbf{x} = \mathbf{b}$.

Of course, forms (1.1) and (1.3) are equivalent. For *theoretical* discussions, however, (1.3) will be the more convenient one.

Unless otherwise stated, it should be noted that, throughout, we will concern ourselves only with real numbers and real functions.

From a computer point of view, it is desirable to know that system (1.1) has *one and only one* solution before one attempts to solve it. Numerical computations for systems which have more than one solution usually yield meaningless results. Numerical computations for systems which have no solutions are always meaningless. The fundamental theorem which assures such existence and uniqueness and is proved in introductory algebra courses is stated now for completeness.

THEOREM 1.1. System (1.1) has one and only one solution if and only if the determinant of A, denoted by |A|, is different from zero.

Theoretically, when $|A| \neq 0$, the solution of (1.1) for given A and b can be given constructively as the quotient of determinants by Cramer's rule. For example, for system

$$x_1 + x_2 - x_3 = 2$$
$$x_1 - x_2 + x_3 = 0$$
$$-x_1 + x_2 + x_3 = 0 \, ,$$

one has, by Cramer's rule,

$$x_1 = \frac{\begin{vmatrix} 2 & 1 & -1 \\ 0 & -1 & 1 \\ 0 & 1 & 1 \end{vmatrix}}{\begin{vmatrix} 1 & 1 & -1 \\ 1 & -1 & 1 \\ -1 & 1 & 1 \end{vmatrix}}, \quad x_2 = \frac{\begin{vmatrix} 1 & 2 & -1 \\ 1 & 0 & 1 \\ -1 & 0 & 1 \end{vmatrix}}{\begin{vmatrix} 1 & 1 & -1 \\ 1 & -1 & 1 \\ -1 & 1 & 1 \end{vmatrix}}, \quad x_3 = \frac{\begin{vmatrix} 1 & 1 & 2 \\ 1 & -1 & 0 \\ -1 & 1 & 0 \end{vmatrix}}{\begin{vmatrix} 1 & 1 & -1 \\ 1 & -1 & 1 \\ -1 & 1 & 1 \end{vmatrix}},$$

so that

$$x_1 = 1 \, , \qquad x_2 = 1 \, , \qquad x_3 = 0 \, .$$

Cramer's rule, however, though reasonable for n=2,3 and 4, becomes readily intractable for increasing values of n because the determinants are difficult to evaluate, and other methods must be used. Since we will be interested often in relatively large values of n, let us introduce next some matrix properties which are common in many applied problems and which will enable us to solve (1.1) quickly and efficiently. In general, the more structure which is imposed on A, the easier it will be to solve (1.1). One must be sure, however, when studying an applied problem, that the structure which has been imposed is consistent with the physical constraints of the problem.

DEFINITION 1.1. System (1.1) is said to be *diagonally dominant* if and only if

(1.4) $$|a_{ii}| \geq \sum_{\substack{j=1 \\ j \neq i}}^{n} |a_{ij}| \, , \qquad i = 1, 2, \ldots, n,$$

with strict inequality valid for at least one value of i.

EXAMPLE 1. The following system is diagonally dominant:

$$-4x_1+2x_2 + x_3 + x_4 = 2$$
$$-x_1-5x_2 + x_3 + x_4 = -1$$
$$3x_1 + x_2 - 6x_3 + 2x_4 = 0$$
$$x_1 + x_2 + 2x_3 - 4x_4 = -1.$$

EXAMPLE 2. The following system is not diagonally dominant, but interchange of the second and third equations yields a system which is:

$$-4x_1+2x_2 + x_3 + x_4 = 2$$
$$3x_1 + x_2 - 6x_3 + 2x_4 = 0$$
$$-x_1-5x_2 + x_3 + x_4 = -1$$
$$x_1 + x_2 + 2x_3 - 4x_4 = -1.$$

EXAMPLE 3. The following system is not diagonally dominant, nor is any system which results by reordering the equations:

$$-x_1+x_2+x_3+x_4 = -1$$
$$x_1 - x_2 + x_3 + x_4 = -1$$
$$x_1 + x_2 - x_3 + x_4 = -1$$
$$x_1 + x_2 + x_3 - x_4 = -1.$$

DEFINITION 1.2. System (1.1) is said to be *tridiagonal* if and only if all elements of A are zero except a_{ii}, $a_{j,j+1}$, $a_{j+1,j}$, i=1,2,...,n; j=1,2,...,n–1, and none of these is zero.

EXAMPLE. The following system is tridiagonal:

$$-3x_1+ x_2 \qquad\qquad = 1$$
$$x_1-2x_2 +x_3 \qquad\quad = -1$$
$$x_2-2x_3 +x_4 \quad = 11$$
$$x_3-2x_4 +x_5 = 2$$
$$x_4-2x_5 = -3.$$

The term tridiagonal, in the last definition, is most appropriate because, in matrix form, **A** has the particular representation

$$
A = \begin{bmatrix}
a_{11} & a_{12} & & & & & & 0 \\
a_{21} & a_{22} & a_{23} & & & & & \\
& a_{32} & a_{33} & a_{34} & & & & \\
& & \cdot & \cdot & \cdot & & & \\
& & & \cdot & \cdot & \cdot & & \\
& & & & \cdot & \cdot & \cdot & \\
& & & & & a_{n-1,n-2} & a_{n-1,n-1} & a_{n-1,n} \\
0 & & & & & & a_{n,n-1} & a_{n,n}
\end{bmatrix},
$$

in which all elements are zero except those on three diagonals: the main diagonal, the superdiagonal (just above the main diagonal), and the subdiagonal (just below the main diagonal).

We turn now to practical methods of solution.

1.3 GAUSS ELIMINATION

If one knows only that $|A| \neq 0$, it may be difficult to solve (1.1) for large n. The method often applied first is an elementary one called Gauss elimination, which is reviewed next by means of an example.

Consider the system

(1.5)　　　　$4x_1 - x_2 + 2x_3 - x_4 = 2$

(1.6)　　　　$x_1 + 4x_2 - x_3 + x_4 = 2$

(1.7)　　　　$x_1 - 2x_2 - 3x_3 + x_4 = 4$

(1.8)　　　　　　$x_2 \quad - 4x_4 = 0$.

It is verified easily that the determinant of the system has the value 290, so that the solution exists and is unique. Next, we add suitable multiples of (1.5) to (1.6), (1.7) and (1.8) to eliminate x_1 in these equations. In this way (1.6)-(1.8) reduce to

(1.6') $\frac{17}{4}x_2 - \frac{3}{2}x_3 + \frac{5}{4}x_4 = \frac{3}{2}$

(1.7') $-\frac{7}{4}x_2 - \frac{7}{2}x_3 + \frac{5}{4}x_4 = \frac{7}{2}$

(1.8') $x_2 \qquad - 4x_4 = 0$.

Next, add suitable multiples of (1.6') to each of (1.7') and (1.8') to eliminate x_2 in these equations. In this way, (1.7') and (1.8') reduce to

(1.7'') $-\frac{70}{17}x_3 + \frac{30}{17}x_4 = \frac{70}{17}$

(1.8'') $\frac{6}{17}x_3 - \frac{73}{17}x_4 = -\frac{6}{17}$.

Next, add a suitable multiple of (1.7'') to (1.8'') to eliminate x_3 in (1.8''). In this way, (1.8'') reduces to

(1.8''') $-\frac{29}{7}x_4 = 0$.

Thus, system (1.5)-(1.8) has been transformed into the equivalent system

(1.5) $4x_1 - x_2 + 2x_3 \; - x_4 = 2$

(1.6') $\frac{17}{4}x_2 - \frac{3}{2}x_3 + \frac{5}{4}x_4 = \frac{3}{2}$

(1.7'') $-\frac{70}{17}x_3 + \frac{30}{17}x_4 = \frac{70}{17}$

(1.8''') $-\frac{29}{7}x_4 = 0$.

Finally, the latter system is solved by backward substitution, that is, from (1.8''') one has $x_4=0$; substitution of $x_4=0$ into (1.7'') yields $x_3=-1$; substitution of $x_4=0$ and $x_3=-1$ into (1.6') yields $x_2=0$; and substitution of $x_4=0$, $x_3=-1$, $x_2=0$ into (1.5) yields $x_1=1$, and the original system is solved.

With a little thought and a few self-generated examples, one can see readily how to direct a digital computer to perform Gauss elimination. The mathematical recipe, or set of mathematical directions, to do this is a particular example of what is called formally an *algorithm.* To develop a Gauss elimination algorithm, one must formalize the *elimination* and the *backward* substitution steps, which are the basic elements of the method. Now, with regard to the elimination step, we must have a means to indicate how the element a_{ij} in the i-th row and j-th column changes during the process. This is accomplished by using the symbol $a_{ij}^{(k)}$, where k is a positive integer and will indicate that the original a_{ij} has been adjusted (k–1) times. With these abbreviations and notations in mind, we next observe that the elimination step can be stated very precisely as follows. Set $a_{ij}^{(1)} = a_{ij}$, $b_i^{(1)} = b_i$. Then, for k=1,2,...,n–1, generate $a_{ij}^{(k+1)}$ and $b_i^{(k+1)}$ recursively by

$$(1.9) \qquad a_{ij}^{(k+1)} = a_{ij}^{(k)} - \frac{a_{ik}^{(k)}}{a_{kk}^{(k)}} a_{kj}^{(k)}, \qquad\qquad b_i^{(k+1)} = b_i^{(k)} - \frac{a_{ik}^{(k)}}{a_{kk}^{(k)}} b_k^{(k)},$$

for i,j=k+1,k+2,...,n–1,n. The system one then has is

$$a_{11}^{(1)}x_1 + a_{12}^{(1)}x_2 + a_{13}^{(1)}x_3 + \ldots + a_{1n}^{(1)}x_n = b_1^{(1)}$$
$$a_{22}^{(2)}x_2 + a_{23}^{(2)}x_3 + \ldots + a_{2n}^{(2)}x_n = b_2^{(2)}$$
$$a_{33}^{(3)}x_3 + \ldots + a_{3n}^{(3)}x_n = b_3^{(3)}$$
$$\begin{matrix} \cdot & & \cdot & & \cdot \\ & \cdot & & \cdot & & \cdot \\ & & \cdot & & \cdot & & \cdot \end{matrix}$$
$$a_{nn}^{(n)}x_n = b_n^{(n)} .$$

Finally, the backward substitution step is

$$(1.10) \qquad x_n = \frac{b_n^{(n)}}{a_{nn}^{(n)}}, \qquad x_i = \frac{b_i^{(i)} - \sum_{j=i+1}^{n} a_{ij}^{(i)} x_j}{a_{ii}^{(i)}}, \qquad i=n-1,n-2,\ldots,2,1,$$

and the algorithm is complete.

In Gauss elimination, if any of the elements $a_{kk}^{(k)}$, which are called the *pivot* elements, vanishes or, in absolute value, becomes very small compared to the other elements $a_{ik}^{(k)}$, i>k, then we attempt to rearrange the remaining rows so as to attain a nonvanishing pivot or to avoid multiplication by a large number. Specifically, for each k we choose j, the smallest integer for which $|a_{jk}^{(k)}| = \max_{i>k} |a_{ik}^{(k)}|$ and interchange rows k and j. This strategy is called *pivoting*. If **A** is diagonally dominant then no pivoting is necessary.

EXAMPLE. Consider the system

$$(1.11) \qquad x_2 \quad -4x_4 = 0$$
$$(1.12) \qquad x_1 - 2x_2 - 3x_3 + x_4 = 4$$
$$(1.13) \qquad x_1 + 4x_2 - x_3 + x_4 = 2$$
$$(1.14) \qquad 4x_1 - x_2 + 2x_3 - x_4 = 2 \;.$$

In this case, since $a_{11}^{(1)} = 0$, there is no multiple of (1.11) which can be added to (1.12), (1.13) and (1.14) to eliminate x_1 in these equations. However, since $|a_{41}^{(1)}| = \max|a_{i1}^{(1)}|$, interchange of equations (1.11) and (1.14) leads to the following equivalent system

$$(1.14) \qquad 4x_1 - x_2 + 2x_3 - x_4 = 2$$
$$(1.12) \qquad x_1 - 2x_2 - 3x_3 + x_4 = 4$$
$$(1.13) \qquad x_1 + 4x_2 - x_3 + x_4 = 2$$
$$(1.11) \qquad x_2 \quad -4x_4 = 0 \;.$$

Now, for k=1, use of (1.9) yields

$$(1.12') \qquad -\frac{7}{4}x_2 - \frac{7}{2}x_3 + \frac{5}{4}x_4 = \frac{7}{2}$$
$$(1.13') \qquad \frac{17}{4}x_2 - \frac{3}{2}x_3 + \frac{5}{4}x_4 = \frac{3}{2}$$
$$(1.11') \qquad x_2 \quad -4x_4 = 0 \;.$$

Next, since the pivot element $a_{22}^{(2)} = -\frac{7}{4}$ in (1.12') is smaller in absolute value than $a_{32}^{(2)} = -\frac{17}{4}$, we interchange (1.12') with (1.13'):

$$(1.13') \qquad \frac{17}{4}x_2 - \frac{3}{2}x_3 + \frac{5}{4}x_4 = \frac{3}{2}$$
$$(1.12') \qquad -\frac{7}{4}x_2 - \frac{7}{2}x_3 + \frac{5}{4}x_4 = \frac{7}{2}$$
$$(1.11') \qquad x_2 \quad -4x_4 = 0$$

so that, for k=2, (1.9) yields

(1.12") $-\frac{70}{17}x_3 + \frac{30}{17}x_4 = \frac{70}{17}$

(1.11") $\frac{6}{17}x_3 - \frac{73}{17}x_4 = -\frac{6}{17}$.

Next, since system (1.12")-(1.11") is diagonally dominant, it no longer requires pivoting and (1.9) applies directly to yield

(1.11"') $-\frac{29}{7}x_4 = 0$.

Thus, system (1.11)-(1.14) has been transformed into the equivalent system

(1.14) $4x_1 - x_2 + 2x_3 - x_4 = 2$

(1.13') $\frac{17}{4}x_2 - \frac{3}{2}x_3 + \frac{5}{4}x_4 = \frac{3}{2}$

(1.12") $-\frac{70}{17}x_3 + \frac{30}{17}x_4 = \frac{70}{17}$

(1.11"') $-\frac{29}{7}x_4 = 0$.

Finally, the latter system is solved by backward substitution by means of (1.10) to yield $x_4=0$, $x_3=-1$, $x_2=0$, $x_1=1$, and the example is complete.

As a high speed computer technique for general linear systems, Gauss elimination has several shortcomings, most notable of which is the rate at which roundoff error accumulates. Not only would coefficients like $-\frac{70}{17}$ in (1.12") be rounded by the computer, but the actual arithmetic calculations would also be rounded, the effects of which can accumulate so rapidly that, in general, Gauss elimination is relatively inefficient for general systems with more than about 400 equations.

1.4 TRIDIAGONAL SYSTEMS

When additional structure is imposed on coefficient matrix **A**, methods similar to Gauss elimination can be developed which are applicable if n is as large as 2000. In particular, this is valid for "certain systems" which are tridiagonal and diagonally dominant, where, by "certain systems" we mean naturally occurring systems in which the main diagonal elements are *negative* and the subdiagonal and the superdiagonal elements are *positive*. We will show first that, for such systems, $|A| \neq 0$ and then show how to solve them easily and accurately.

In anticipation of the proof that $|A| \neq 0$, we first give some preliminary results.

LEMMA 1.1. If $x = a_1 x_1 + a_2 x_2 + \ldots + a_k x_k$ for k positive numbers $a_1, a_2, \ldots, a_k$ which satisfy $a_1 + a_2 + \ldots + a_k = 1$, then

$$(1.15) \qquad \min_i(x_i) \leq x \leq \max_i(x_i) ,$$

where the equal signs apply if and only if $x_1 = x_2 = \ldots = x_k$.

PROOF. For $j = 1, 2, \ldots, k$, one has $a_j > 0$ and $x_j \leq \max(x_i)$. Thus

$$(1.16) \qquad x \leq a_1(\max x_i) + a_2(\max x_i) + \ldots + a_k(\max x_i)$$
$$= (a_1 + a_2 + \ldots + a_k)(\max x_i) = \max(x_i) ,$$

where equality is valid throughout if and only if $x_1 = x_2 = \ldots = x_k$. Similarly, since $a_j > 0$ and since $x_j \geq \min(x_i)$, one has

$$(1.17) \qquad x \geq a_1(\min x_i) + a_2(\min x_i) + \ldots + a_k(\min x_i)$$
$$= (a_1 + a_2 + \ldots + a_k)(\min x_i) = \min(x_i) ,$$

where, again, equality follows throughout if and only if $x_1 = x_2 = \ldots = x_k$.

LEMMA 1.2. If $x = a_1 x_1 + a_2 x_2 + \ldots + a_k x_k$ for k positive numbers $a_1, a_2, \ldots, a_k$ which satisfy $a_1 + a_2 + \ldots + a_k < 1$, then

$$(1.18) \qquad \min(0, x_1, x_2, \ldots, x_k) \leq x \leq \max(0, x_1, x_2, \ldots, x_k) ,$$

where the equal signs apply if and only if $x_1 = x_2 = \ldots = x_k = 0$.

PROOF. By setting $b = 1 - (a_1 + a_2 + \ldots + a_k)$, one has

$$(1.19) \qquad x = a_1 x_1 + a_2 x_2 + \ldots + a_k x_k + b \cdot 0 .$$

Thus, since $a_j > 0$ and since $b > 0$, Lemma 1.1 implies (1.18).

We are now ready for a major theorem.

THEOREM 1.2. Let system (1.1) be tridiagonal, diagonally dominant, and satisfy

(1.20) $a_{ii} < 0,$ $i = 1, 2, \ldots, n$

(1.21) $a_{j,j+1} > 0,$ $j = 1, 2, \ldots, n-1$

(1.22) $a_{j+1,j} > 0,$ $j = 1, 2, \ldots, n-1.$

Then the solution of the system exists and is unique.

PROOF. To show that the solution of the above system exists and is unique, we need only show that $|A| \neq 0$. This can be done by showing that the only solution of the related homogeneous system

$$
\begin{aligned}
a_{11}x_1 + a_{12}x_2 &= 0 \\
a_{21}x_1 + a_{22}x_2 + a_{23}x_3 &= 0 \\
a_{32}x_2 + a_{33}x_3 + a_{34}x_4 &= 0 \\
&\;\;\vdots \\
a_{n,n-1}x_{n-1} + a_{nn}x_n &= 0
\end{aligned}
$$

(1.23)

is $x_1 = x_2 = \ldots = x_n = 0$. If we assume, then, that system (1.23) has a nonzero solution and show that this must lead to a contradiction, the validity of the theorem will follow.

Assume that (1.23) has a solution for which at least one of $x_1, x_2, \ldots, x_n$ is not zero. Without loss of generality, assume that one of these values is positive (a completely analogous proof follows under the assumption that one is negative). Now, if one of $x_1, x_2, \ldots, x_n$ is positive, then there is a maximum positive number in this set, since there are only n numbers under consideration. Hence, let x_k satisfy

(1.24) $x_k \geq x_i,$ $i = 1, 2, \ldots, n,$

and consider the various possibilities $k=1$, $k=n$ and $1<k<n$.

If $k=1$, then (1.24) becomes

$$x_1 \geq x_i, \qquad i=1,2,\ldots,n.$$

But then the first equation in (1.23) implies

(1.25) $$x_2 = -\frac{a_{11}}{a_{12}} x_1 .$$

However, $-(a_{11}/a_{12})>0$ by (1.20) and (1.21), so that $x_2>0$. But, further, by diagonal dominance, $-(a_{11}/a_{12})\geq 1$, so that $x_2 \geq x_1$. However, $x_1 \geq x_2$, so that $x_2=x_1$ is also maximal and positive and

(1.26) $$-a_{11} = a_{12} .$$

Next, consider the case k=n. Then a completely analogous argument yields $x_{n-1}=x_n$ and

(1.27) $$-a_{nn} = a_{n,n-1} .$$

Consider now the case 1<k<n. Then

$$x_k \geq x_i, \qquad\qquad i=1,2,\ldots,n.$$

But the k-th equation in (1.23) implies

(1.28) $$x_k = -\frac{a_{k,k-1}}{a_{kk}} x_{k-1} - \frac{a_{k,k+1}}{a_{kk}} x_{k+1} .$$

However, by (1.20)-(1.22) $-(a_{k,k-1}/a_{kk})>0$ and $-(a_{k,k+1}/a_{kk})>0$. Moreover, by diagonal dominance,

(1.29) $$-\frac{a_{k,k-1}}{a_{kk}} - \frac{a_{k,k+1}}{a_{kk}} \leq 1 .$$

Note now that strict inequality cannot apply in (1.29) since, otherwise, Lemma 1.2, applied to (1.28), would imply that $x_k \leq \max(0, x_{k-1}, x_{k+1})$ where the equal sign applies only if $x_{k-1} = x_{k+1} = 0$. But this is not possible since $x_k = \max(x_1, x_2, \ldots, x_n) > 0$. Consequently we have

$$(1.30) \qquad -a_{kk} = a_{k,k-1} + a_{k,k+1} .$$

Moreover, Lemma 1.1 applied to (1.28) yields $x_k \leq \max(x_{k-1}, x_{k+1})$, and since x_k is maximal and positive by hypothesis, we have

$$(1.31) \qquad x_k = x_{k-1} = x_{k+1} .$$

Thus, x_{k-1} and x_{k+1} are also maximal and positive. Consequently, the above analysis can be repeated for x_j, $j = k-1, k-2, \ldots, 2$, and $j = k+1, k+2, \ldots, n-1$, so that

$$(1.32) \qquad -a_{jj} = a_{j,j-1} + a_{j,j+1} , \qquad j = 2, 3, \ldots, n-1.$$

From the above discussion, it follows that $x_1 = x_2 = \ldots = x_n$. Thus (1.26), (1.27) and (1.32) are valid, which implies

$$(1.33) \qquad -a_{ii} = \sum_{\substack{j=1 \\ j \neq i}}^{n} a_{ij} , \qquad i = 1, 2, \ldots, n.$$

But, $a_{ii} < 0$ implies $-a_{ii} = |a_{ii}|$, while $a_{ij} \geq 0$, $i \neq j$ implies $a_{ij} = |a_{ij}|$. Thus (1.33) can be written as

$$(1.34) \qquad |a_{ii}| = \sum_{\substack{j=1 \\ j \neq i}}^{n} |a_{ij}| , \qquad i = 1, 2, \ldots, n.$$

Finally, we note that (1.34) contradicts the assumption of diagonal dominance, since strict inequality must be valid for at least one of $i = 1, 2, \ldots, n$. Thus, by contradiction, system (1.23) has only the zero solution and $|A| \neq 0$.

Though we will turn next to the actual solution of tridiagonal systems, one should note immediately that Theorem 1.2 will serve to guide our intuition in later discussions. It will prompt us to seek diagonal dominance whenever possible and to have negative elements on the main diagonal and nonnegative elements elsewhere.

Let us now turn to the actual solution of tridiagonal systems which satisfy the conditions of Theorem 1.2. The method to be developed will be a two-step method based on the following observations. Note first that if a linear algebraic system has a unique solution and if the coefficient matrix is nonzero only on the main and the superdiagonal, then it can be solved easily, as can a system in which the coefficient matrix is nonzero only on the main and subdiagonal.

EXAMPLE 1. The system

$$\begin{bmatrix} -3 & 1 & 0 & 0 \\ 0 & -3 & 1 & 0 \\ 0 & 0 & -3 & 1 \\ 0 & 0 & 0 & -3 \end{bmatrix} \begin{bmatrix} x_1 \\ x_2 \\ x_3 \\ x_4 \end{bmatrix} = \begin{bmatrix} 2 \\ -4 \\ -5 \\ -3 \end{bmatrix}$$

is equivalent to

$$-3x_1 + x_2 = 2$$
$$-3x_2 + x_3 = -4$$
$$-3x_3 + x_4 = -5$$
$$-3x_4 = -3 ,$$

which is solved readily by *backward* substitution. Thus, from the last equation, $x_4 = 1$, while from the third equation $x_3 = 2$, from the second $x_2 = 2$, and from the first $x_1 = 0$.

EXAMPLE 2. The system

$$\begin{bmatrix} -3 & 0 & 0 & 0 \\ 1 & -3 & 0 & 0 \\ 0 & 1 & -3 & 0 \\ 0 & 0 & 1 & -3 \end{bmatrix} \begin{bmatrix} z_1 \\ z_2 \\ z_3 \\ z_4 \end{bmatrix} = \begin{bmatrix} 0 \\ 6 \\ 4 \\ 7 \end{bmatrix}$$

is equivalent to

$$-3z_1 = 0$$
$$z_1 - 3z_2 = 6$$
$$z_2 - 3z_3 = 4$$
$$z_3 - 3z_4 = 7 ,$$

which is solved readily by *forward* substitution. Thus, from the first equation, $z_1=0$, while from the second $z_2=-2$, from the third $z_3=-2$, and from the fourth $z_4=-3$.

The method to be developed will consist of two steps. First, it will require solving a system in which only the main diagonal and subdiagonal elements are nonzero. Then it will require solving a system in which only the main diagonal and superdiagonal elements are nonzero. To understand the details let us consider the system

$$
\begin{aligned}
a_{11}x_1+a_{12}x_2 &= b_1 \\
a_{21}x_1+a_{22}x_2+a_{23}x_3 &= b_2 \\
a_{32}x_2+a_{33}x_3+a_{34}x_4 &= b_3 \\
&\ \ \vdots \\
a_{n,n-1}x_{n-1}+a_{nn}x_n &= b_n ,
\end{aligned}
$$

(1.35)

or, in matrix form,

(1.36) $\mathbf{A\,x = b}$,

where

(1.37) $\mathbf{A} = \begin{bmatrix} a_{11} & a_{12} & & & & \\ a_{21} & a_{22} & a_{23} & & \text{\Large 0} & \\ & a_{32} & a_{33} & a_{34} & & \\ & & \cdot & \cdot & \cdot & \\ & & & \cdot & \cdot & \cdot \\ \text{\Large 0} & & & & a_{n,n-1} & a_{n,n} \end{bmatrix}$, $\mathbf{x} = \begin{bmatrix} x_1 \\ x_2 \\ x_3 \\ \cdot \\ \cdot \\ \cdot \\ x_n \end{bmatrix}$, $\mathbf{b} = \begin{bmatrix} b_1 \\ b_2 \\ b_3 \\ \cdot \\ \cdot \\ \cdot \\ b_n \end{bmatrix}$.

Consider first the possibility of factoring **A**, that is, writing it as a product of matrices:

(1.38) $A = L \cdot U$,

where L is a matrix with nonzero elements on the main diagonal and the subdiagonal only, while U is a matrix with nonzero elements on the main diagonal and the superdiagonal only. The letter L indicates that the matrix is *lower* triangular, while the letter U indicates that the matrix is *upper* triangular. Rewriting A in the form (1.38) is called matrix decomposition. Specifically, let

$$(1.39) \qquad L = \begin{bmatrix} p_1 & 0 & & & & & 0 \\ a_{21} & p_2 & 0 & & & & \\ & a_{32} & p_3 & 0 & & & \\ & & \cdot & \cdot & \cdot & & \\ & & & \cdot & \cdot & \cdot & \\ & & & & \cdot & \cdot & \cdot \\ 0 & & & & & a_{n,n-1} & p_n \end{bmatrix} ,$$

$$(1.40) \qquad U = \begin{bmatrix} 1 & q_1 & & & & & 0 \\ 0 & 1 & q_2 & & & & \\ & 0 & 1 & q_3 & & & \\ & & \cdot & \cdot & \cdot & & \\ & & & \cdot & \cdot & \cdot & \\ & & & & \cdot & \cdot & \cdot \\ 0 & & & & & 0 & 1 \end{bmatrix} ,$$

where $p_1, p_2, p_3, \ldots, p_n$, and $q_1, q_2, q_3, \ldots, q_{n-1}$ are parameters to be determined. Then, (1.39) and (1.40) imply

$$(1.41) \quad LU = \begin{bmatrix} p_1 & p_1q_1 & & & & 0 \\ a_{21} & a_{21}q_1+p_2 & p_2q_2 & & & \\ & a_{32} & a_{32}q_2+p_3 & p_3q_3 & & \\ & & \cdot & \cdot & \cdot & \\ & & & \cdot & \cdot & \cdot \\ & & & & \cdot & \cdot \\ 0 & & & & a_{n,n-1} & a_{n,n-1}q_{n-1}+p_n \end{bmatrix},$$

so that, from (1.38),

$$p_1 = a_{11},$$

$$q_1 = a_{12}/p_1, \qquad\qquad p_2 = a_{22}-a_{21}q_1,$$

$$q_2 = a_{23}/p_2, \qquad\qquad p_3 = a_{33}-a_{32}q_2,$$

$$\cdot \qquad\qquad\qquad\qquad \cdot$$
$$\cdot \qquad\qquad\qquad\qquad \cdot$$
$$\cdot \qquad\qquad\qquad\qquad \cdot$$

$$q_{n-1} = a_{n-1,n}/p_{n-1}, \qquad p_n = a_{nn}-a_{n,n-1}q_{n-1},$$

or, more compactly,

$$(1.42) \quad \begin{cases} p_1 = a_{11} \\ q_{j-1} = a_{j-1,j}/p_{j-1}, \\ p_j = a_{jj}-a_{j,j-1}q_{j-1}, \qquad j=2,3,\ldots,n. \end{cases}$$

Given matrix $\mathbf{A}$, formulas (1.42) yield the required p's and q's in a recursive fashion.

EXAMPLE. Let

$$\mathbf{A} = \begin{bmatrix} -2 & 1 & 0 & 0 & 0 \\ 1 & -2 & 1 & 0 & 0 \\ 0 & 1 & -2 & 1 & 0 \\ 0 & 0 & 1 & -2 & 1 \\ 0 & 0 & 0 & 1 & -2 \end{bmatrix}.$$

Then,

$$a_{ii} = -2 , \qquad i=1,2,3,4,5,$$

$$a_{j,j+1} = 1 , \qquad j=1,2,3,4,$$

$$a_{j+1,j} = 1 , \qquad j=1,2,3,4,$$

so that

$$p_1 = -2$$
$$q_1 = -1/2 , \qquad p_2 = -3/2$$
$$q_2 = -2/3 , \qquad p_3 = -4/3$$
$$q_3 = -3/4 , \qquad p_4 = -5/4$$
$$q_4 = -4/5 , \qquad p_5 = -6/5 .$$

Finally, the decomposition (1.38) is given by

$$
\begin{bmatrix}
-2 & 1 & 0 & 0 & 0 \\
1 & -2 & 1 & 0 & 0 \\
0 & 1 & -2 & 1 & 0 \\
0 & 0 & 1 & -2 & 1 \\
0 & 0 & 0 & 1 & -2
\end{bmatrix}
=
\begin{bmatrix}
-2 & 0 & 0 & 0 & 0 \\
1 & -3/2 & 0 & 0 & 0 \\
0 & 1 & -4/3 & 0 & 0 \\
0 & 0 & 1 & -5/4 & 0 \\
0 & 0 & 0 & 1 & -6/5
\end{bmatrix}
\begin{bmatrix}
1 & -1/2 & 0 & 0 & 0 \\
0 & 1 & -2/3 & 0 & 0 \\
0 & 0 & 1 & -3/4 & 0 \\
0 & 0 & 0 & 1 & -4/5 \\
0 & 0 & 0 & 0 & 1
\end{bmatrix} ,
$$

and the example is complete.

The way we will solve

(1.43) $\mathbf{A} \, \mathbf{x} = \mathbf{b}$

will be as follows. Determine $\mathbf{L}$ and $\mathbf{U}$ such that

(1.44) $\mathbf{A} = \mathbf{L} \, \mathbf{U} ,$

and (1.43) can be written as

(1.45) $\quad$ **L U x = b** .

If, next, we set

(1.46) $\quad$ **U x = z** ,

then

(1.47) $\quad$ **L z = b** .

We will solve (1.47) simply for $z_1, z_2, \ldots, z_n$ by forward substitution. Using these values in (1.46), we will then solve (1.46) by backward substitution to yield $x_1, x_2, \ldots, x_n$, which will solve the given system (1.43). Let us now carry out the details of this two step process.

Consider first (1.47). This system is

$$
\begin{bmatrix}
p_1 & 0 & & & & & 0 \\
a_{21} & p_2 & 0 & & & & \\
& a_{32} & p_3 & 0 & & & \\
& & \cdot & \cdot & \cdot & & \\
& & & \cdot & \cdot & \cdot & \\
0 & & & & \cdot & \cdot & \cdot \\
& & & & & a_{n,n-1} & p_n
\end{bmatrix}
\begin{bmatrix}
z_1 \\ z_2 \\ z_3 \\ \cdot \\ \cdot \\ \cdot \\ z_n
\end{bmatrix}
=
\begin{bmatrix}
b_1 \\ b_2 \\ b_3 \\ \cdot \\ \cdot \\ \cdot \\ b_n
\end{bmatrix}
$$

or

$$
\begin{aligned}
p_1 z_1 &= b_1 \\
a_{21} z_1 + p_2 z_2 &= b_2 \\
a_{32} z_2 + p_3 z_3 &= b_3 \\
&\;\vdots \\
a_{n,n-1} z_{n-1} + p_n z_n &= b_n ,
\end{aligned}
$$

the solution of which, by forward substitution, is

$$z_1 = b_1/p_1$$
$$z_2 = (b_2 - a_{21}z_1)/p_2$$
$$z_3 = (b_3 - a_{32}z_2)/p_3$$

$$\vdots$$

$$z_n = (b_n - a_{n,n-1}z_{n-1})/p_n \,.$$

In compact form, this solution is given by

(1.48) $$z_1 = \frac{b_1}{p_1} \; ; \qquad z_j = \frac{b_j - a_{j,j-1}z_{j-1}}{p_j} , \qquad j=2,3,\ldots,n.$$

Now that the z's are known, we solve (1.46), or,

$$
\begin{bmatrix}
1 & q_1 & & & & & \large{0} \\
0 & 1 & q_2 & & & & \\
 & 0 & 1 & q_3 & & & \\
 & & \cdot & \cdot & \cdot & & \\
 & & & \cdot & \cdot & \cdot & \\
 & \large{0} & & & \cdot & \cdot & \cdot \\
 & & & & & 0 & 1
\end{bmatrix}
\begin{bmatrix}
x_1 \\ x_2 \\ x_3 \\ \cdot \\ \cdot \\ \cdot \\ x_n
\end{bmatrix}
=
\begin{bmatrix}
z_1 \\ z_2 \\ z_3 \\ \cdot \\ \cdot \\ \cdot \\ z_n
\end{bmatrix}
$$

or,

$$x_1 + q_1 x_2 = z_1$$
$$x_2 + q_2 x_3 = z_2$$

$$\vdots$$

$$x_{n-1} + q_{n-1}x_n = z_{n-1}$$
$$x_n = z_n \,.$$

The solution of this system, by backward substitution, is

$$x_n = z_n$$
$$x_{n-1} = z_{n-1} - x_n q_{n-1}$$
$$\cdot$$
$$\cdot$$
$$\cdot$$
$$x_2 = z_2 - x_3 q_2$$
$$x_1 = z_1 - x_2 q_1$$

or, in compact form,

(1.49) $x_n = z_n$; $x_j = z_j - x_{j+1} q_j$, $j = n-1, n-2, \ldots, 1.$

In summary, then, for any tridiagonal system of n equations in $x_1, x_2, x_3, \ldots, x_n$ which satisfy the assumptions of Theorem 1.2, one need only proceed as follows. First, determine the p's and q's from (1.42). Next, determine the z's from (1.48). Finally, determine the x's from (1.49) and the system is solved.

EXAMPLE. Consider the system

$$-2x_1 + x_2 \qquad\qquad = 1$$
$$x_1 - 2x_2 + x_3 \qquad = 0$$
$$x_2 - 2x_3 + x_4 \qquad = 0$$
$$x_3 - 2x_4 + x_5 = 0$$
$$x_4 - 2x_5 = 0 .$$

Then,

$$a_{ii} = -2 , \qquad\qquad i = 1,2,3,4,5,$$
$$a_{j,j+1} = 1 , \qquad\qquad j = 1,2,3,4,$$
$$a_{j+1,j} = 1 , \qquad\qquad j = 1,2,3,4.$$

From the preceding example, we have

$$p_1 = -2, \qquad p_2 = -\frac{3}{2}, \qquad p_3 = -\frac{4}{3}, \qquad p_4 = -\frac{5}{4}, \qquad p_5 = -\frac{6}{5},$$

$$q_1 = -\frac{1}{2}, \qquad q_2 = -\frac{2}{3}, \qquad q_3 = -\frac{3}{4}, \qquad q_4 = -\frac{4}{5}.$$

Moreover, $b_1 = 1$, $b_2 = b_3 = b_4 = b_5 = 0$. Hence, from (1.48),

$$z_1 = -\frac{1}{2}, \qquad z_2 = -\frac{1}{3}, \qquad z_3 = -\frac{1}{4}, \qquad z_4 = -\frac{1}{5}, \qquad z_5 = -\frac{1}{6}.$$

Finally, from (1.49),

$$x_5 = -\frac{1}{6}, \qquad x_4 = -\frac{1}{3}, \qquad x_3 = -\frac{1}{2}, \qquad x_2 = -\frac{2}{3}, \qquad x_1 = -\frac{5}{6},$$

and the system is solved.

Tridiagonal systems which do not satisfy the assumptions of Theorem 1.2 do occur. Often these can be solved by the method of this section or by the method of the next section. One such system which will be of interest later is described in the corollary of the next theorem.

THEOREM 1.3. **Let Ax=b be a diagonally dominant, linear algebraic system in which strict inequality holds for each value of i in (1.4). Then $|A| \neq 0$.**

PROOF. To show that $|A| \neq 0$, we need only prove that the zero solution is the unique solution of **Ax=0**. Assume then that **Ax=0** has a nonzero solution $x_1, x_2, \ldots, x_n$. Then, at least one of $x_1, x_2, \ldots, x_n$, say x_k, is maximum in absolute value. Thus,

$$|x_k| \geq |x_i|, \qquad\qquad i = 1, 2, \ldots, n.$$

Consider now the k-th equation of **Ax=0**:

$$\sum_{j=1}^{n} a_{kj} x_j = 0, \qquad \text{or} \qquad -a_{kk} x_k = \sum_{\substack{j=1 \\ j \neq k}}^{n} a_{kj} x_j.$$

Thus,

$$|a_{kk}|\ |x_k| = |\sum_{j \neq k} a_{kj} x_j| \leq |x_k| \sum_{j \neq k} |a_{kj}| \ .$$

Since $|x_k| > 0$,

$$|a_{kk}| \leq \sum_{j \neq k} |a_{kj}| \ ,$$

which contradicts the strict inequality assumption and the theorem is proved.

Corollary 1.3. **Let Ax=b be a diagonally dominant, tridiagonal linear algebraic system in which strict inequality holds for each value of i in (1.4). Then $|A| \neq 0$.**

The proof follows directly from Theorem 1.3

1.5 THE GENERALIZED NEWTON'S METHOD

The methods of the last two sections are called direct methods because, barring roundoff error accumulation, they lead to a solution in a fixed, predetermined number of steps. When applicable, they are very useful. In general, however, they do *not* apply to the solution of *nonlinear* algebraic and transcendental systems of equations. In this section, we will develop a very powerful method [H.M.Lieberstein (1959)], which is qualitatively different from the ones developed thus far. It is an iterative method and is called the generalized Newton's method because it is an extension of the usual Newton's method. It can be applied to *both* linear and nonlinear systems, but only with some computer artistry. It is not direct in that it requires some judicious initial guesswork and the number of steps it requires is not fixed but varies with the accuracy one demands of the answer. Because it is cyclic in nature, it has the capacity of damping out roundoff error accumulation.

Let us begin by recalling the classical Newton's method for solving one equation in one unknown, say,

(1.50) $f(x) = 0$.

For intuition, we first draw the graph of $y=f(x)$, as shown in Figure 1.1. Now, the problem of determining the real roots of (1.50) is equivalent to that of finding the zeros of $y=f(x)$, that is, of finding where the graph intersects the x-axis. Let us try to accomplish the latter as follows.

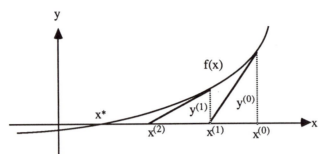

Figure 1.1

As shown in Figure 1.1, let x^* be a zero of $f(x)$. Since x^* is, in general, unknown, let us make an initial guess at it, say $x^{(0)}$. As will be discussed later, it is at times possible to determine $x^{(0)}$ so that it is a "good" approximation to x^*. Now, if $f(x^{(0)})=0$, then, of course, the problem is solved. However, invariably, $f(x^{(0)})\neq 0$, in which case we try to generate a new approximation, $x^{(1)}$, which is better than $x^{(0)}$, as follows. Set $y^{(0)}=f(x^{(0)})$. At the point $(x^{(0)},y^{(0)})$, the slope of the tangent line $L^{(0)}$ to $f(x)$ is $f'(x^{(0)})$ and the equation of $L^{(0)}$ is

(1.51) $y-f(x^{(0)}) = f'(x^{(0)})(x-x^{(0)})$.

If $f'(x^{(0)})\neq 0$, then $L^{(0)}$ meets the x-axis in a point $(x^{(1)},0)$. From (1.51), then,

$$0-f(x^{(0)}) = f'(x^{(0)})(x^{(1)}-x^{(0)}) ,$$

or, equivalently,

(1.52) $x^{(1)} = x^{(0)} - \dfrac{f(x^{(0)})}{f'(x^{(0)})}$.

If $f(x^{(1)})=0$, then the problem is solved. If $f(x^{(1)})\neq 0$, we try to construct an improvement $x^{(2)}$ of $x^{(1)}$ in the same spirit as $x^{(1)}$ was an improvement of $x^{(0)}$. Thus, set $y^{(1)}=f(x^{(1)})$, and let $L^{(1)}$ be the tangent line to $f(x)$ at $(x^{(1)},y^{(1)})$. If $f'(x^{(1)})\neq 0$, then $L^{(1)}$ meets the x-axis in a point $(x^{(2)},0)$, so that

$$0-f(x^{(1)}) = f'(x^{(1)})(x^{(2)}-x^{(1)}) ,$$

or, equivalently,

$$(1.53) \qquad x^{(2)} = x^{(1)} - \frac{f(x^{(1)})}{f'(x^{(1)})} .$$

Again, if $f(x^{(2)})=0$, then the problem is solved. If $f(x^{(2)})\neq 0$, then we construct $x^{(3)}$, $x^{(4)}$, ..., in the same spirit as $x^{(1)}$ and $x^{(2)}$ were constructed. After k+1 steps the real number $x^{(k+1)}$ is determined by the formula

$$(1.54) \qquad x^{(k+1)} = x^{(k)} - \frac{f(x^{(k)})}{f'(x^{(k)})} , \qquad\qquad f'(x^{(k)}) \neq 0 .$$

The iterative procedure defined by recursion formula (1.54) is called Newton's method and (1.54), itself, is called Newton's formula. A solution of (1.50) is a value $x^{(K)}$ which (1.54) self-reproduces, that is $x^{(K+1)}=x^{(K)}$.

EXAMPLE 1. Let us approximate a root of

$$(1.55) \qquad x^3 + \sqrt{3}\, x^2 - 2x - 2\sqrt{3} = 0$$

by Newton's method. To do this, set

$$(1.56) \qquad f(x) = x^3 + \sqrt{3}\, x^2 - 2x - 2\sqrt{3} ,$$

so that

$$f'(x) = 3x^2 + 2\sqrt{3}\, x - 2 ,$$

and Newton's formula becomes

$$(1.57) \qquad x^{(k+1)} = x^{(k)} - \frac{(x^{(k)})^3 + \sqrt{3}\, (x^{(k)})^2 - 2(x^{(k)}) - 2\sqrt{3}}{3(x^{(k)})^2 + 2\sqrt{3}(x^{(k)}) - 2} .$$

To start the iteration, we first assign a value to $x^{(0)}$. In general, it is advantageous to choose $x^{(0)}$ close to a root, if we can so determine $x^{(0)}$. For equations like (1.55), this can usually be accomplished by evaluating (1.56) at, for example, x=0,1,2,3,.... Thus,

$$f(0) = -2\sqrt{3}$$

$$f(1) = 1+\sqrt{3} -2-2\sqrt{3} = -1-\sqrt{3}$$

$$f(2) = 8+4\sqrt{3} -4-2\sqrt{3} = 4+2\sqrt{3}$$

$$f(3) = 27+9\sqrt{3} -6-2\sqrt{3} = 21+7\sqrt{3}$$

.
.
.

Observing that $f(1)<0$ and $f(2)>0$ implies immediately that a root lies in the range $1<x<2$, so that a reasonable guess is $x^{(0)}=1$ or $x^{(0)}=2$. Substitution of $x^{(0)}=2$ in (1.57) with $k=0$ yields

$$x^{(1)} = 2-\frac{(2)^3+\sqrt{3}\,(2)^2-2(2)-2\sqrt{3}}{3(2)^2+2\sqrt{3}\,(2)-2},$$

and it is now apparent that we must also decide on an approximation for $\sqrt{3}$. For simplicity, we will use a one decimal place approximation and then calculate $x^{(1)}$ to one decimal place (in general we would use the maximum number of places allowed on a computer when doing the calculation). Hence, $x^{(1)}=1.6$. Setting $k=1,2,3,4,\ldots$, then yields $x^{(2)}=1.4$, $x^{(3)}=1.4$, $x^{(4)}=1.4$, ..., from which it is clear that the approximate solution is $x^{(2)}=1.4$, since $x^{(2)}=x^{(3)}=x^{(4)}=\ldots=1.4$. It is also clear that once one finds that $x^{(3)}=x^{(2)}$, then it is a waste of effort to continue the iteration, because all further iterates are also equal to $x^{(2)}$.

In general, the criterion for terminating Newton's method is

(1.58) $x^{(K+1)} = x^{(K)}$

for some integer K. However, from the computer point of view, (1.58) is not an exact equality. It means that $x^{(K+1)}$ and $x^{(K)}$ agree to the number of decimal places in one's computation. Another way of stating (1.58) is as follows. Given a positive tolerance ε, which fixes the accuracy of one's computation, then

(1.59) $|x^{(K+1)}-x^{(K)}| < \varepsilon$.

Note that the use of an equals sign in (1.58), even though $x^{(K+1)}$ and $x^{(K)}$ are not exactly equal, is a common practice throughout mathematics. For example, in trigonometry we write

$$\sin 27° = 0.45399$$

when, in fact, what we really mean is

$$|0.45399 - \sin 27°| < 10^{-5}.$$

EXAMPLE 2. Consider the problem in Example 1, again. To make it difficult, we set $x^{(0)}=100000$ and rounded to nine decimal places. In 33 iterations, Newton's method yielded $x^{(33)}=x^{(32)}=1.414213562$, which is a nine decimal place approximation of the real root $x=\sqrt{2}$.

For very complicated equations which will be studied later, it will be of value to have a method which yields a root in fewer iterations than those of Newton's method. For this reason, instead of constructing the line $L^{(0)}$ shown in Figure 1.1, let us try to determine a line through $(x^{(0)},y^{(0)})$ which is *close* to $L^{(0)}$, but which intersects the x-axis in a point x^+, where $x^*<x^+<x^{(1)}$, so that x^+ is a *better* approximation than is $x^{(1)}$. Such a line would have an equation of the form

(1.60) $$y-f(x^{(0)}) = \tau f'(x^{(0)})(x-x^{(0)}) ,$$

where τ is close to unity. Setting y=0 and $x=x^{(1)}=x^+$ in (1.60) yields

(1.61) $$x^{(1)} = x^{(0)} - \frac{f(x^{(0)})}{\tau f'(x^{(0)})} , \qquad f'(x^{(0)}) \neq 0 ,$$

and just as Newton's formula (1.54) was developed by considering first (1.52), so from (1.61) follows the recursion formula

(1.62) $$x^{(k+1)} = x^{(k)} - \frac{f(x^{(k)})}{\tau f'(x^{(k)})} , \qquad f'(x^{(k)}) \neq 0 .$$

For notational simplicity, we set $\omega=1/\tau$ and note that since τ is close to unity, so is ω. Indeed, by ω being close to unity we will, in particular, mean $0<\omega<2$, so that (1.62) becomes

(1.63) $x^{(k+1)} = x^{(k)} - \omega \dfrac{f(x^{(k)})}{f'(x^{(k)})}$, $f'(x^{(k)}) \neq 0$, $0 < \omega < 2$.

Formula (1.63) is called the **generalized Newton's formula** and ω is called an over-relaxation factor. The modification of Newton's method which uses (1.63) in place of (1.48) is called the **generalized Newton's method.** Of course, Newton's formula is a particular case of (1.63) when $\omega=1$.

EXAMPLE. Consider again approximating a root of (1.55), that is,

$$x^3 + \sqrt{3}\, x^2 - 2x - 2\sqrt{3} = 0 ,$$

by the generalized Newton's method. In place of (1.57), the generalized Newton's formula is

(1.64) $x^{(k+1)} = x^{(k)} - \omega \dfrac{(x^{(k)})^3 + \sqrt{3}\,(x^{(k)})^2 - 2(x^{(k)}) - 2\sqrt{3}}{3(x^{(k)})^2 + 2\sqrt{3}(x^{(k)}) - 2}$.

Choosing $\omega=1.3$, and, as in Example 1, calculating only to one decimal place only with $x^{(0)}=2$, we find from (1.64) that $x^{(1)}=1.4$ and $x^{(2)}=1.4$. Thus, one has again that the approximate solution is 1.4, but one has done only two iterations, whereas the previous example required three iterations. Though this saving seems meager at present, we will see in later examples that the resulting savings can be quite substantial.

        Suppose next that one wishes to solve a system of two equations in two unknowns, say,

(1.65) $f(x,y) = 0$

(1.66) $g(x,y) = 0$.

Since this system, as given, does not take advantage of the subscripting capability of computers, we first rewrite it as

(1.67) $f_1(x_1,x_2) = 0$

(1.68) $f_2(x_1,x_2) = 0$,

where $f_1=f$, $f_2=g$, $x_1=x$, $x_2=y$. A natural generalization of (1.63) which we shall use for

(1.67)-(1.68) is

$$(1.69) \qquad x_1^{(k+1)} = x_1^{(k)} - \omega \frac{f_1(x_1^{(k)}, x_2^{(k)})}{\partial f_1(x_1^{(k)}, x_2^{(k)})/\partial x_1} \, , \qquad \frac{\partial f_1}{\partial x_1} \neq 0 \, , \qquad 0 < \omega < 2$$

$$(1.70) \qquad x_2^{(k+1)} = x_2^{(k)} - \omega \frac{f_2(x_1^{(k+1)}, x_2^{(k)})}{\partial f_2(x_1^{(k+1)}, x_2^{(k)})/\partial x_2} \, , \qquad \frac{\partial f_2}{\partial x_2} \neq 0 \, , \qquad 0 < \omega < 2 \, .$$

Note that (1.69) is related directly to x_1 and f_1, while (1.70) is related directly to x_2 and f_2. Also note that the result $x_1^{(k+1)}$ is used in (1.70), that is, new data is being used just as soon as it becomes available.

EXAMPLE 1. Consider the very simple linear system

$$2x_1 - x_2 = -3$$

$$x_1 - 2x_2 = -3 \, ,$$

the exact solution of which is $x_1 = -1$, $x_2 = 1$. Rewriting the system as

$$(1.71) \qquad 2x_1 - x_2 + 3 = 0$$

$$(1.72) \qquad x_1 - 2x_2 + 3 = 0 \, ,$$

we take

$$f_1(x_1, x_2) = 2x_1 - x_2 + 3$$

$$f_2(x_1, x_2) = x_1 - 2x_2 + 3 \, .$$

These choices for f_1 and f_2 are, at present, arbitrary, since there is no apparent reason why $2x_1 - x_2 + 3$ should not be $f_2(x_1, x_2)$. Later, in the study of differential equations, we will see that there is a natural way, when given two equations like (1.71)-(1.72), to choose the f_1 and f_2 functions. The iteration formulas (1.69) and (1.70) take the forms

(1.73) $$x_1^{(k+1)} = x_1^{(k)} - \omega \frac{2x_1^{(k)} - x_2^{(k)} + 3}{2}$$

(1.74) $$x_2^{(k+1)} = x_2^{(k)} - \omega \frac{x_1^{(k+1)} - 2x_2^{(k)} + 3}{-2} .$$

To implement the iteration, we must guess $x_1^{(0)}$ and $x_2^{(0)}$ and also choose ω. For simplicity, let us choose $\omega=1$. Moreover, since there is no longer a simple method for approximating $x_1^{(0)}$ and $x_2^{(0)}$ so that they are close to the exact solution, we simply set $x_1^{(0)}=x_2^{(0)}=0$. Thus, (1.73) and (1.74) imply

$$x_1^{(1)} = -3/2, \qquad x_2^{(1)} = 3/4 ,$$

$$x_1^{(2)} = -9/8 , \qquad x_2^{(2)} = 15/16 ,$$

$$x_1^{(3)} = -33/32 , \qquad x_2^{(3)} = 63/64 ,$$

$$x_1^{(4)} = -129/128 , \qquad x_2^{(4)} = 255/256 ,$$

$$\vdots \qquad\qquad \vdots$$

from which it is apparent directly from the x_1 iterates that they are converging to -1, and from the x_2 iterates that they are converging to 1, which are the correct answers.

EXAMPLE 2. Consider the nonlinear system

$$x_1 - 3x_2 + e^{x_1} = 3$$

$$-2x_1 + x_2 + e^{x_2} = -1 ,$$

the exact solution of which is not known. Set

$$f_1(x_1, x_2) = x_1 - 3x_2 + e^{x_1} - 3$$

$$f_2(x_1, x_2) = -2x_1 + x_2 + e^{x_2} + 1 .$$

For this system, the generalized Newton's formulas reduce to

$$x_1^{(k+1)} = x_1^{(k)} - \omega \frac{x_1^{(k)} - 3x_2^{(k)} + e^{x_1^{(k)}} - 3}{e^{x_1^{(k)}} + 1}$$

$$x_2^{(k+1)} = x_2^{(k)} - \omega \frac{-2x_1^{(k+1)} + x_2^{(k)} + e^{x_2^{(k)}} + 1}{e^{x_2^{(k)}} + 1} .$$

For $x_1^{(0)} = x_0^{(0)} = 0$ and $\omega = 1.5$, it follows that

$$x_1^{(1)} = 0 - 1.5 \frac{0 - 3(0) + e^0 - 3}{e^0 + 1} = 1.5$$

$$x_2^{(1)} = 0 - 1.5 \frac{-2(1.5) + 0 + e^0 + 1}{e^0 + 1} = 0.75 .$$

The iteration would then continue until for some positive integer K:

$$x_1^{(K+1)} = x_1^{(K)}, \qquad\qquad x_2^{(K+1)} = x_2^{(K)}.$$

Finally one would have to check that $x_1^{(K)}$ and $x_2^{(K)}$ were actually solutions of the given system, since we did not know at the outset whether or not the system even had a solution.

Now let us extend (1.69) and (1.70) to the most general system which can occur. Suppose one has to solve the system

$$f_1(x_1, x_2, x_3, \ldots, x_n) = 0$$
$$f_2(x_1, x_2, x_3, \ldots, x_n) = 0$$
$$f_3(x_1, x_2, x_3, \ldots, x_n) = 0$$

(1.75)
$$.$$
$$.$$
$$.$$

$$f_n(x_1, x_2, x_3, \ldots, x_n) = 0 .$$

Then the generalized Newton's formulas to be used are

$$x_1^{(k+1)} = x_1^{(k)} - \omega \frac{f_1(x_1^{(k)}, x_2^{(k)}, x_3^{(k)}, \ldots, x_n^{(k)})}{\partial f_1(x_1^{(k)}, x_2^{(k)}, x_3^{(k)}, \ldots, x_n^{(k)})/\partial x_1}$$

$$x_2^{(k+1)} = x_2^{(k)} - \omega \frac{f_2(x_1^{(k+1)}, x_2^{(k)}, x_3^{(k)}, \ldots, x_n^{(k)})}{\partial f_2(x_1^{(k+1)}, x_2^{(k)}, x_3^{(k)}, \ldots, x_n^{(k)})/\partial x_2}$$

(1.76) $$x_3^{(k+1)} = x_3^{(k)} - \omega \frac{f_3(x_1^{(k+1)}, x_2^{(k+1)}, x_3^{(k)}, \ldots, x_n^{(k)})}{\partial f_3(x_1^{(k+1)}, x_2^{(k+1)}, x_3^{(k)}, \ldots, x_n^{(k)})/\partial x_3}$$

$$\cdot$$
$$\cdot$$
$$\cdot$$

$$x_n^{(k+1)} = x_n^{(k)} - \omega \frac{f_n(x_1^{(k+1)}, x_2^{(k+1)}, \ldots, x_{n-1}^{(k+1)}, x_n^{(k)})}{\partial f_n(x_1^{(k+1)}, x_2^{(k+1)}, \ldots, x_{n-1}^{(k+1)}, x_n^{(k)})/\partial x_n} \; .$$

As a special case, consider the so-called *mildly nonlinear* systems which have the following form:

$$a_{11}x_1 + a_{12}x_2 + a_{13}x_3 + \ldots + a_{1n}x_n + f_1(x_1) = 0$$

$$a_{21}x_1 + a_{22}x_2 + a_{23}x_3 + \ldots + a_{2n}x_n + f_2(x_2) = 0$$

$$a_{31}x_1 + a_{32}x_2 + a_{33}x_3 + \ldots + a_{3n}x_n + f_3(x_3) = 0$$

(1.77)

$$a_{n1}x_1 + a_{n2}x_2 + a_{n3}x_3 + \ldots + a_{nn}x_n + f_n(x_n) = 0 \, ,$$

where $f_i(x)$, i=1,2,...,n, are *nonlinear* functions in only one variable. The preceding Example 2 was an example of such a system in which $f_1(x_1) = e^{x_1} - 3$ and $f_2(x_2) = e^{x_2} + 1$. Next we state a theorem which assures the existence and the uniqueness of the solution of mildly nonlinear systems.

THEOREM 1.4. Let the linear part of system (1.77) be tridiagonal, diagonally dominant, and satisfy

$$a_{ii} < 0, \qquad\qquad i=1,2,\dots,n$$

$$a_{j,j+1} > 0, \qquad\qquad j=1,2,\dots,n-1$$

$$a_{j+1,j} > 0, \qquad\qquad j=1,2,\dots,n-1.$$

If, in addition, $f_i(x)$, $i=1,2,\dots,n$, are differentiable and $f_i'(x){\leq}0$ for all x, then system (1.77) has one and only one solution.

PROOF. Assume that $x_1,x_2,\dots,x_n$ and $y_1,y_2,\dots,y_n$ are two solutions of (1.77). Then,

$$(1.78) \qquad \sum_{j=1}^{n} a_{ij}x_j + f_i(x_i) = 0, \qquad\qquad i=1,2,\dots,n,$$

and

$$(1.79) \qquad \sum_{j=1}^{n} a_{ij}y_j + f_i(y_i) = 0, \qquad\qquad i=1,2,\dots,n.$$

Then, by subtracting (1.78) from (1.79), one finds

$$(1.80) \qquad \sum_{j=1}^{n} a_{ij}(y_j - x_j) + [f_i(y_i) - f_i(x_i)] = 0, \qquad\qquad i=1,2,\dots,n.$$

Since $f_i(x)$ is differentiable one has, by the Mean Value Theorem,

$$f_i(y_i) - f_i(x_i) = f_i'(\xi_i)(y_i - x_i),$$

where ξ_i lies between x_i and y_i. By setting $e_i = y_i - x_i$, equations (1.80) become

$$(1.81) \qquad \sum_{j=1}^{n} a_{ij}e_j + f_i'(\xi_i)e_i = 0, \qquad\qquad i=1,2,\dots,n.$$

Note, now, that equations (1.81) constitute a linear system with unknowns $e_1,e_2,\dots,e_n$. Moreover, since $f_i'(\xi_i){\leq}0$, this system satisfies the conditions of Theorem 1.2, and hence its unique solution is $e_1=e_2=\ldots=e_n=0$. Thus, $x_1=y_1$, $x_2=y_2$, ..., $x_n=y_n$.

The proof that system (1.77) does have a solution is beyond the present scope, since it requires topological methods which are relatively advanced [J.M.Ortega and W.C.Rheinboldt (1970)]. However, from a practical point of view, we will use the following relatively simplistic approach to extablish existence. We will generate a solution, when possible, by iteration and verify that it is a solution by direct substitution.

For mildly nonlinear systems the generalized Newton's formulas (1.76) reduce to

$$x_1^{(k+1)} = x_1^{(k)} - \omega \frac{a_{11}x_1^{(k)}+a_{12}x_2^{(k)}+a_{13}x_3^{(k)}+\ldots+a_{1n}x_n^{(k)}+f_1(x_1^{(k)})}{a_{11}+f_1'(x_1^{(k)})}$$

$$x_2^{(k+1)} = x_2^{(k)} - \omega \frac{a_{21}x_1^{(k+1)}+a_{22}x_2^{(k)}+a_{23}x_3^{(k)}+\ldots+a_{2n}x_n^{(k)}+f_2(x_2^{(k)})}{a_{22}+f_2'(x_2^{(k)})}$$

$$(1.82) \qquad x_3^{(k+1)} = x_3^{(k)} - \omega \frac{a_{31}x_1^{(k+1)}+a_{32}x_2^{(k+1)}+a_{33}x_3^{(k)}+\ldots+a_{3n}x_n^{(k)}+f_3(x_3^{(k)})}{a_{33}+f_3'(x_3^{(k)})}$$

$$\cdot$$
$$\cdot$$
$$\cdot$$

$$x_n^{(k+1)} = x_n^{(k)} - \omega \frac{a_{n1}x_1^{(k+1)}+a_{n2}x_2^{(k+1)}+\ldots+a_{n,n-1}x_{n-1}^{(k+1)}+a_{nn}x_n^{(k)}+f_n(x_n^{(k)})}{a_{nn}+f_n'(x_n^{(k)})} \, .$$

In particular, if $f_i(x)$, $i=1,2,\ldots,n$, are constants, then system (1.77) is linear, and the corresponding generalized Newton's method is called Successive Over-Relaxation, which is abbreviated **SOR**.

1.6 REMARKS ON THE GENERALIZED NEWTON'S METHOD

In general, the choice one should make for ω in the generalized Newton's formulas is difficult to determine *a priori*. However, with some artful computing, a good choice can be made in the following fashion. Heuristically, Newton's method works as follows. From an initial guess $x^{(0)}$, the first iterates behave erratically, while later ones, when they do converge, converge very rapidly. Hence, suppose we are considering

$$(1.83) \qquad f(x) = 0$$

and would like to apply the iteration formula

$$x^{(k+1)} = x^{(k)} - \omega \frac{f(x^{(k)})}{f'(x^{(k)})}.$$

We choose several values of ω in the range $0<\omega<2$ and then examine only the first ten iterates for each ω. For example, suppose we choose $\omega=0.5$, 1.0, 1.5 and 1.75, and find, as shown in Table 1.1, the listed iterates. Examination of the iterates for each of the ω's indicates that the choice $\omega=1.5$ is the most promising, because the convergence seems to be present from $x^{(4)}$ onward. Thus, $\omega=1.5$ is the choice made and the calculations would then proceed using more decimal places than those used in generating the table.

When the system one is trying to solve has a particular form, then convergence often can be guaranteed *a priori*. One such case is when the system is linear, tridiagonal and satisfies the assumptions of Theorem 1.2. In this case it is known [R.S.Varga (1962)] that the generalized Newton's method converges for *all* initial guesses and for *all* ω in the range $0<\omega<2$. Often ω in the range $1.3<\omega<1.8$ is excellent and can reduce a fifty minute computation executed with $\omega=1$ to a computation which requires less than one or two minutes. When the system to be solved is mildly nonlinear and satisfies the assumptions of Theorem 1.4, then the generalized Newton's method converges for all initial guesses and all ω in a subrange of $0<\omega<2$ [S.Schechter (1962)].

Table 1.1

	$\omega=0.5$	$\omega=1.0$	$\omega=1.5$	$\omega=1.75$
$x^{(0)}$	0.00	0.00	0.00	0.00
$x^{(1)}$	10.21	10.24	10.24	9.31
$x^{(2)}$	3.22	19.21	17.36	4.22
$x^{(3)}$	−4.31	32.70	19.24	−15.37
$x^{(4)}$	17.23	16.70	26.43	−67.22
$x^{(5)}$	27.19	19.70	23.42	−55.22
$x^{(6)}$	57.23	23.16	25.40	−40.20
$x^{(7)}$	96.13	25.19	24.17	7.34
$x^{(8)}$	−87.11	22.75	24.33	9.37
$x^{(9)}$	40.32	27.66	24.39	10.25
$x^{(10)}$	11.15	24.63	24.38	23.01

It is important for two reasons that the answer always be checked. First, one may not know *a priori,* as indicated at the end of Section 1.5, that (1.83), for example, even has a solution, and the check provides an *a posteriori* proof that it does. Second, the check provides a degree of confidence that the computer program being used is correct.

Finally, note that, unlike other Newtonian iteration methods for systems, our formulas do not require matrix inversion. These other methods can be analyzed readily from the theoretical point of view, but are not easily implementable on a computer, the reason being that matrix inversion is usually difficult to accomplish with significant accuracy for all but relatively small systems.

1.7 EIGENVALUES AND EIGENVECTORS

The state of a physical system, especially an atomic or a molecular system, can be characterized by the energy of the system. Algebraically, this characterization is often achieved by means of the eigenvalues and eigenvectors of an associated matrix. For this reason, we turn now to the fundamental eigenvalue-eigenvector problem for matrices.

For a nonsingular, n×n matrix A, the algebraic eigenvalue-eigenvector problem is that of finding a set of n constants $\lambda_1,\lambda_2,\ldots,\lambda_n$ and a corresponding set of n *nonzero* vectors $v_1,v_2,\ldots,v_n$ which satisfy

$$(1.84) \qquad A\,v_i = \lambda_i v_i\,, \qquad\qquad i=1,2,\ldots,n.$$

The constants λ_i are called the eigenvalues of A and the vectors v_i are called eigenvectors.

If (1.84) is rewritten as

$$(1.85) \qquad (A-\lambda_i I)v_i = 0\,, \qquad i=1,2,\ldots,n,$$

where I is the unit matrix, then, since the v_i are to be nonzero, (1.85) implies the necessity of the determinantal equation

$$(1.86) \qquad |A-\lambda_i I| = 0\,.$$

However, (1.86) is equivalent to a polynomial of degree n in λ_i, so that A has exactly n eigenvalues, which may be real or complex.

The problem of determining the eigenvalues and eigenvectors of a given matrix is, in general, one of exceptional difficulty. A large number of numerical methods, each applicable to a particular class of matrices, have been developed for this purpose. [For good surveys of the methods available, see A.Ralston and P.Rabinowitz (1978), Chapter 9; D.M.Young and R.T.Gregory (1973), volume II, Chapter 14.] However, since it is of practical value to know the *spectral radius*, that is, the eigenvalue of *maximum* absolute value, of a given matrix, and since, very often, this can be accomplished easily, we will explore a method which is often used for this purpose and which produces a corresponding eigenvector simultaneously. It is called the *power method*.

Assume that A is an $n \times n$ matrix with n eigenvalues $\lambda_1, \lambda_2, ..., \lambda_n$ in which λ_1 is real and

$$(1.87) \qquad |\lambda_1| > |\lambda_j|, \qquad j = 2, 3, ..., n.$$

In addition, assume that A has n linearly independent eigenvectors $v_1, v_2, ..., v_n$ which satisfy (1.84) and which are normalized in the sense that the maximum component in absolute value is unity. Now, let vector $x^{(0)}$ be any initial guess and form the sequence of vectors $x^{(1)}, x^{(2)}, x^{(3)}, ...,$ from

$$(1.88) \qquad x^{(k)} = A\, x^{(k-1)}, \qquad k = 1, 2, 3, ...$$

Then if

$$x^{(0)} = c_1 v_1 + c_2 v_2 + ... + c_n v_n, \qquad c_1 \neq 0,$$

it follows that

$$x^{(1)} = Ax^{(0)} = c_1 \lambda_1 v_1 + c_2 \lambda_2 v_2 + ... + c_n \lambda_n v_n$$
$$x^{(2)} = Ax^{(1)} = c_1 \lambda_1^2 v_1 + c_2 \lambda_2^2 v_2 + ... + c_n \lambda_n^2 v_n$$
$$x^{(3)} = Ax^{(2)} = c_1 \lambda_1^3 v_1 + c_2 \lambda_2^3 v_2 + ... + c_n \lambda_n^3 v_n$$

$$\cdot$$
$$\cdot$$
$$\cdot$$

$$x^{(k)} = Ax^{(k-1)} = c_1 \lambda_1^k v_1 + c_2 \lambda_2^k v_2 + ... + c_n \lambda_n^k v_n.$$

Thus,

(1.89) $x^{(k)} = \lambda_1^k [c_1 v_1 + c_2(\lambda_2/\lambda_1)^k v_2 + \ldots + c_n(\lambda_n/\lambda_1)^k v_n]$

and, as $k \rightarrow \infty$, (1.87) and (1.89) imply

(1.90) $x^{(k)} \sim \lambda_1^k c_1 v_1$

(1.91) $x^{(k+1)} \sim \lambda_1^{k+1} c_1 v_1$.

Hence the ratios of the corresponding, but nonzero, components of $x^{(k+1)}$ and $x^{(k)}$ converge to λ_1.

The result above is easily modified and formalized into a convenient algorithm as follows. Guess $x^{(0)}$. Generate the sequence of vectors $z^{(k)}$, $x^{(k)}$, $k=1,2,3,\ldots$, by

(1.92) $z^{(k)} = (1/m_{k-1})x^{(k-1)}$, $x^{(k)} = A z^{(k)}$, $k=1,2,3,\ldots,$

where m_{k-1} is the component of $x^{(k-1)}$ of maximum absolute value. Then, under the assumptions given above, the sequence of values m_k converges to λ_1 and $z^{(k)}$ converges to an eigenvector v_1 of λ_1.

EXAMPLE. Consider the simple matrix

$$\begin{bmatrix} 4 & 3 & 0 \\ 0 & 3 & 0 \\ 1 & 1 & 2 \end{bmatrix}.$$

Since the matrix is only 3×3, let us first determine all the eigenvalues directly. These are given by the equation

$$\begin{vmatrix} 4-\lambda & 3 & 0 \\ 0 & 3-\lambda & 0 \\ 1 & 1 & 2-\lambda \end{vmatrix} = 0 ,$$

or, equivalently, by

$$(4-\lambda)(3-\lambda)(2-\lambda) = 0 .$$

Hence, the eigenvalues are $\lambda_1=4$, $\lambda_2=3$, $\lambda_3=2$. To determine an eigenvector corresponding to $\lambda_1=4$, we have, from (1.84),

$$
\begin{bmatrix} 4 & 3 & 0 \\ 0 & 3 & 0 \\ 1 & 1 & 2 \end{bmatrix} \begin{bmatrix} v_{1,1} \\ v_{1,2} \\ v_{1,3} \end{bmatrix} = 4 \begin{bmatrix} v_{1,1} \\ v_{1,2} \\ v_{1,3} \end{bmatrix}
$$

which is equivalent to the system

$$
\begin{aligned}
4v_{1,1} + 3v_{1,2} \qquad &= 4v_{1,1} \\
3v_{1,2} \qquad &= 4v_{1,2} \\
v_{1,1} + v_{1,2} + 2v_{1,3} &= 4v_{1,3} \,,
\end{aligned}
$$

which, in turn, is equivalent to

$$
\begin{aligned}
3v_{1,2} \qquad &= 0 \\
3v_{1,2} \qquad &= 4v_{1,2} \\
v_{1,1} + v_{1,2} - 2v_{1,3} &= 0 \,.
\end{aligned}
$$

Thus, $v_{1,2}=0$, $v_{1,1}=2v_{1,3}$. Setting $v_{1,1}=1$ yields $v_{1,3}=1/2$ and an eigenvector is

$$
\mathbf{v}_1 = \begin{bmatrix} 1 \\ 0 \\ 1/2 \end{bmatrix}.
$$

All other eigenvectors are proportional to this one. Let us now see how λ_1 and $\mathbf{v}_1$ are generated numerically by (1.92). Let us guess

$$
\mathbf{x}^{(0)} = \begin{bmatrix} 3 \\ 0 \\ 0 \end{bmatrix}.
$$

Then, calculating to two decimal places, one finds

$$z^{(1)} = \frac{1}{3}\begin{bmatrix} 3 \\ 0 \\ 0 \end{bmatrix} = \begin{bmatrix} 1 \\ 0 \\ 0 \end{bmatrix}, \qquad x^{(1)} = \begin{bmatrix} 4 & 3 & 0 \\ 0 & 3 & 0 \\ 1 & 1 & 2 \end{bmatrix}\begin{bmatrix} 1 \\ 0 \\ 0 \end{bmatrix} = \begin{bmatrix} 4 \\ 0 \\ 1 \end{bmatrix}$$

$$z^{(2)} = \frac{1}{4}\begin{bmatrix} 4 \\ 0 \\ 1 \end{bmatrix} = \begin{bmatrix} 1 \\ 0 \\ 0.25 \end{bmatrix}, \qquad x^{(2)} = \begin{bmatrix} 4 & 3 & 0 \\ 0 & 3 & 0 \\ 1 & 1 & 2 \end{bmatrix}\begin{bmatrix} 1 \\ 0 \\ 0.25 \end{bmatrix} = \begin{bmatrix} 4 \\ 0 \\ 1.50 \end{bmatrix}$$

$$z^{(3)} = \frac{1}{4}\begin{bmatrix} 4 \\ 0 \\ 1.50 \end{bmatrix} = \begin{bmatrix} 1 \\ 0 \\ 0.38 \end{bmatrix}, \qquad x^{(3)} = \begin{bmatrix} 4 & 3 & 0 \\ 0 & 3 & 0 \\ 1 & 1 & 2 \end{bmatrix}\begin{bmatrix} 1 \\ 0 \\ 0.38 \end{bmatrix} = \begin{bmatrix} 4 \\ 0 \\ 1.76 \end{bmatrix}$$

$$z^{(4)} = \frac{1}{4}\begin{bmatrix} 4 \\ 0 \\ 1.76 \end{bmatrix} = \begin{bmatrix} 1 \\ 0 \\ 0.44 \end{bmatrix}, \qquad x^{(4)} = \begin{bmatrix} 4 & 3 & 0 \\ 0 & 3 & 0 \\ 1 & 1 & 2 \end{bmatrix}\begin{bmatrix} 1 \\ 0 \\ 0.44 \end{bmatrix} = \begin{bmatrix} 4 \\ 0 \\ 1.88 \end{bmatrix}$$

$$z^{(5)} = \begin{bmatrix} 1 \\ 0 \\ 0.47 \end{bmatrix}, \qquad x^{(5)} = \begin{bmatrix} 4 \\ 0 \\ 1.94 \end{bmatrix}$$

$$z^{(6)} = \begin{bmatrix} 1 \\ 0 \\ 0.49 \end{bmatrix}, \qquad x^{(6)} = \begin{bmatrix} 4 \\ 0 \\ 1.98 \end{bmatrix}$$

$$z^{(7)} = \begin{bmatrix} 1 \\ 0 \\ 0.50 \end{bmatrix}, \qquad x^{(7)} = \begin{bmatrix} 4 \\ 0 \\ 2.00 \end{bmatrix}$$

$$z^{(8)} = \begin{bmatrix} 1 \\ 0 \\ 0.50 \end{bmatrix}, \qquad x^{(8)} = \begin{bmatrix} 4 \\ 0 \\ 2.00 \end{bmatrix}.$$

Since $z^{(8)} = z^{(7)}$ and $x^{(8)} = x^{(7)}$, it follows that $\lambda_1 = 4$ and $v_1 = z^{(7)}$, which are identical to the results first derived algebraically.

EXERCISES

Basic Exercises

1. Solve for x_2 and x_6 by Cramer's rule and then give your opinion of this technique for finding the complete solution.

$$-9x_1-x_2+x_3+x_4+x_5+3x_6 = 2$$
$$2x_1-7x_2-x_3+x_4+x_5 + x_6 = -12$$
$$x_1+2x_2-9x_3-x_4+x_5+3x_6 = -33$$
$$x_1+x_2+2x_3-7x_4-x_5 + x_6 = -29$$
$$x_1+x_2+x_3+2x_4-9x_5-3x_6 = 21$$
$$x_1+x_2+x_3 + x_4+2x_5-7x_6 = -13 .$$

2. Which of the following are, or by reordering are then, diagonally dominant?

(a)
$$-4x_1 = 0$$
$$-5x_2 = 1$$
$$6x_3 = 2$$
$$-7x_4 = 3$$

(b)
$$-6x_1+2x_2 = 1$$
$$x_1-7x_2+3x_3 = 3$$
$$x_2-8x_3+4x_4 = -2$$
$$x_3-9x_4 = 4$$

(c)
$$-3x_1 +x_2 +x_3 +x_4 = 1$$
$$x_1-3x_2 +x_3 +x_4 = 0$$
$$x_1 +x_2-3x_3 +x_4 = -2$$
$$x_1 +x_2 +x_3-3x_4 = -4$$

(d)
$$-4x_1 +x_2+x_3+x_4 = 7$$
$$-4x_2 +x_3-x_4 = 1$$
$$-4x_3-x_4 = -1$$
$$-4x_4 = 3$$

(e)
$$-x_1+x_2 = 10$$
$$x_1-x_2+x_3 = 6$$
$$x_2-x_3+x_4 = -2$$
$$x_3-x_4+x_5 = 0$$
$$x_4-x_5 = 3$$

(f)
$$x_1-4x_2 +x_3 -x_4 = 1$$
$$2x_1 +x_2-8x_3 +x_4 = 2$$
$$x_1-x_2+3x_3-6x_4 = -1$$
$$x_1 +x_2 +x_3 + x_4 = 0 .$$

3. Which of the systems in Exercise 2 are tridiagonal?

4. Solve by Gauss elimination:

$$-x_1+x_2+x_3+x_4 = -2$$
$$x_1-x_2+x_3+x_4 = -2$$
$$x_1+x_2-x_3+x_4 = 2$$
$$x_1+x_2+x_3-x_4 = 2 .$$

5. Solve the system in Exercise 1 by Gauss elimination. Is your result exact or does it contain error? If there is error, explain its source.

6. Show that the determinant of each of the following matrices is nonzero. Then, write each in the special product form (1.38), that is, as $A=LU$.

(a) $\begin{bmatrix} -2 & 1 & 0 & 0 \\ 1 & -2 & 1 & 0 \\ 0 & 1 & -2 & 1 \\ 0 & 0 & 1 & -2 \end{bmatrix}$
(b) $\begin{bmatrix} -3 & 1 & 0 & 0 \\ 2 & -3 & 1 & 0 \\ 0 & 2 & -3 & 1 \\ 0 & 0 & 2 & -3 \end{bmatrix}$

(c) $\begin{bmatrix} 10 & -3 & 0 & 0 \\ 4 & 10 & -3 & 0 \\ 0 & 4 & 10 & -3 \\ 0 & 0 & 4 & 10 \end{bmatrix}$
(d) $\begin{bmatrix} 5 & -2 & 0 & 0 & 0 \\ -2 & 5 & -2 & 0 & 0 \\ 0 & -2 & 5 & -2 & 0 \\ 0 & 0 & -2 & 5 & -2 \\ 0 & 0 & 0 & -2 & 5 \end{bmatrix}$

(e) $\begin{bmatrix} -4 & 3 & 0 & 0 & 0 \\ 1 & -4 & 3 & 0 & 0 \\ 0 & 1 & -4 & 3 & 0 \\ 0 & 0 & 1 & -4 & 3 \\ 0 & 0 & 0 & 1 & -4 \end{bmatrix}$
(f) $\begin{bmatrix} -4 & 3 & 0 & 0 & 0 \\ 2 & -5 & 3 & 0 & 0 \\ 0 & 3 & -6 & 3 & 0 \\ 0 & 0 & 4 & -7 & 3 \\ 0 & 0 & 0 & 5 & -8 \end{bmatrix} .$

7. Under the assumptions of Theorem 1.2, show that none of $p_1, p_2, p_3, \ldots, p_n$ in (1.39) can be zero.

8. Show that each of the following tridiagonal systems has a unique solution. Then solve and check the answers.

(a) $\begin{aligned} -2x_1 +x_2 \quad\quad &= 0 \\ x_1-4x_2 +x_3 \quad &= 0 \\ x_2-3x_3 +x_4 &= -7 \\ x_3-2x_4 &= 14 \end{aligned}$

(b) $\begin{aligned} 5x_1-2x_2 \quad\quad &= 1 \\ -2x_1+5x_2-2x_3 &= 2 \\ -2x_2+5x_3-2x_4 &= 3 \\ -2x_3+5x_4-2x_5 &= 4 \\ -2x_4+5x_5 &= 17 \end{aligned}$

(c) $\begin{aligned} -2x_1 +x_2 \quad\quad &= 5 \\ x_1-2x_2 +x_3 \quad &= -4 \\ x_{j-1}-2x_j+x_{j+1} &= 0, \quad j=3,4,5,\ldots,97,98 \\ x_{98}-2x_{99}+x_{100} &= -8 \\ x_{99}-2x_{100} &= 13 \end{aligned}$

(d) $\begin{aligned} 3x_1 +x_2 \quad\quad\quad\quad\quad &= 2 \\ x_1+3x_2-x_3 \quad\quad\quad\quad &= -3 \\ -x_2+3x_3 +x_4 \quad\quad\quad &= 3 \\ x_3+3x_4 +x_5 \quad\quad &= -1 \\ x_4+3x_5 + x_6 &= 1 \\ x_5+3x_6 &= -2 \;. \end{aligned}$

9. With the aid of the generalized Newton's method, find all the roots of

$$x^5+31.00x^4-30.75x^3+39.00x^2-31.75x+8.00 = 0 \;.$$

10. Use the generalized Newton's method with $\omega=1$ and $\omega=1.5$ to solve to four decimal places the algebraic system in Exercise 1. Compare the relative efficiencies of the two choices of ω.

11. Using the generalized Newton's method, solve to four decimal places the tridiagonal system of Exercise 8(c) and check the result.

12. Using the generalized Newton's method, solve to four decimal places each of the following nonlinear systems and check the results.

(a) $\quad -10y_1 +5y_2 \qquad\qquad -e^{y_1} = 0$

$\qquad 5y_1 -10y_2 +5y_3 \qquad -e^{y_2} = 0$

$\qquad\qquad 5y_2 -10y_3 +5y_4 -e^{y_3} = 0$

$\qquad\qquad\qquad 5y_3 -10y_4 -e^{y_4} = 0$

(b) $\quad \dfrac{x_1}{\sqrt{1+(x_1/0.2)^2}} - \dfrac{[\cos^2(\pi/16)](x_2-x_1)}{\sqrt{1+[\cos^2(\pi/16)][(x_2-x_1)/0.2]^2}} = 0$

$\qquad \dfrac{[\cos^2(\pi/16)](x_2-x_1)}{\sqrt{1+[\cos^2(\pi/16)][(x_2-x_1)/0.2]^2}} - \dfrac{[\cos^2(\pi/8)](x_3-x_2)}{\sqrt{1+[\cos^2(\pi/8)][(x_3-x_2)/0.2]^2}} = 0$

$\qquad \dfrac{[\cos^2(\pi/8)](x_3-x_2)}{\sqrt{1+[\cos^2(\pi/8)][(x_3-x_2)/0.2]^2}} - \dfrac{[\cos^2(3\pi/16)]x_3}{\sqrt{1+[\cos^2(3\pi/16)](x_3/0.2)^2}} = 0 \ .$

13. Using the power method, find to two decimal places the eigenvalue of maximum absolute value and the corresponding eigenvector of each of the following:

(a) $\begin{bmatrix} 0.20 & 0.90 & 1.32 \\ -11.20 & 22.28 & -10.72 \\ -5.80 & 9.45 & -1.94 \end{bmatrix}$
　　　　(b) $\begin{bmatrix} 10 & 7 & 8 & 7 \\ 7 & 5 & 6 & 5 \\ 8 & 6 & 10 & 9 \\ 7 & 5 & 9 & 10 \end{bmatrix}$

(c) $\begin{bmatrix} 2.0 & -0.7 & 0.3 & -0.2 \\ -0.7 & 0.5 & 0.1 & 0.4 \\ 0.3 & 0.1 & 0.3 & 0.1 \\ -0.2 & 0.4 & 0.1 & 0.2 \end{bmatrix}.$

14. Solve by any method and check:

$$y = (3.075)(10)^{-2}e^{-(7.570)10^{-9}(2x-7440)^2}$$

$$-(1.840)10^{-4} = -\frac{8}{x} + \frac{10^{12}y}{(1+y)(2x)^{1+y}} + \frac{4}{(x^2+2760^2)^{1/2}} \ .$$

Supplementary Exercises

15. Determine which of the following systems has no solution and which has an infinite number of solutions:

(a)
$$-x_1+3x_2 +x_3 = 5$$
$$2x_1 -x_2-2x_3 = 3$$
$$x_1+4x_2 -x_3 = 6$$

(b)
$$x_1+3x_2 +x_3 = 5$$
$$2x_1 -x_2+2x_3 = 3$$
$$x_1+4x_2 +x_3 = 6 .$$

16. Solve by Gauss elimination:

$$6x_1 + 3x_2 + 2x_3 = 6$$
$$6x_1 + 4x_2 + 3x_3 = 0$$
$$20x_1+15x_2+12x_3= 0 .$$

17. Consider the system

$$10x_1+7x_2 +8x_3 +7x_4 = a_1$$
$$7x_1+5x_2 +6x_3 + 5x_4 = a_2$$
$$8x_1+6x_2+10x_3 +9x_4 = a_3$$
$$7x_1+5x_2 +9x_3+10x_4 = a_4 .$$

Show that if $a_1=32$, $a_2=23$, $a_3=33$, $a_4=31$, the exact solution is $(1,1,1,1)$. Next show that if $a_1=32.1$, $a_2=22.9$, $a_3=32.9$, $a_4=31.1$, the solution is $(6,-7.2,2.9,-0.1)$. Explain why such small changes in the a's yield such large changes in the x's.

18. Show that the exact solution of the tridiagonal system

$$-2x_1+ x_2 = -1$$
$$x_{i-1}-2x_i+x_{i+1} = 0 , \qquad i=2,3,\ldots,n-1$$
$$x_{n-1}-2x_n = 0$$

is

$$x_i = \frac{n+1-i}{n+1}, \qquad i=1,2,\ldots,n.$$

19. Approximate the real root of

$$x^3 - 3.7x^2 + 6.25x - 4.069 = 0 .$$

20. Find, to two decimal places, the smallest root of each of the following:

(a) $x^x = 6-2x$,   (b) $\sin(x) = e^{-x}$.

21. Approximate, to three decimal places, the real root of

$$x^2 + 4\sin(x) = 0 .$$

22. Approximate, to three decimal places, the real root of

$$3x - \cos(x) = 1 .$$

23. Find a real root of each of the following polynomial equations and check each answer:

(a) $x^7 - x^6 + x^5 - x^4 + x^3 - x^2 + x - 1 = 0$

(b) $x^7 - 2x^6 + 3x^5 - 4x^4 + 5x^3 - 6x^2 + 7x - 8 = 0$

(c) $x^7 + 2x^6 + 3x^5 + 4x^4 + 5x^3 + 6x^2 + 7x + 8 = 0 .$

24. Approximate a solution of the following system:

$$x_1^4 + x_2^4 = 67$$
$$x_1^3 - 3x_1 x_2 = -35$$

near $x_1 = 2$, $x_2 = 3$.

25. Approximate a solution to each of the following systems:

(a) $x_1^2 + x_2^2 - x_3 = 2$

$\log(x_1) - x_2 - x_3 = 1$

$x_1 + e^{x_2} + x_3 = 0$

(b) $x_1 + x_2 + x_3 + x_4 = 0$

$\log(x_1) + x_2 + \log(x_3) + x_4 = -2$

$x_1 x_2 x_3 - x_4 = 0$

$x_1 - e^{x_2} + x_3 + e^{x_4} = 2$.

26. Approximate, to two decimal places, a solution of

$x_2 - \sin(x_1) = 1.32$

$x_1 - \cos(x_2) = 0.85$.

27. Let $x = x^*$ be a root of $f(x) = 0$. Rewrite the equation in the form $x = F(x)$. Assume that $|F'(x)| \le L < 1$ on an interval $a < x^* < b$. Let $x^{(0)}$ be any value in the range $a < x^{(0)} < b$. Then, prove that the sequence defined by $x^{(k+1)} = F(x^{(k)})$, $k = 0, 1, 2, \ldots$, converges to x^*.

28. Using the power method, find the maximum eigenvalue and the corresponding eigenvector for

$$\begin{bmatrix} 25 & -41 & 10 & -6 \\ -41 & 68 & -17 & 10 \\ 10 & -17 & 5 & -3 \\ -6 & 10 & -3 & 2 \end{bmatrix} .$$

2

Approximation

2.1 INTRODUCTION

Before the development of digital computers, one did not have the facility to store and manipulate large data sets easily. The powerful, classical methodologies developed at that time for such data sets were founded on concepts and techniques from the calculus. Basically, these enable one to approximate discrete data by a continuous function and then to analyze the resulting continuous function. In this chapter, we will study these classical methods, but, whenever possible, we will indicate how even these can be improved by supplemental computer application.

2.2 DISCRETE FUNCTIONS

Any experimental or observational data set consists of a *finite* number of numbers. Such a set is represented easily mathematically by a *discrete function*, which will be defined formally in this section.

For two real numbers a and b that satisfy a<b, let [a,b] denote the interval of real numbers $a \leq x \leq b$. Subdivide [a,b] into n equal parts, each of length h=(b−a)/n, by the n+1 linearly ordered points $a=x_0<x_1<x_2<\ldots<x_n=b$. Then, the set $x_0,x_1,x_2,\ldots,x_n$ is called an R_{n+1} set. For example, the interval [0,2] is divided into five equal parts, each of length h=0.4, by the R_6 set $x_0=0$, $x_1=0.4$, $x_2=0.8$, $x_3=1.2$, $x_4=1.6$, $x_5=2$.

A given R_{n+1} set is also called a set of *grid points* and the positive value h is called a *grid size*. The points x_0 and x_n are called *end* grid points, while $x_1,x_2,\ldots,x_{n-1}$ are called *interior* grid points. This terminology will have special value in Chapter IV.

We have elected to subdivide [a,b] into equal parts only for mathematical simplicity. The discussion which follows can be extended readily to apply to the case where [a,b] is divided into unequal parts. Note also that the notation used for the points of an R_{n+1} set is entirely consistent with the subscripting capability of modern computers.

DEFINITION 2.1. A function y=f(x) that is defined only on an R_{n+1} set is called a *discrete function.*

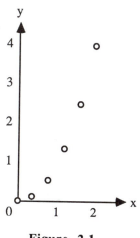

Figure 2.1

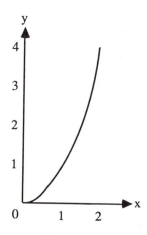

Figure 2.2

The notation $y_i=f(x_i)$, in which x and y are given the same subscript, will be especially convenient when dealing with discrete functions.

EXAMPLE 1. Let [0,2] be divided into five equal parts by the R_6 set $x_0=0$, $x_1=0.4$, $x_2=0.8$, $x_3=1.2$, $x_4=1.6$, $x_5=2$. Then the equation $y_i=x_i^2$, i=0,1,2,3,4,5, defines a discrete function on R_6. Also, $y_i=x_i^2$, $x_i\in R_6$, defines the same function, where, of course "$x_i\in R_6$" is used in the customary mathematical way to mean "x_i is an element in the set R_6".

EXAMPLE 2. Let [0,2] be divided into five equal parts by the R_6 set $x_0=0$, $x_1=0.4$, $x_2=0.8$, $x_3=1.2$, $x_4=1.6$, $x_5=2$. Then the list $f(x_0)=0$, $f(x_1)=1$, $f(x_2)=-1$, $f(x_3)=0$, $f(x_4)=2$, $f(x_5)=-3$ defines a discrete function on R_6.

DEFINITION 2.2. If $y_i=f(x_i)$, $x_i\in R_{n+1}$, is a discrete function, then the set of points (x_i,y_i), i=0,1,2,...,n is called the *graph* of the function.

EXAMPLE 1. Let [0,2] be divided into five equal parts by the R_6 set $x_0=0$, $x_1=0.4$, $x_2=0.8$, $x_3=1.2$, $x_4=1.6$, $x_5=2$. The graph of the discrete function $y_i=x_i^2$, $x_i\in R_6$, is shown in Figure 2.1.

EXAMPLE 2. Let [0,2] be divided into 10^5 equal parts by the R_{100001} set $x_i=(2i)/(10)^5$, i=0,1,2,...,10^5. The graph of the discrete function $y_i=x_i^2$, $x_i\in R_{100001}$, is shown in Figure 2.2.

With regard to Example 2, above, note that the graph of the discrete function appears to the naked eye to be identical with the graph of $y=x^2$, $0 \leq x \leq 2$. The idea of "packing" a large, *but finite,* number of points sufficiently close so that each is physically indistinguishable from others nearby is, of course, the basis of television, and suggests that finite sets might serve one's purposes just as well as infinite ones. The availability of large memory banks in modern computers even makes such a finite approach feasible. However, the proper manipulation of such data sets requires extensive knowledge from areas of the computer sciences, and, therefore, will not be developed here.

2.3 PIECEWISE LINEAR INTERPOLATION

Suppose one is given a discrete function $y_i=f(x_i)$ on an R_{n+1} set. Then, in general, *interpolation* is the process of approximating y at a point x which is between two points x_i and x_{i+1} of the R_{n+1} set. A general procedure for obtaining such an approximation is as follows. First, approximate the given discrete function by a continuous function which is defined on [a,b]. Then, use the value of this continuous function at the abscissa of interest.

There are many ways to implement this approach to interpolation and we shall study some of those which are particularly useful. We begin quite simply.

DEFINITION 2.3. Let $y_i=f(x_i)$, $i=0,1,2,...,n$, be a discrete function that is defined on a given R_{n+1} set. Then the piecewise linear interpolating function L(x) of y_i is the function which is continuous on [a,b] and whose graph consists of straight line segments joining consecutive pairs of points (x_i,y_i), (x_{i+1},y_{i+1}), for $i=0,1,2,...$ n–1, that is,

$$(2.1) \qquad L(x) = y_i + \frac{y_{i+1}-y_i}{h} (x-x_i), \qquad x_i \leq x \leq x_{i+1}, \qquad i=0,1,2,...,n-1.$$

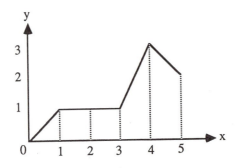

Figure 2.3.

EXAMPLE. On the R_6 set $x_0=0$, $x_1=1$, $x_2=2$, $x_3=3$, $x_4=4$, $x_5=5$, let the discrete function $y_i=f(x_i)$, $i=0,1,2,3,4,5$, be defined by $y_0=0$, $y_1=y_2=y_3=1$, $y_4=3$, $y_5=2$. Then the piecewise linear interpolating function $L(x)$ of y_i is defined on $[0,5]$ by

$$L(x)=\begin{cases} x, & 0\leq x\leq 1 \\ 1, & 1\leq x\leq 2 \\ 1, & 2\leq x\leq 3 \\ 2x-5, & 3\leq x\leq 4 \\ -x+7, & 4\leq x\leq 5, \end{cases}$$

and its graph is given in Figure 2.3. An approximation of y at x=0.7 by means of L yields $y(0.7)=L(0.7)=0.7$. An approximation of y at x=3.5 by means of L yields $y(3.5)=L(3.5)=2(3.5)-5=2$.

If additional knowledge about a data set implies that piecewise linear interpolation is not adequate, then a piecewise parabolic interpolating function can be used. This is described next.

2.4 PIECEWISE PARABOLIC INTERPOLATION

Let us begin by examining the problem of determining a parabolic arc through three points. For this purpose, consider, first, three *noncollinear* points (x_0,y_0), (x_1,y_1), (x_2,y_2), where x_0, x_1, x_2 belong to an R_{n+1} set and, without loss of generality, assume $x_1=0$. (Of course, this is not a restriction because the translation of axes $x^*=x-x_1$ yields $x_1^*=0$.) Hence, $x_0=-h$ and $x_2=h$. Assume the three points lie on a parabolic arc whose equation is of the quadratic form

$$(2.2) \qquad y = a+bx+cx^2, \qquad x_0\leq x\leq x_2,$$

where a, b, c are parameters to be determined. The axis of any parabola given by (2.2) is vertical. Now, if each of the given points is to lie on the graph of (2.2), then

$$y_0 = a-bh+ch^2$$

$$y_1 = a$$

$$y_2 = a+bh+ch^2,$$

the unique solution of which, for a, b, and c, is

$$(2.3) \qquad a = y_1$$

$$(2.4) \qquad b = \frac{y_2 - y_0}{2h}$$

$$(2.5) \qquad c = \frac{y_2 - 2y_1 + y_0}{2h^2} \,.$$

Thus, a solution of the problem is given by (2.2), where a, b, c are determined from (2.3)-(2.5). That is,

$$y = y_1 + \frac{y_2 - y_0}{2h} x + \frac{y_2 - 2y_1 + y_0}{2h^2} x^2 \,.$$

Moreover, $c \neq 0$. To prove this, we need only show that the numerator of the fraction in (2.5) is not zero. This numerator can be written as

$$y_2 - 2y_1 + y_0 = [(y_2 - y_1) - (y_1 - y_0)] \,.$$

However, $(y_2 - y_1) \neq (y_1 - y_0)$ because the points are not collinear. Thus, the numerator, and hence c, are not zero.

When considering a data set, we have no reason to assume that (x_0, y_0), (x_1, y_1), (x_2, y_2), are not collinear, as was assumed above. In the special case of linearity, there is, of course, no parabolic arc with an equation of the form (2.2) through the points. However, formulas (2.2)-(2.5) are still valid, but with c=0, so that (2.2) reduces to a linear form. When discussing parabolic interpolation, we *allow* this degeneracy in the case of three collinear points. We now define piecewise parabolic interpolation.

DEFINITION 2.4. For an R_{n+1} set $x_0, x_1, x_2, \ldots, x_n$, let n be *even*. Then the piecewise parabolic interpolating function P(x) of a given discrete function $y_i = f(x_i)$, $x_i \in R_{n+1}$, is defined to be that function which is continuous on $[x_0, x_n]$ and whose graph consists of piecewise parabolic arcs, each with vertical axis, through consecutive triples of points (x_{i-1}, y_{i-1}), (x_i, y_i), (x_{i+1}, y_{i+1}), $i = 1, 3, 5, 7, \ldots, n-1$, that is,

$$(2.6) \qquad P(x) = y_i + \frac{y_{i+1} - y_{i-1}}{2h} (x - x_i) + \frac{y_{i+1} - 2y_i + y_{i-1}}{2h^2} (x - x_i)^2,$$

$$x_{i-1} \leq x \leq x_{i+1}, \qquad i = 1, 3, 5, \ldots, n-1.$$

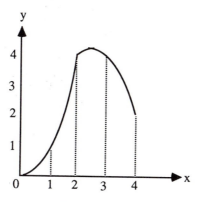

Figure 2.4.

EXAMPLE. On the R_5 set $x_0=0$, $x_1=1$, $x_2=2$, $x_3=3$, $x_4=4$, let the discrete function $y_i=f(x_i)$ be defined by $y_0=0$, $y_1=1$, $y_2=4$, $y_3=4$, $y_4=2$. Then, with $i=1$, $x_0=0$, $x_1=1$, $x_2=2$, $y_0=0$, $y_1=1$, $y_2=4$, substitution in (2.6) yields

$$P(x)=x^2, \qquad 0 \leq x \leq 2 \ .$$

Proceeding to the interval [2,4], with $i=3$, $x_2=2$, $x_3=3$, $x_4=4$, and $y_2=4$, $y_3=4$, $y_4=2$, (2.6) yields

$$P(x) =-2+5x-x^2, \qquad 2 \leq x \leq 4 \ .$$

Thus,

$$P(x)=\begin{cases} x^2, & 0 \leq x \leq 2 \\ -2+5x-x^2, & 2 \leq x \leq 4 \end{cases}$$

is the piecewise parabolic interpolating function for the given discrete function. The graph of P(x) is given in Figure 2.4.

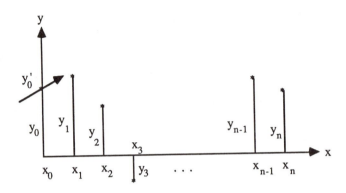

Figure 2.5.

*2.5 CUBIC SPLINE INTERPOLATION

Piecewise linear and piecewise parabolic interpolation have two practical limitations. First, they are not usually applicable for *extrapolation,* that is, for approximation of function values outside the interval of data collection. Extrapolation is particularly valuable when it yields accurate predictions about future events. Second, they can yield implications which are physically unrealistic. For example, if the data sets in Figures 2.3 or 2.4 were to represent positions of a satellite in motion, then the derivatives of $L(x)$ and $P(x)$ would correspond to the satellite's velocity. However, the sharp corners of the graphs are points of discontinuity of the first derivative, which correspond to erratic jumps in the satellite's velocity. Physically, the velocity of a satellite changes slowly and smoothly.

To a certain extent, both the above shortcomings can be overcome by the use of *cubic spline* interpolation, which is motivated and developed as follows. Let us suppose that we wish to analyze a very large set of satellite tracking data. We will in this case make three assumptions. First, as in parabolic interpolation, we assume that n is even. This is not a severe assumption when a data set is very large, since it may require, at worst, dropping a single reading. Second, we assume that a time scale has been chosen in which the initial time x_0 is zero, so that $x_1 = h$, $x_2 = 2h$, and so forth. Of course, any time scale can be transformed into one in which $x_0 = 0$ by a translation of axes. Finally, we assume that, in addition to the discrete function $y_i = f(x_i)$, $x_i \in R_{n+1}$, we know also y_0'. Physically, this corresponds to having measured the satellite's initial velocity. A typical data set is then shown graphically in Figure 2.5.

Let us determine first a continuous function through the points (x_0, y_0), (x_1, y_1) and (x_2, y_2) which has the given derivative value y_0'. To do this, we need four parameters, and a general cubic expression is a natural choice, since it has four undetermined coefficients. Consider, then,

(2.7) $y = a+bx+cx^2+dx^3.$

Since $x_0=0$, it follows from (2.7) that

(2.8) $y_0 = a$.

Also,

(2.9) $y_1 = a+bh+ch^2+dh^3$
(2.10) $y_2 = a+2bh+4ch^2+8dh^3.$

Moreover, since

$$y'= b+2cx+3dx^2,$$

then

(2.11) $y_0' = b$.

The unique solution of system (2.8)-(2.11) is

(2.12) $a = y_0$
(2.13) $b = y_0'$
(2.14) $c = \dfrac{8y_1-7y_0-y_2-6hy_0'}{4h^2}$
(2.15) $d = \dfrac{3y_0-4y_1+y_2+2hy_0'}{4h^3}$.

The resulting formula, called a *cubic spline* interpolating formula, is then

(2.16) $y(x) = y_0+y_0'x+ \dfrac{8y_1-7y_0-y_2-6hy_0'}{4h^2}x^2+ \dfrac{3y_0-4y_1+y_2+2hy_0'}{4h^3}x^3.$

Moreover, from (2.16) one has

$$y'(x) = y_0' + \frac{8y_1 - 7y_0 - y_2 - 6hy_0'}{2h^2} x + 3 \frac{3y_0 - 4y_1 + y_2 + 2hy_0'}{4h^3} x^2,$$

so that

(2.17) $\qquad y_2' = y_0' + 2 \dfrac{y_2 - 2y_1 + y_0}{h}.$

      The formulas (2.16) and (2.17) are the basic ones and can be applied to determine a piecewise cubic spline interpolating function $S(x)$ on the entire R_{n+1} set. Specifically, on the sequence of intervals $[x_{i-1}, x_{i+1}]$, $i=1,3,5,7,\ldots,n-1$, $S(x)$ is given by

(2.18) $\qquad S(x) = y_{i-1} + y_{i-1}'(x - x_{i-1}) + \dfrac{8y_i - 7y_{i-1} - y_{i+1} - 6hy_{i-1}'}{4h^2}(x - x_{i-1})^2$

$$+ \frac{3y_{i-1} - 4y_i + y_{i+1} + 2hy_{i-1}'}{4h^3}(x - x_{i-1})^3,$$

with

(2.19) $\qquad y_{i+1}' = y_{i-1}' + 2 \dfrac{y_{i+1} - 2y_i + y_{i-1}}{h}.$

EXAMPLE. On the R_5 set $x_0=0$, $x_1=1$, $x_2=2$, $x_3=3$, $x_4=4$, let the discrete function $y_i = f(x_i)$ be defined by $y_0=2$, $y_1=3$, $y_2=2$, $y_3=1$, $y_4=2$. Also assume that $y_0'=1$. Then, on $0 \le x \le 2$, substitution into (2.18) with $i=1$ yields

(2.20) $\qquad y = 2 + x + \dfrac{x^2}{2} - \dfrac{x^3}{2},$ $\qquad\qquad 0 \le x \le 2$.

Moreover, (2.19) yields

(2.21) $\qquad y_2' = -3$.

Proceeding to the interval $2 \leq x \leq 4$ with i=3 by using (2.21), one has from (2.18)

(2.22) $y = 2 - 3(x-2) + \frac{5}{2}(x-2)^2 - \frac{1}{2}(x-2)^3,$ $2 \leq x \leq 4$.

Though (2.22) is a perfectly acceptable form for the answer, one could multiply out and recombine to find, equivalently,

(2.23) $y = 22 - 19x + \frac{11}{2}x^2 - \frac{1}{2}x^3,$ $2 \leq x \leq 4$.

Thus, a piecewise cubic spline interpolating function for the given data set is

$$S(x) = \begin{cases} 2 + x + \dfrac{x^2}{2} - \dfrac{x^3}{2}, & 0 \leq x \leq 2 \\[2mm] 22 - 19x + \dfrac{11}{2}x^2 - \dfrac{1}{2}x^3, & 2 \leq x \leq 4 . \end{cases}$$

Finally, note that the derivative of $S(x)$ at x=2 exists and is continuous. Indeed, as a check, one has

$$\frac{d}{dx}[2 + x + \frac{1}{2}x^2 - \frac{1}{2}x^3]\big|_{x=2} = [1 + x - \frac{3}{2}x^2]\big|_{x=2} = -3$$

and

$$\frac{d}{dx}[22 - 19x + \frac{11}{2}x^2 - \frac{1}{2}x^3]\big|_{x=2} = [-19 + 11x - \frac{3}{2}x^2]\big|_{x=2} = -3 ,$$

and the example is complete.

   A large variety of other types of spline functions and spline problems have been and are being studied. For example, if one knows both y_0' and y_0'', then *cubic spline* interpolation can be developed on the interval between a consecutive *pair* of points of an R_{n+1} set rather than on an interval which contains a consecutive *triple* of points. Assuming $x_0 = 0$, one need merely determine a, b, c, d in (2.9) from the given data $y(x_0) = y(0) = y_0$, $y(x_1) = y(h) = y_1$, $y'(x_0) = y'(0) = y_0'$, and $y''(x_0) = y''(0) = y_0''$. Thus, from (2.9),

$$y_0 = a , \qquad y_0' = b , \qquad y_0'' = 2c , \qquad y_1 = a+bh+ch^2+dh^3,$$

so that, on $0 \leq x \leq x_1$, $y(x)$ is given by

$$y(x) = y_0+y_0'x+ \frac{y_0''}{2}x^2+ \frac{y_1-y_0-hy_0'-\frac{h^2}{2}y_0''}{h^3}x^3.$$

One can now generate, recursively, a piecewise cubic spline on the entire R_{n+1} set so that at each of the points $x_1,x_2,\ldots,x_{n-1}$ *both* the *first* and the *second* derivatives exist and are continuous. Specifically, on each subsequent interval $[x_i,x_{i+1}]$, $i=1,2,\ldots,n-1$, one first determines y_i' and y_i'' by

(2.24)

$$y_i' = y_{i-1}'+hy_{i-1}''+ 3\, \frac{y_i-y_{i-1}-hy_{i-1}'-\frac{h^2}{2}y_{i-1}''}{h}$$

$$y_i'' = y_{i-1}''+6\, \frac{y_i-y_{i-1}-hy_{i-1}'-\frac{h^2}{2}y_{i-1}''}{h^2} ,$$

and then $y(x)$ is given by

(2.25) $$y(x) = y_i+y_i'(x-x_i) + \frac{y_i''}{2}(x-x_i)^2 + \frac{y_{i+1}-y_i-hy_i'-\frac{h^2}{2}y_i''}{h^3}(x-x_i)^3.$$

Interestingly enough, piecewise cubic splines in which both the first and the second derivatives exist and are continuous at $x_1,x_2,\ldots,x_{n-1}$ can also be generated on the entire R_{n+1} set in a *single* step by solving a tridiagonal system of linear algebraic equations. Let us show how to do this for a problem of wide interest, and, indeed, even drop the assumption that $x_0=0$.

Let $y_i=f(x_i)$, $x_i \in R_{n+1}$, be a discrete function. We wish to generate on $[x_0,x_n]$ an associated piecewise cubic spline function $S(x)$, such that

(2.26) $$S(x) = S_i(x) , \qquad x_{i-1} \leq x \leq x_i , \qquad i=1,2,\ldots,n,$$

where $S_i(x)$ is a cubic expression. Moreover, in order for $S(x)$ to be an interpolating function one must have

(2.27) $S(x_i) = y_i$, $i=0,1,2,\ldots,n$.

Finally, the continuity of the first and second derivatives is expressed by

(2.28) $S_i'(x_i) = S_{i+1}'(x_i)$

$$i=1,2,\ldots,n-1.$$

(2.29) $S_i''(x_i) = S_{i+1}''(x_i)$

For this purpose, let z_0 and z_n be given constants and let

$$z_i = S_i''(x_i) = S_{i+1}''(x_i), \qquad i=1,2,\ldots,n-1.$$

Then consider the function

(2.30) $S_i''(x) = z_{i-1}\dfrac{x_i-x}{h} + z_i\dfrac{x-x_{i-1}}{h}$, $x_{i-1} \leq x \leq x_i$.

At x_{i-1}, $S_i''(x)$ has the value z_{i-1}, while at x_i it has the value z_i. From (2.30), by integration, one has

$$S_i'(x) = -z_{i-1}\frac{(x_i-x)^2}{2h} + z_i\frac{(x-x_{i-1})^2}{2h} + c_1,$$

(2.31) $S_i(x) = z_{i-1}\dfrac{(x_i-x)^3}{6h} + z_i\dfrac{(x-x_{i-1})^3}{6h} + c_1 x + c_2$.

Since $y_i = S(x_i)$, then, from (2.31),

$$y_i = (1/6)z_ih^2 + c_1x_i + c_2$$

(2.32)

$$y_{i-1} = (1/6)z_{i-1}h^2 + c_1x_{i-1} + c_2 .$$

Solving (2.32) for c_1 and c_2 yields

(2.33) $$c_1 = \frac{(y_i - y_{i-1}) - (z_i - z_{i-1})h^2/6}{h}$$

(2.34) $$c_2 = \frac{(x_iy_{i-1} - x_{i-1}y_i) - (x_iz_{i-1} - x_{i-1}z_i)h^2/6}{h} .$$

Substitution of (2.33) and (2.34) into (2.31) yields

(2.35) $$S_i(x) = z_{i-1}\frac{(x_i-x)[(x_i-x)^2-h^2]}{6h} + z_i\frac{(x-x_{i-1})[(x-x_{i-1})^2-h^2]}{6h} + \frac{y_{i-1}(x_i-x)+y_i(x-x_{i-1})}{h} .$$

Differentiation of (2.35) yields

(2.36) $$S_i'(x) = z_{i-1}\frac{h^2-3(x_i-x)^2}{6h} + z_i\frac{3(x-x_{i-1})^2-h^2}{6h} + \frac{y_i-y_{i-1}}{h} .$$

Thus, if we require (2.28), then

(2.37) $$\frac{z_{i-1}h}{6} + \frac{z_ih}{3} + \frac{y_i-y_{i-1}}{h} = -\frac{z_ih}{3} - \frac{z_{i+1}h}{6} + \frac{y_{i+1}-y_i}{h} ,$$

or, equivalently,

(2.38) $$\frac{1}{6}z_{i-1} + \frac{2}{3}z_i + \frac{1}{6}z_{i+1} = \frac{y_{i+1}-2y_i+y_{i-1}}{h^2} , \qquad i=1,2,\dots,n-1.$$

Since z_0 and z_n are given, this is a tridiagonal system of $n-1$ linear equations in $z_1, z_2, \dots, z_{n-1}$, which satisfies the conditions of Corollary 1.3. Thus, the system has a unique solution and hence $S(x)$ in (2.26) is generated directly on $[x_0, x_n]$ by solving system (2.38) and substituting into (2.35).

EXAMPLE. Given the discrete function $y(0)=1$, $y(1)=2$, $y(2)=4$, $y(3)=0$, $y(4)=-10$, let us construct the piecewise cubic spline function $S(x)$ on $[0,4]$ which satisfies

$$z_0 = z_4 = 0$$

$$S(x_i) = y(x_i) , \qquad i=0,1,2,3,4$$

$$S(x) = S_i(x) , \qquad x_{i-1} \le x \le x_i , \qquad i=1,2,3,4$$

$$S_i'(x_i) = S_{i+1}'(x_i) , \qquad i=1,2,3$$

$$S_i''(x_i) = S_{i+1}''(x_i) , \qquad i=1,2,3.$$

Direct substitution into (2.38) for i=1,2,3, yields

$$(1/6)z_0 + (2/3)z_1 + (1/6)z_2 = 1$$

$$(1/6)z_1 + (2/3)z_2 + (1/6)z_3 = -6$$

$$(1/6)z_2 + (2/3)z_3 + (1/6)z_4 = -6 ,$$

or, since $z_0 = z_4 = 0$,

$$4z_1 + z_2 \qquad = 6$$

$$z_1 + 4z_2 + z_3 = -36$$

$$z_2 + 4z_3 = -36 .$$

The unique solution of this system is

$$z_1 = 99/28 , \qquad z_2 = -57/7 , \qquad z_3 = -195/28 .$$

Thus, $S(x)$, which is given piecewise by (2.35), is defined by

$$S_1(x) = \frac{99}{28} \frac{x(x^2-1)}{6} + (1-x) + 2x , \qquad\qquad 0 \leq x \leq 1,$$

$$S_2(x) = \frac{99}{28} \frac{(2-x)[(2-x)^2-1]}{6} - \frac{57}{7} \frac{(x-1)[(x-1)^2-1]}{6} + 2(2-x) + 4(x-1) , \qquad 1 \leq x \leq 2,$$

$$S_3(x) = -\frac{57}{7} \frac{(3-x)[(3-x)^2-1]}{6} + \frac{195}{28} \frac{(x-2)[(x-2)^2-1]}{6} + 4(3-x) , \qquad 2 \leq x \leq 3,$$

$$S_4(x) = -\frac{195}{28} \frac{(4-x)[(4-x)^2-1]}{6} - 10(x-3) , \qquad\qquad 3 \leq x \leq 4.$$

2.6 LAGRANGE INTERPOLATION

In Sections 2.3 and 2.4, it was shown how to construct a piecewise linear function through consecutive pairs of points and a piecewise quadratic function through consecutive triplets of points. One can proceed in this spirit to construct a piecewise cubic through consecutive quadruplets of points, a piecewise quartic through consecutive quintuplets of points and so forth. Instead of treating each such possibility individually, we will explore next an approach developed by Lagrange which will enable one to write down, almost at once, the desired result for any given number of points.

To understand the Lagrange method, let us begin again with $x_0, x_1, x_2 \in R_{n+1}$ and consider the unique parabolic arc, with vertical axis, through the three points (x_0, y_0), (x_1, y_1), (x_2, y_2). If the equation is

$$(2.39) \qquad y = a + b(x-x_1) + c(x-x_1)^2,$$

then a, b and c are given uniquely by (2.3)-(2.5). Lagrange devised a rather simple method for writing the same solution, but in a different form. In particular, he assumed the solution was of the form

$$y(x) = y_0 A(x) + y_1 B(x) + y_2 C(x) ,$$

where $A(x)$, $B(x)$ and $C(x)$ are polynomials of degree two. Since it was necessary to have

$$y(x_0) = y_0 , \qquad\qquad y(x_1) = y_1 , \qquad\qquad y(x_2) = y_2 ,$$

it was sufficient to choose $A(x)$, $B(x)$ and $C(x)$ so that

$$A(x_0) = 1, \qquad A(x_1) = 0, \qquad A(x_2) = 0$$

$$B(x_0) = 0, \qquad B(x_1) = 1, \qquad B(x_2) = 0$$

$$C(x_0) = 0, \qquad C(x_1) = 0, \qquad C(x_2) = 1.$$

Lagrange observed that simple quadratic functions which satisfied these conditions are

$$A = \frac{(x-x_1)(x-x_2)}{(x_0-x_1)(x_0-x_2)}, \qquad B = \frac{(x-x_0)(x-x_2)}{(x_1-x_0)(x_1-x_2)}, \qquad C = \frac{(x-x_0)(x-x_1)}{(x_2-x_0)(x_2-x_1)}.$$

Thus, his solution was

(2.40)
$$y = y_0 \frac{(x-x_1)(x-x_2)}{(x_0-x_1)(x_0-x_2)} + y_1 \frac{(x-x_0)(x-x_2)}{(x_1-x_0)(x_1-x_2)} + y_2 \frac{(x-x_0)(x-x_1)}{(x_2-x_0)(x_2-x_1)}.$$

Of course, (2.40) and (2.39), with a, b, c given by (2.3)-(2.5), are mathematically identical. To see this, one need only multiply out and recombine the terms of (2.40). However, the structure of (2.40) is more lucid than (2.39). Indeed, this structure can be characterized precisely as follows. Consider, for example, the first term

(2.41)
$$y_0 \frac{(x-x_1)(x-x_2)}{(x_0-x_1)(x_0-x_2)}.$$

This term is associated with the first point (x_0,y_0) of the three points (x_0,y_0), (x_1,y_1), (x_2,y_2). The denominator of (2.41) is constructed from x_0 by subtracting from it all other possible choices of x_i and by multiplying the resulting differences. The numerator is constructed from the denominator by replacing x_0 by x. The coefficient y_0 is the value of y corresponding to x_0. In the same spirit, it follows that corresponding to (x_1,y_1) we would have a term whose denominator is $(x_1-x_0)(x_1-x_2)$, whose numerator is $(x-x_0)(x-x_2)$, and whose coefficient is y_1; corresponding to (x_2,y_2) one has a term whose denominator is $(x_2-x_0)(x_2-x_1)$, whose numerator is $(x-x_0)(x-x_1)$, and whose coefficient is y_2. The sum of the three resulting terms is (2.40).

Suppose now that one wishes to construct a cubic polynomial through four points (x_0,y_0), (x_1,y_1), (x_2,y_2), (x_3,y_3), where x_0,x_1,x_2,x_3 belong to an R_{n+1} set. If one would wish to guess the answer from the above discussion, the reasoning would be as follows. There must be four terms. The first one would correspond to (x_0,y_0). Its denominator would be

$$(x_0-x_1)(x_0-x_2)(x_0-x_3) ,$$

its numerator would be

$$(x-x_1)(x-x_2)(x-x_3) ,$$

and its coefficient would be y_0. Thus, the first term would be

$$y_0 \frac{(x-x_1)(x-x_2)(x-x_3)}{(x_0-x_1)(x_0-x_2)(x_0-x_3)} .$$

The second term, corresponding to (x_1,y_1), would be

$$y_1 \frac{(x-x_0)(x-x_2)(x-x_3)}{(x_1-x_0)(x_1-x_2)(x_1-x_3)} .$$

The third term would correspond to (x_2,y_2), and would be

$$y_2 \frac{(x-x_0)(x-x_1)(x-x_3)}{(x_2-x_0)(x_2-x_1)(x_2-x_3)} .$$

The fourth term, corresponding to (x_3,y_3), would be

$$y_3 \frac{(x-x_0)(x-x_1)(x-x_2)}{(x_3-x_0)(x_3-x_1)(x_3-x_2)} .$$

Thus, the guess would be

$$(2.42) \qquad y = y_0 \frac{(x-x_1)(x-x_2)(x-x_3)}{(x_0-x_1)(x_0-x_2)(x_0-x_3)} + y_1 \frac{(x-x_0)(x-x_2)(x-x_3)}{(x_1-x_0)(x_1-x_2)(x_1-x_3)}$$

$$+ y_2 \frac{(x-x_0)(x-x_1)(x-x_3)}{(x_2-x_0)(x_2-x_1)(x_2-x_3)} + y_3 \frac{(x-x_0)(x-x_1)(x-x_2)}{(x_3-x_0)(x_3-x_1)(x_3-x_2)} \, .$$

The fact that (2.42) is actually the correct result follows because each term is cubic and because substitution of $x=x_0,x_1,x_2,x_3$ implies easily $y=y_0,y_1,y_2,y_3$, respectively. If one wishes to multiply (2.42) out and rewrite it in the form

$$y = a+bx+cx^2+dx^3,$$

one can do so. However, from the computer point of view, both forms are relatively simple and acceptable.

EXAMPLE. The cubic polynomial whose graph contains the four points (0,1), (1,2), (2,0) and (3,–2) is constructed as follows. Setting $x_0=0$, $x_1=1$, $x_2=2$, $x_3=3$, $y_0=1$, $y_1=2$, $y_2=0$, $y_3=-2$ and substituting in (2.42) yields

$$(2.43) \qquad y = -\frac{1}{6}(x-1)(x-2)(x-3)+x(x-2)(x-3)-\frac{1}{3}x(x-1)(x-2) \, .$$

Although not necessary, multiplication and recombination of (2.43) yields the equivalent form

$$y = \frac{1}{2}x^3-3x^2+\frac{7}{2}x+1 \, .$$

The special forms (2.40) and (2.42) are called Lagrange interpolation forms. The *general Lagrange interpolation* formula of degree k, denoted by $p_k(x)$, is given as follows. Let $x_0,x_1,x_2,\ldots,x_k$ belong to an R_{n+1} set, where n is a multiple of k. Consider the distinct points (x_0,y_0), (x_1,y_1), ..., (x_k,y_k). Define $Q_j(x)$ by

$$Q_j(x) = \frac{(x-x_0)(x-x_1)\ldots(x-x_{j-1})(x-x_{j+1})\ldots(x-x_{k-1})(x-x_k)}{(x_j-x_0)(x_j-x_1)\ldots(x_j-x_{j-1})(x_j-x_{j+1})\ldots(x_j-x_{k-1})(x_j-x_k)} \, .$$

Then, $p_k(x)$ is given by

$$(2.44) \qquad p_k(x) = \sum_{j=0}^{k} y_j Q_j(x) ,$$

where $p_k(x)$ is, *except* for degenerate cases, equivalent to a polynomial of degree k. The formula (2.44) is the basic one and can be applied to determine a piecewise Lagrange interpolating function of degree k on the entire R_{n+1} set by using translation of axes.

2.7 LEAST SQUARES

The graphs of all the continuous, approximating functions considered thus far contained *all* the points of a given discrete function. We now seek to construct approximating functions whose graphs need not necessarily contain the points of a given discrete function, but which "fit the function closely". This can be accomplished very easily by the method of least squares, which is described first by means of a simple example.

EXAMPLE. On the R_4 set $x_0=0$, $x_1=1$, $x_2=2$, $x_3=3$, consider the discrete function $y_0=-1$, $y_1=1$, $y_2=3$, $y_3=4$. The graph of this function consists of the four points in Figure 2.6. Consider now the problem of finding the equation of the straight line in the figure which, in some sense, fits the data well. Let the equation of the line be

$$(2.45) \qquad y = a_1 + a_2 x .$$

Then, at x_0, x_1, x_2, x_3 the *vertical* distances between the points and the line are, respectively,

$$(2.46) \qquad \varepsilon_0 = a_1 + a_2 x_0 - y_0$$

$$(2.47) \qquad \varepsilon_1 = a_1 + a_2 x_1 - y_1$$

$$(2.48) \qquad \varepsilon_2 = a_1 + a_2 x_2 - y_2$$

$$(2.49) \qquad \varepsilon_3 = a_1 + a_2 x_3 - y_3 .$$

We have taken vertical distances because it is far simpler to calculate these than to calculate perpendicular distances. The values ε_0, ε_1, ε_2, ε_3 are, in a sense, errors, since they are a measure of how far the line misses the points. Our objective is to *minimize* these errors. One way to do this would be to minimize the sum

$$(2.50) \qquad E = \varepsilon_0 + \varepsilon_1 + \varepsilon_2 + \varepsilon_3 .$$

However, this would cause distasteful problems. For example, if, as seen in Figure 2.6, one were to have $\varepsilon_0=0.1$, $\varepsilon_1=-0.1$, $\varepsilon_2=-0.2$, $\varepsilon_3=0.2$, then E=0 in (2.50) and it is somewhat misleading to conclude that the total error is zero. In place of (2.50), it would be more reasonable to minimize the sum

(2.51) $E = |\varepsilon_0|+|\varepsilon_1|+|\varepsilon_2|+|\varepsilon_3|$.

However, this may be quite difficult to do because the usual method for minimization requires setting the derivative of the function equal to zero, and it should be recalled that even simple functions like y=|x| are not always differentiable. An easy way out is to minimize the function

(2.52) $E = \varepsilon_0^2+\varepsilon_1^2+\varepsilon_2^2+\varepsilon_3^2$,

for it is always differentiable, its individual terms are always nonnegative, and, because of its specific parabolic structure, it always has a unique minimum. From (2.46)-(2.49), then, (2.52) becomes

$$E = [a_1+1]^2+[(a_1+a_2)-1]^2+[(a_1+2a_2)-3]^2+[(a_1+3a_2)-4]^2 .$$

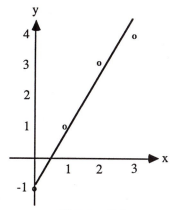

Figure 2.6

To minimize E, we simply consider the system of equations

$$\frac{\partial E}{\partial a_1} = 0 , \qquad \frac{\partial E}{\partial a_2} = 0 ,$$

which reduces to

$$4a_1 + 6a_2 = 7$$
$$6a_1 + 14a_2 = 19 .$$

The solution of the system is $a_1 = -0.8$, $a_2 = 1.7$. The desired fit, called the *least squares* fit because (2.52) has been minimized, is

$$y = -0.8 + 1.7x ,$$

and the example is complete.

      The method of least squares is described generally as follows. Let $y_i = f(x_i)$, $i = 0,1,2,\ldots,n$, be a given discrete function on an R_{n+1} set. Let $y = g(x)$, $x_0 \le x \le x_n$, be a continuous function chosen to fit the given data by least squares. Assume $g(x)$ contains m independent parameters $a_1, a_2, \ldots, a_m$. Then one minimizes

$$E = [g(x_0) - y_0]^2 + [g(x_1) - y_1]^2 + \ldots + [g(x_n) - y_n]^2$$

by solving the algebraic system

$$\frac{\partial E}{\partial a_1} = 0 , \qquad \frac{\partial E}{\partial a_2} = 0 , \qquad \ldots , \qquad \frac{\partial E}{\partial a_m} = 0$$

to yield the desired values of $a_1, a_2, \ldots, a_m$.

      Let us now consider a second example, which will reveal another subtle aspect of the method of least squares.

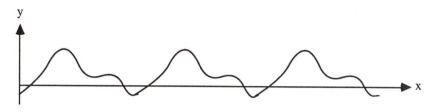

Figure 2.7

EXAMPLE. On an oscilloscope, periodic behavior is observed in the form shown in Figure 2.7. We are confronted now with the basic "decision problem" in the least squares method. We have to choose the function y=g(x) to fit the data. Most applied researchers who use the least squares method use the linear fit

$$y = a_1 + a_2 x$$

without saying so explicitly. It would be foolish to use the linear fit, or the quadratic fit

$$y = a_1 + a_2 x + a_3 x^2,$$

or the cubic fit

$$y = a_1 + a_2 x + a_3 x^2 + a_4 x^3$$

for the data shown in the figure. Indeed, the choice of the function to be used must depend on some knowledge of the behavior of the phenomenon under observation. For the oscilloscope function, it would be more reasonable, for example, to try first a periodic function like

(2.53) $y = a_1 \sin(x) + a_2 \cos(x)$.

If one actually carries out measurements and determines, approximately, that the points $(0,-0.9)$, $(\pi/4,1.5)$, $(\pi/2,3.1)$, $(3\pi/4,3)$, $(\pi,1.1)$ are on the graph shown in Figure 2.7, then the least squares fit with (2.52) yields

$$E = [a_1\sin(0)+a_2\cos(0)+0.9]^2+[a_1\sin(\pi/4)+a_2\cos(\pi/4)-1.5]^2$$
$$+[a_1\sin(\pi/2)+a_2\cos(\pi/2)-3.1]^2+[a_1\sin(3\pi/4)+a_2\cos(3\pi/4)-3]^2$$
$$+[a_1\sin(\pi)+a_2\cos(\pi)-1.1]^2 .$$

In this case the system

$$\frac{\partial E}{\partial a_1} = 0 , \qquad\qquad \frac{\partial E}{\partial a_2} = 0$$

implies

$$2a_1 = \frac{4.5}{2}\sqrt{2}+3.1$$

$$3a_2 = -2-\frac{1.5}{2}\sqrt{2} .$$

One then has, approximately, $a_1=3.14$, $a_2=-1.02$, and the resulting least squares periodic fit is

$$y = 3.14\sin(x)-1.02\cos(x) .$$

EXERCISES

Basic Exercises

1. Let [0,2] be divided into ten equal parts by the R_{11} set $x_i=(2i)/10$, $i=0,1,2,\ldots,10$. Draw the graph of the discrete function $y_i=f(x_i)$, $x_i \in R_{11}$, for each of the following.

(a) $f(x_i) = x_i$,   (b) $f(x_i) = x_i^2$,

(c) $f(x_i) = x_i^3-1$,   (d) $f(x_i) = \sin(\pi x_i)$,

(e) $f(x_i) = e^{x_i}$,   (f) $y_0 = y_1 = y_3 = y_5 = y_7 = y_8 = y_{10} = 2$,

$\qquad\qquad\qquad\qquad\qquad\qquad$ $y_2=y_4=y_6=-2$, $y_9=0$.

2. Let [0,2] be divided into 1000 equal parts by the R_{1001} set $x_i=(2i)/1000$, $i=0,1,2,\ldots,1000$. Plot the discrete function

$$y_i = \frac{(5x_i-0.5)(12x_i-1.2)}{(x_i^2-9)(x_i+1)}, \quad x_i \in R_{1001}.$$

3. For each of the following discrete functions, find and graph the piecewise linear interpolating function $L(x)$.

(a) $y_i = x_i^3$ on the R_5 set $x_i=2i$, $i=0,1,2,3,4$

(b) $y_i = x_i^3$ on the R_7 set $x_i=i/10$, $i=0,1,2,3,4,5,6$

(c) $y_i = \cos(\pi x_i)$ on the R_{11} set $x_i=i/10$, $i=0,1,2,\ldots,10$

(d) $f(x_i)$ is defined on the R_5 set $x_i=i$, $i=0,1,2,3,4$, by
$f(x_0)=8$, $f(x_1)=-3$, $f(x_2)=2$, $f(x_3)=0$, $f(x_4)=0$

(e) $f(x_i)$ is defined on the R_7 set $x_i=i$, $i=0,1,2,3,4,5,6$, by
$y_0=0$, $y_1=2$, $y_2=3$, $y_3=5$, $y_4=3$, $y_5=0$, $y_6=-1$.

4. For each of the discrete functions in Exercise 3, find and graph the piecewise parabolic interpolating function $P(x)$.

5. Assuming that $y_0'=0$, find the piecewise cubic spline function $S(x)$ of form (2.18) for each discrete function in Exercise 3. Show that $S'(x)$ exists and is continuous at all interior grid points of the form x_i, where i takes on even values only.

6. Assuming $z_0=z_4=0$, determine the piecewise cubic spline function $S(x)$ of form (2.26) for each discrete function in Exercise 3.

7. On the R_3 set $x_0=0$, $x_1=1$, $x_2=2$, a discrete function is defined by $f(0)=1$, $f(1)=2$, $f(2)=5$. Determine the Lagrange form of the parabolic interpolating function for the given discrete function.

8. On the R_7 set $x_i=i$, $i=0,1,2,3,4,5,6$, a discrete function is defined by $f(0)=0$, $f(1)=1$, $f(2)=2$, $f(3)=4$, $f(4)=3$, $f(5)=0$, $f(6)=1$. Determine the Lagrange form of the unique polynomial of degree six for the given discrete function.

9. By least squares, fit the points (0,0), (1,2), (2,4), (3,5), (4,6.1) with a straight line $y=a_1+a_2x$. Plot both the data points and the linear least square fit.

10. By least squares, fit the points (−3,−0.72), (−2,−0.01), (−1,0.50), (0,0.82), (1,0.89), (2,0.81), (3,0.50) with a parabola $y=a_1+a_2x+a_3x^2$. Plot both the data points and the resulting quadratic least square fit.

11. By least squares, fit the points (0,1), (1,0), (2,−7), (3,−26) with a polynomial of the form $y=a_1+a_2x+a_3x^2+a_4x^3$ and show that the result is identical with the cubical polynomial through the given points.

12. A discrete function of period 2π is given by the data set

x	0	$\pi/4$	$\pi/2$	$3\pi/4$	π	$5\pi/4$	$3\pi/2$	$7\pi/4$	2π
y	−6	2	5	−1	−2	1	3	−5	−6 .

By least squares, find a fit of the form

$$y = a_1\sin(x)+a_2\cos(x)+a_3\sin(2x)+a_4\cos(2x) .$$

13. By least squares, fit the function $y=a_1e^x+a_2e^{2x}$ to the data

x	0	1	2	3
y	−1.1	−7.5	−22.0	−60.1 .

14. Given a discrete function $y_i=f(x_i)$, i=0,1,2,...,n, develop the formulas for least square fit of the form $y=a_1+a_2x$, where the ε_i are the perpendicular distances from the (x_i,y_i) to the line.

15. Chebyschev approximation utilizes special polynomials $C_n(x)$ of degree n in x which are defined by

$$C_n(x) = \cos[n \; \arccos(x)] , \qquad n=0,1,2,3,... .$$

Show that $C_0(x)=1$, $C_1(x)=x$, $C_2(x)=2x^2-1$, $C_3(x)=4x^3-3x$, $C_4(x)=8x^4-8x^2+1$. Then prove that

$$C_{n+1}(x) = 2xC_n(x)-C_{n-1}(x), \qquad n=1,2,3,\dots .$$

16. Using least squares, fit each of $\alpha=ARe^{-BR}$, $\alpha=ARe^{-BR^2}$, $\alpha=Ae^{-B(R-C)^2}$ to the data set

α	.02802	.03040	.03247	.02850	.02900	.02497	.02447	.02856	.02566
R	2910	1860	1380	1100	910	780	680	600	540

Supplementary Exercises

17. Find the polynomial $p(x)$ of degree less than or equal to 3 which minimizes

$$\int_{-1}^{1} [e^x-p(x)]^2dx .$$

18. Using least squares, fit the following data set with a straight line:

x	0.5	1.0	1.5	2.0	2.5	3.0
y	0.31	0.82	1.29	1.85	2.51	3.02

19. Using least squares, fit the data set

x	0.0	0.1	0.2	0.3	0.4	0.5	0.6	0.7	0.8	0.9
y	13.195	3.230	3.253	3.261	3.252	3.228	3.181	3.127	3.059	2.976

with $y=a+bx+cx^2$.

20. Using least squares, fit the following data set with a straight line:

x	1	2	3	4	5	6	7	8	9	10	11	12	13	14	15
y	1.7	-.5	.3	.9	1.1	-.7	-.6	.7	.4	.3	.1	-.2	.5	-.3	.2 .

21. Using least squares, fit the data set

x	0	2	4	6	8	10	12	14	16
y	2.0	1.4	0.9	1.0	0.7	0.5	0.3	0.3	0.1

with $y = ae^{bx}$.

22. Using least squares, fit the data set

x	1	2	3	4
y	7	11	17	27

with $y = ae^{bx}$.

23. Find the unique cubic polynomial through the points $(1,-7)$, $(3,5)$, $(4,8)$, $(6,14)$.

24. Find the unique polynomial of degree 5 through the points $(-2,-27)$, $(-1,2)$, $(0,1)$, $(1,0)$, $(2,29)$, $(4,1017)$.

25. Using least squares, fit the data set

x	0.78	1.56	2.34	3.12	3.81
y	2.50	1.20	1.12	2.25	4.28

with $y = a + bx + cx^2$.

26. By least squares, fit the data set

$$\begin{array}{c|cccc} x & 2.2 & 2.7 & 3.5 & 4.0 \\ \hline y & 65 & 60 & 53 & 50 \end{array}$$

with $y=ax^b$.

27. In analogy with Lagrange's interpolation formula, the Hermite periodic interpolation formula is

$$y = \frac{\sin(x-x_1)\sin(x-x_2)\ldots\sin(x-x_k)}{\sin(x_0-x_1)\sin(x_0-x_2)\ldots\sin(x_0-x_k)}y_0$$

$$+\frac{\sin(x-x_0)\sin(x-x_2)\ldots\sin(x-x_k)}{\sin(x_1-x_0)\sin(x_1-x_2)\ldots\sin(x_1-x_k)}y_1$$

.
.
.

$$+\frac{\sin(x-x_0)\sin(x-x_1)\ldots\sin(x-x_{k-1})}{\sin(x_k-x_0)\sin(x_k-x_1)\ldots\sin(x_k-x_{k-1})}y_k.$$

Using this formula with k=4, approximate y at x=0.65 (radians, of course) from the data set

$$\begin{array}{c|ccccc} x & 0.4 & 0.5 & 0.6 & 0.7 & 0.8 \\ \hline y & 0.098 & 0.009 & -0.079 & -0.158 & -0.219 \end{array}.$$

3

Approximate Integration and Differentiation

3.1 INTRODUCTION

The basic rule for finding the constant value of the definite integral

$$(3.1) \qquad \int_a^b f(x)dx$$

is as follows. Given a function $f(x)$ which is continuous on $[a,b]$, find an antiderivative $F(x)$ of $f(x)$, that is, a function which satisfies $F'(x) \equiv f(x)$, and then apply the formula

$$(3.2) \qquad \int_a^b f(x)dx = F(b)-F(a) \ .$$

Unfortunately, too often this rule **cannot** be applied, even though every continuous function $f(x)$ has an antiderivative. To make this possibly ambiguous point clear, consider $f(x)=x^2$, $1 \leq x \leq 2$. Suppose we wish to determine the constant value of

$$\int_1^2 x^2 dx \ .$$

Then,

$$(3.3) \qquad F(x) = \int_0^x t^2 dt$$

is an antiderivative of f(x), since $F'(x) \equiv x^2$, by the Fundamental Theorem of Calculus. Unfortunately, (3.3) is of no practical value when applied in (3.2), since it yields

$$\int_1^2 x^2 dx = \left(\int_0^x t^2 dt \right)\Big|_1^2 = \int_0^2 t^2 dt - \int_0^1 t^2 dt = \int_1^2 t^2 dt \ .$$

On the other hand, if one uses the antiderivative

(3.4) $F(x) = \dfrac{x^3}{3}$,

then (3.2) yields

$$\int_1^2 x^2 dx = \frac{x^3}{3} \Big|_1^2 = \frac{8}{3} - \frac{1}{3} = \frac{7}{3},$$

and the constant value of the integral has been found.

A theoretical antiderivative of the form (3.3) is useful for theoretical purposes. A practical antiderivative of the form (3.4) is *essential* for practical purposes. The theoretical antiderivative always exists. The practical one *rarely* exists, though the opposite impression is given too often in elementary books. Indeed, for any arbitrary continuous function f(x), the probability is *zero* that a practical antiderivative can be produced.

To illustrate the scientific difficulties which can result, consider the particular integral

(3.5) $\int_a^b e^{-x^2} dx$.

The graph of the function $y = e^{-x^2}$ is shown in Figure 3.1 and is called a *normal curve*. It is also described as a *bell shaped* curve. Functions like this are most common in analyzing experimental data when large populations are sampled or tested. The particular integral (3.5) represents that portion of a population between x=a and x=b. Unfortunately, it can be proved that e^{-x^2} does *not* have a practical antiderivative, so that (3.5) cannot be evaluated using (3.2).

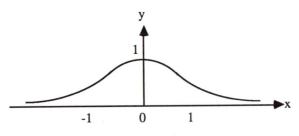

Figure 3.1.

For the reasons given above, we turn now to numerical methods for approximating the value of a given definite integral (3.1). Generally speaking, our approach will be to approximate $f(x)$ on $[a,b]$ by a function $g(x)$ which can be integrated practically, and then use the approximation

$$(3.6) \qquad \int_a^b f(x)dx \approx \int_a^b g(x)dx \ ,$$

where the symbol $\approx$ will be utilized occasionally to emphasize a general formula for approximation. The theory and methods of Chapter II will be used extensively in the construction of $g(x)$.

3.2 THE TRAPEZOIDAL RULE

Let $f(x)$ be continuous on $[a,b]$. Subdivide $[a,b]$ into n equal parts, each of length h, by the R_{n+1} set $a=x_0<x_1<x_2<...<x_n=b$ and let $y_i=f(x_i)$, $i=0,1,2,...,n$. Let $g(x)$ be the piecewise linear interpolating function $L(x)$ for the discrete function $y_i=f(x_i)$, $i=0,1,2,...,n$. Then (3.6) is called the *trapezoidal rule*.

The formulas for the trapezoidal rule can be developed easily as follows. The equation of the straight line determined by (x_0,y_0) and (x_1,y_1) in Figure 3.2 is

$$(3.7) \qquad y = y_0 + \frac{y_1-y_0}{h}(x-x_0) \ .$$

Using (3.7) for $g(x)$ yields

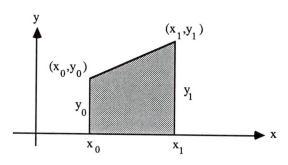

Figure 3.2

(3.8) $\displaystyle\int_{x_0}^{x_1} g(x)dx = \left[y_0 x + \frac{y_1 - y_0}{h}\frac{(x-x_0)^2}{2}\right]\Big|_{x_0}^{x_1} = \frac{h(y_0 + y_1)}{2}$.

If, as shown in Figure 3.2, both y_0 and y_1 have the same sign, then (3.8) is the shaded area of the trapezoid, from which follows the name *trapezoidal rule*.

In general, with f(x) and L(x) as shown in Figure 3.3, it follows from (3.8) that

$$\int_a^b g(x)dx = \int_a^b L(x)dx = \sum_{i=0}^{n-1} \int_{x_i}^{x_{i+1}} L(x)dx = \sum_{i=0}^{n-1}\frac{h(y_i + y_{i+1})}{2} = \frac{h}{2}\sum_{i=0}^{n-1}[f(x_i) + f(x_{i+1})] \ .$$

Thus, on [a,b], the trapezoidal rule takes the form

(3.9) $\displaystyle\int_a^b f(x)dx \approx \frac{h}{2}\sum_{i=0}^{n-1}[f(x_i) + f(x_{i+1})]$.

For programming purposes, however, a form more convenient than (3.9) can be derived as follows:

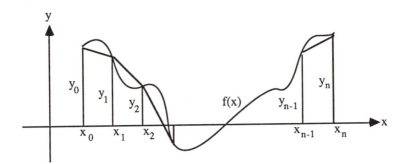

Figure 3.3

$$\int_a^b f(x)dx \approx \frac{h}{2}\sum_{i=0}^{n-1}(y_i+y_{i+1}) = \frac{h}{2}[(y_0+y_1)+(y_1+y_2)+(y_2+y_3)+\ldots+(y_{n-1}+y_n)]$$

$$= \frac{h}{2}[y_0+2y_1+2y_2+2y_3+\ldots+2y_{n-1}+y_n],$$

or, in concise notation,

$$\int_a^b f(x)dx \approx \frac{h}{2}\left[y_0+2\left(\sum_{i=1}^{n-1}y_i\right)+y_n\right].$$

EXAMPLE 1. Let us approximate

(3.10) $$\int_0^{\pi/2} \sin(x)dx$$

by the trapezoidal rule. Of course there is no need to approximate this integral, since it can be evaluated exactly. Indeed, the exact value of the integral (3.10) is 1. We continue, however, for illustrative purposes only. Hence, divide $[0,\pi/2]$ into four equal parts by the R_5 set $x_0=0$, $x_1=\pi/8$, $x_2=\pi/4$, $x_3=3\pi/8$, $x_4=\pi/2$. Then, to five decimal places, $y_0=\sin(0)=0.00000$, $y_1=\sin(\pi/8)=0.38268$, $y_2=\sin(\pi/4)=0.70711$, $y_3=\sin(3\pi/8)=0.92388$, $y_4=\sin(\pi/2)=1.00000$. Thus, the trapezoidal rule yields the approximation

(3.11) $\int_0^{\pi/2} \sin(x)dx \approx \frac{\pi}{16} [0+2(0.38268+0.70711+0.92388)+1] = 0.98712$.

EXAMPLE 2. Let us approximate

$$\int_0^3 e^{-x^2}dx$$

by the trapezoidal rule. For this integral divide [0,3] into three equal parts by the R_4 set $x_0=0$, $x_1=1$, $x_2=2$, $x_3=3$. Then, $y_0=1$, $y_1=e^{-1}$, $y_2=e^{-4}$, $y_3=e^{-9}$, which results in the six decimal place approximation

$$\int_0^3 e^{-x^2}dx \approx \frac{1}{2}[1+2(0.367879+0.018316)+0.000123] = 0.886257 .$$

Example 2 above, presents us with the immediate problem of determining the error in the result. The following theorems are of value for this purpose.

THEOREM 3.1. Let **f(x) be continuous and have continuous, and hence bounded, derivatives through the second order on [a,b]. In applying the trapezoidal rule to approximate (3.1), let the error E_i on $[x_i,x_{i+1}]$ be defined by**

(3.12) $E_i = \int_{x_i}^{x_{i+1}} f(x)dx - \frac{h}{2}[f(x_i) + f(x_{i+1})]$.

Then

(3.13) $|E_i| \leq \frac{h^3}{12} \max_{[x_i,x_{i+1}]} |f''(x)|$.

PROOF. For convenience, let F be an antiderivative of f. Then, (3.12) can be written in the form

(3.14) $E_i = F(x_i+h)-F(x_i) - \frac{h}{2}[f(x_i+h)+f(x_i)]$.

In order to give a *short* proof of the theorem, we resort now to the introduction of a "trick" function $G(t)$, which is defined as follows:

(3.15) $G(t) = F(x_i+t)-F(x_i) - \frac{t}{2}[f(x_i+t)+f(x_i)]-\frac{t^3}{h^3}E_i$.

Then, $G(0)=G(h)=0$, so that, by Rolle's theorem, there exists a value α, $0<\alpha<h$, such that $G'(\alpha)=0$. But,

$$G'(t) = \frac{1}{2}\left[f(x_i+t)-f(x_i) - t\frac{df(x_i+t)}{dt}\right] - 3\frac{t^2}{h^3}E_i ,$$

from which it follows that $G'(0)=0$ also. Thus,

$$G'(0) = G'(\alpha) = 0 , \qquad\qquad 0<\alpha<h,$$

and Rolle's theorem can be applied a second time to yield $G''(\mu)=0$, where $0<\mu<\alpha$. Thus,

$$-\frac{\mu}{2}\frac{d^2f(x_i+t)}{dt^2}\Big|_{t=\mu} - 6\frac{\mu}{h^3}E_i = 0 ,$$

which implies

$$E_i = -\frac{h^3}{12}\frac{d^2f(x_i+t)}{dt^2}\Big|_{t=\mu} .$$

But, for $x=x_i+t$,

$$\frac{df(x_i+t)}{dt} = \frac{df(x)}{dx} = f'(x) ,$$

$$\frac{d^2f(x_i+t)}{dt^2} = \frac{d^2f(x)}{dx^2} = f''(x) ,$$

so that

(3.16) $\qquad E_i = -\frac{h^3}{12} f''(\xi_i) , \qquad\qquad x_i < \xi_i < x_{i+1} .$

Finally, (3.13) follows directly from (3.16).

THEOREM 3.2. **Under the assumptions of Theorem 3.1, let E, the total error, be given by**

$$E = E_0 + E_1 + E_2 + ... + E_{n-1} .$$

Then,

$$\lim_{h \to 0} E = 0 .$$

PROOF. With the aid of (3.16), one has

$$E = \sum_{i=0}^{n-1} E_i = -\frac{h^3}{12} \sum_{i=0}^{n-1} f''(\xi_i) = -\frac{h^2}{12} \sum_{i=0}^{n-1} hf''(\xi_i) .$$

Hence,

$$\lim_{h \to 0} E = -[\lim_{h \to 0} \frac{h^2}{12}] \int_a^b f''(x)dx = 0 .$$

THEOREM 3.3. **Let E be the total error made in approximating (3.1) by the trapezoidal rule. Then, under the assumptions of Theorem 3.1,**

(3.17) $\qquad |E| \le \frac{h^3}{12} \sum_{i=0}^{n-1} (\max_{[x_i, x_{i+1}]} |f''(x)|) .$

PROOF. Since $E=E_0+E_1+E_2+...+E_{n-1}$, by (3.13) one has

$$|E| \leq \sum_{i=0}^{n-1} |E_i| = \frac{h^3}{12} \sum_{i=0}^{n-1} |f''(\xi_i)| \leq \frac{h^3}{12} \sum_{i=0}^{n-1} (\max_{[x_i,x_{i+1}]} |f''(x)|)$$

which is (3.17).

The quantity on the right hand side of inequality (3.17) is called an ***error bound***. It is not the error itself, but it is a positive number which the absolute value of the error cannot exceed.

EXAMPLE 1. Let us find an error bound for the approximation (3.11) and show that it is consistent with the exact error. First note that

(3.18) $|E| = 1 - 0.98712 = 0.01288$.

From (3.13), however, one has

$$|E| \leq \frac{(\pi/8)^3}{12} \sum_{i=0}^{3} (\max_{[x_i,x_{i+1}]} |\sin(x)|) = \frac{(\pi/8)^3}{12} [\sin(\pi/8)+\sin(\pi/4)+\sin(3\pi/8)+1]$$

$$< \frac{\pi^3}{6144} (0.39+0.71+0.93+1) = \frac{3.03\pi^3}{6144} < 0.01616 ,$$

which is completely consistent with (3.18).

Inequality (3.17) also can be applied to determine the number of subdivisions to be used in the trapezoidal rule in order to assure, *a priori,* a desired accuracy.

EXAMPLE 2. Let us determine into how many subdivisions one can divide $[0,\pi/2]$ so that application of the trapezoidal rule to (3.10) will yield a result that is correct to 10^{-3}. For this purpose, let the subdivision consist of n equal parts. Then, since $|\sin(x)| \leq 1$,

$$|E| \leq \frac{h^3}{12} \sum_{i=0}^{n-1} (\max_{[x_i,x_{i+1}]} |\sin(x)|) < \frac{h^3}{12} \sum_{i=0}^{n-1} 1 = \frac{nh^3}{12} .$$

But, $nh=\pi/2$, so that one wishes

$$\frac{nh^3}{12} = \frac{\pi/2}{12}\left(\frac{\pi}{2n}\right)^2 < 10^{-3},$$

which implies

$$n > \sqrt{\frac{(10\pi)^3}{96}}.$$

However, since

$$\sqrt{\frac{(10\pi)^3}{96}} < \sqrt{\frac{10^3(3.1416)^3}{96}} < 17.98,$$

then any choice $n \geq 18$ would suffice.

3.3 SIMPSON'S RULE

In general, for about the same amount of computation required by the trapezoidal rule, one can apply Simpson's rule and greatly improve the accuracy of the approximation. Simpson's rule is defined as follows.

Given $f(x)$ continuous on [a,b], subdivide [a,b] into n equal parts, where n is *even*, by the R_{n+1} set $a=x_0<x_1<x_2<...<x_n=b$ and let $y_i=f(x_i)$, $i=0,1,2,...,n$. Let $g(x)$ be the piecewise parabolic interpolating function $P(x)$ of the discrete function $y_i=f(x_i)$, $i=0,1,2,...,n$. Then (3.6) is called *Simpson's rule*.

The precise formulas for Simpson's rule can be developed as follows. Consider first a particular arrangement of the three values x_0, x_1, x_2 such that $x_0=-h$, $x_1=0$, $x_2=h$. Then, from (2.6), the parabolic function through (x_0,y_0), (x_1,y_1), (x_2,y_2), is

$$g(x) = y_1 + \frac{y_2-y_0}{2h}x + \frac{y_2-2y_1+y_0}{2h^2}x^2.$$

Hence,

(3.19) $\int_{x_0}^{x_2} g(x)dx = \frac{h}{3}(y_0 + 4y_1 + y_2)$.

But, the right hand side of (3.19) is independent of x_0, x_1, x_2, that is, the integral is independent of translation in the x direction, so no loss of generality results from the particular choice $x_1 = 0$.

Extending the application of (3.19) to the entire interval [a,b] implies

(3.20) $\int_a^b f(x)dx \approx \frac{h}{3}[(y_0 + 4y_1 + y_2) + (y_2 + 4y_3 + y_4) + \ldots + (y_{n-2} + 4y_{n-1} + y_n)]$

or

(3.21) $\int_a^b f(x)dx \approx \frac{h}{3}[y_0 + 4y_1 + 2y_2 + 4y_3 + 2y_4 + \ldots + 2y_{n-2} + 4y_{n-1} + y_n]$.

Formula (3.20) can be written, for theoretical purposes, as

(3.22) $\int_a^b f(x)dx \approx \frac{h}{3} \sum_{i=0}^{(n/2)-1} (y_{2i} + 4y_{2i+1} + y_{2i+2})$.

EXAMPLE. Let us again approximate (3.10), but by Simpson's rule. Dividing $[0, \pi/2]$ into four equal parts by the R_5 set $x_0 = 0$, $x_1 = \pi/8$, $x_2 = \pi/4$, $x_3 = 3\pi/8$, $x_4 = \pi/2$, we have $y_0 = 0$, $y_1 = 0.38268$, $y_2 = 0.70711$, $y_3 = 0.92388$, $y_4 = 1$. Simpson's rule then yields

$$\int_0^{\pi/2} \sin(x)dx \approx \frac{\pi}{24}[0 + 4(0.38268) + 2(0.70711) + 4(0.92388) + 1] = 1.0001 ,$$

which is far more accurate than the result (3.11), which was obtained using the trapezoidal rule on the very same R_5 set.

The improvement in accuracy, in general, of Simpson's rule when compared to the trapezoidal rule is reflected by the following analogue of Theorem 3.3.

THEOREM 3.4. Let f(x) be continuous and have continuous derivatives through the fourth order on [a,b]. In applying Simpson's rule to approximate (3.1), let the error E_{2i}, i=0,1,2,...,(n/2)–1, be defined by

$$E_{2i} = \int_{x_{2i}}^{x_{2i+2}} f(x)dx \; - \frac{h[f(x_{2i}) + 4f(x_{2i+1}) + f(x_{2i+2})]}{3}.$$

Then the total error E made in approximating the integral satisfies the inequality

$$(3.23) \qquad |E| \leq \frac{h^5}{90} \sum_{i=0}^{(n/2)-1} \left(\max_{[x_{2i},x_{2i+2}]} \left| \frac{d^4f(x)}{dx^4} \right| \right).$$

PROOF. The proof follows in a fashion completely analogous to that of Theorems 3.1-3.3, if one introduces the "trick" function

$$G(t) = F(x_{2i+1}+t) - F(x_{2i+1}-t) - \frac{t}{3}[f(x_{2i+1}+t) + 4f(x_{2i+1}) + f(x_{2i+1}-t)] - \frac{t^5}{h^5}E_{2i}.$$

Of course, under the assumptions of Theorem 3.4, one has, as in Theorem 3.2, that

$$\lim_{h \to 0} E = 0 \,,$$

with the proof being entirely similar to that of Theorem 3.2.

Intuitively, error bounds (3.17) and (3.23) reveal why Simpson's rule is, in general, more accurate than the trapezoidal rule, for, if h is small, then $h^5/90$ is *much* smaller than $h^3/12$. Indeed, Simpson's rule has proved to be of such practical value that it is usually recommended as the first approximation formula to apply in any problem requiring numerical integration.

3.4 ROMBERG INTEGRATION

When applicable, a very popular and useful higher order method for numerical integration is Romberg integration. It utilizes the results of the trapezoidal rule for two different grid sizes and then projects what the exact result, that is, the result for zero grid size, should be.

In order to develop this type of integration, recall first (3.16). Then

$$E = \sum_{i=0}^{n-1} E_i = -\frac{h^3}{12} \sum_{i=0}^{n-1} f''(\xi_i) .$$

Now, let M denote the mean value of $f''(\xi_i)$, i=0,1,2,...,n–1, that is

$$(3.24) \qquad M = \frac{1}{n} \sum_{i=0}^{n-1} f''(\xi_i) ,$$

so that

$$E = -\frac{h^3}{12} nM = -\frac{h^3}{12} \frac{b-a}{h} M = -\frac{h^2(b-a)}{12} M .$$

Next, assume that for grid size h_1, I_1 is the approximation by the trapezoidal rule and that the error is E_1. Thus,

$$E_1 = -\frac{h_1^2(b-a)}{12} M_1 .$$

For grid size $h_2 \neq h_1$, let I_2 be the approximation and let E_2 be the error, so that

$$E_2 = -\frac{h_2^2(b-a)}{12} M_2 .$$

Hence, if I is the exact integral, then

$$(3.25) \qquad I = I_1 + E_1 = I_1 - \frac{h_1^2(b-a)}{12} M_1 ,$$

$$(3.26) \qquad I = I_2 + E_2 = I_2 - \frac{h_2^2(b-a)}{12} M_2 ,$$

Now, if M_1 and M_2 are approximately equal, then (3.25) and (3.26) imply

$$\frac{I-I_1}{h_1^2} = \frac{I-I_2}{h_2^2},$$

so that, finally,

(3.27) $$I = \frac{I_1 h_2^2 - I_2 h_1^2}{h_2^2 - h_1^2}.$$

The result (3.27), which approximates the exact solution from numerical solutions with grid sizes h_1 and h_2, is called "extrapolation to zero grid size".

EXAMPLE. Consider the integral (3.10). Division of $[0,\pi/2]$ into four equal parts and application of the trapezoidal rule yields, as shown in (3.11), $I_1 = 0.98712$. Division of $[0,\pi/2]$ into eight equal parts yields $I_2 = 0.99678$. Hence, using five decimal place accuracy, we have

$$I = \frac{(0.98712)(\pi/16)^2 - (0.99678)(\pi/8)^2}{(\pi/16)^2 - (\pi/8)^2} = \frac{-0.11566}{-0.11566} = 1.00000,$$

which is the exact answer.

In practice, the assumption that M_1 and M_2 are approximately equal may be difficult to verify. But, one can expect it to be valid if h_1 and h_2 are close to zero, and if $f''(x)$ is continuous on $[a,b]$. In fact, from (3.24), as $h \to 0$, we get

$$M = \frac{1}{n} \sum_{i=0}^{n-1} f''(\xi_i) = \frac{1}{b-a} \sum_{i=0}^{n-1} h f''(\xi_i) \to \frac{1}{b-a} \int_a^b f''(x) dx = C \quad \text{(constant)}.$$

Thus, as $h \to 0$, one can expect $M_1 \to C$ and $M_2 \to C$, so that M_1 and M_2 will be, approximately, equal.

3.5 Remarks About Numerical Integration

There exist numerous other techniques for numerical integration [see, e.g., P.J.Davis and P.Rabinowitz (1967)]. These include Gaussian, Chebyschev, Newton-Cotes, and Monte Carlo integration. Better types of error bounds than those given in section 3.2 and 3.3 have been and are being developed using the methods of functional analysis [M.Golomb and H.F.Weinberger (1959)]. It should be noted also that almost no major techniques are available for the approximation of multiple integrals except those of Monte Carlo methods, and these have been highly restrictive in their applicability.

3.6 Numerical Differentiation

One must develop methods of numerical integration because it may be difficult, if not impossible, to find a practical antiderivative of a given continuous function. The situation is not at all the same with respect to differentiation. Given a function which is differentiable, one can always differentiate it readily. However, equations with derivatives, that is differential equations, are rarely solvable. Because of the importance of these equations in science and technology, we will want to solve them numerically, and, for this reason, we develop now the essential numerical differentiation formulas.

Intuitively, one could begin quite simply as follows. From the definition of a derivative, one has

$$f'(x) = \lim_{\Delta x \to 0} \frac{f(x+\Delta x)-f(x)}{\Delta x}.$$

However, on a computer, Δx cannot be made arbitrarily small, so that deleting $\lim_{\Delta x \to 0}$ in the above equation yields the approximation

$$f'(x) \approx \frac{f(x+\Delta x)-f(x)}{\Delta x}.$$

If $\Delta x > 0$, the approximation is called a forward difference approximation. If $\Delta x < 0$, it is called a backward difference approximation. However, because these and a large variety of other formulas for derivatives can be derived using a very general Taylor expansion approach, we will in this section develop and apply this more general approach.

It is worth noting, before proceeding, that Δx in the definition of a derivative can be positive or negative. However, the grid size h, as defined in Section 2.2, is always positive.

Throughout, let f(x) have continuous derivatives up to and including order four on interval [a,b]. Let [a,b] be divided into n equal parts by the R_{n+1} set $x_0, x_1, x_2, \ldots, x_n$. Let x_i

be a typical *interior* grid point, that is, a value from $x_1, x_2, \ldots, x_{n-1}$, and let us develop first three approximations for $f'(x_i)$.

Consider

(3.28) $f'(x_i) = \alpha f(x_i) + \beta f(x_{i+1})$,

where α and β are parameters. Inserting into (3.28) a finite Taylor expansion for $f(x_{i+1})$ about x_i yields

$$f'(x_i) = \alpha f(x_i) + \beta [f(x_i) + h f'(x_i) + \frac{h^2}{2} f''(\xi)] , \qquad x_i < \xi < x_{i+1},$$

from which one has

(3.29) $f'(x_i) = (\alpha + \beta) f(x_i) + \beta h f'(x_i) + \beta \frac{h^2}{2} f''(\xi)$.

Setting corresponding coefficients equal in (3.29) yields

$$\alpha + \beta = 0$$

$$\beta h = 1 ,$$

from which one has

$$\beta = \frac{1}{h} , \qquad \alpha = -\frac{1}{h} ,$$

so that (3.28) implies

(3.30) $f'(x_i) \approx \dfrac{f(x_{i+1}) - f(x_i)}{h}$, while (3.29) implies that the error in approximation (3.30) is

(3.31) $E = \dfrac{h}{2} f''(\xi)$, $x_i < \xi < x_{i+1}$.

Since approximation (3.30) utilizes the two points x_i and x_{i+1}, it is called a two-point, *forward* difference approximation for $f'(x_i)$.

In a completely analogous fashion, it follows that a two-point, *backward* difference approximation for $f'(x_i)$ is

(3.32) $f'(x_i) \approx \dfrac{f(x_i) - f(x_{i-1})}{h}$,

and the error in (3.32) is

(3.33) $E = -\dfrac{h}{2} f''(\xi)$, $x_{i-1} < \xi < x_i$.

Next, let us develop a three-point, central difference approximation for $f'(x_i)$, that is, an approximation of the form

(3.34) $f'(x_i) = \alpha f(x_i) + \beta f(x_{i+1}) + \gamma f(x_{i-1})$,

Substitution of Taylor expansions into (3.34) implies that

(3.35) $f'(x_i) = \alpha f(x_i) + \beta[f(x_i) + hf'(x_i) + \dfrac{h^2}{2} f''(x_i) + \dfrac{h^3}{6} f'''(\xi_1)]$

$+ \gamma[f(x_i) - hf'(x_i) + \dfrac{h^2}{2} f''(x_i) - \dfrac{h^3}{6} f'''(\xi_2)]$,

where $x_i < \xi_1 < x_{i+1}$ and $x_{i-1} < \xi_2 < x_i$. Recombination of terms in (3.35) yields

(3.36) $\quad f'(x_i) = (\alpha+\beta+\gamma)f(x_i)+h(\beta-\gamma)f'(x_i)+\dfrac{h^2}{2}(\beta+\gamma)f''(x_i)$

$$+\dfrac{h^3}{6}[\beta f'''(\xi_1)-\gamma f'''(\xi_2)] \ .$$

From the system

$$\alpha+\beta+\gamma = 0$$

$$h(\beta-\gamma) = 1$$

$$\beta+\gamma = 0 \ ,$$

one finds

$$\alpha = 0 \ , \qquad \beta = \frac{1}{2h} \ , \qquad \gamma = -\frac{1}{2h} \ ,$$

so that the desired three-point approximation is

(3.37) $\quad f'(x_i) \approx \dfrac{f(x_{i+1})-f(x_{i-1})}{2h} \ ,$

and the error E in the approximation is

(3.38) $\quad E = \dfrac{h^2}{12}[f'''(\xi_1)+f'''(\xi_2)] \ .$

Of course, the three-point formula (3.37) is, in fact, only a two-point formula, so that, in the future, we will simply call it the **central** difference formula.

Finally, let us develop a single, very basic, formula for approximating $f''(x_i)$. For this purpose, let

(3.39) $\quad f''(x_i) = \alpha f(x_i)+\beta f(x_{i+1})+\gamma f(x_{i-1}) \ .$

Then, insertion of finite Taylor expansions through order four into (3.39) yields

$$f''(x_i) = (\alpha+\beta+\gamma)f(x_i)+h(\beta-\gamma)f'(x_i)+\frac{h^2}{2}(\beta+\gamma)f''(x_i)$$

$$+\frac{h^3}{6}(\beta-\gamma)f'''(x_i)+\frac{h^4}{24}[\beta f^{iv}(\xi_1)+\gamma f^{iv}(\xi_2)]$$

where $x_i<\xi_1<x_{i+1}$ and $x_{i-1}<\xi_2<x_i$. Solution of

$$\alpha+\beta+\gamma = 0$$

$$\beta-\gamma = 0$$

$$h^2(\beta+\gamma) = 2 ,$$

yields

$$\alpha = -\frac{2}{h^2}, \qquad \beta = \frac{1}{h^2}, \qquad \gamma = \frac{1}{h^2} .$$

Thus, a three-point, central difference approximation is

(3.40) $$f''(x_i) \approx \frac{f(x_{i+1})-2f(x_i)+f(x_{i-1})}{h^2} ,$$

and the error E in approximation (3.40) is

(3.41) $$E = \frac{h^2}{24}[f^{iv}(\xi_1)+f^{iv}(\xi_2)] .$$

　　　In our study of the numerical solution of differential equations, we will need only the four formulas (3.30), (3.32), (3.37) and (3.40). Note, finally, that (3.30) can be applied at x_0 when $f(x)$ is differentiable at $x=x_0$, while (3.32) can be applied at x_n when $f(x)$ is differentiable at $x=x_n$.

EXERCISES

Basic Exercises

1. Divide $[0,1]$ into six equal parts and, using six decimal places, approximate each of the following by the trapezoidal rule:

   (a) $\displaystyle\int_0^1 x^3 dx$ (b) $\displaystyle\int_0^1 \cos(\pi x)dx$ (c) $\displaystyle\int_0^1 \frac{1}{1+x}dx$.

2. Find both the exact error and, by Theorem 3.2, a bound for $|E|$ for each result of Exercise 1.

3. Divide $[0,1]$ into six equal parts and, using six decimal places, approximate

   $$\int_0^1 e^{-x^2}dx$$

   by the trapezoidal rule. Then determine an error bound for the result.

4. Using the trapezoidal rule, approximate, correct to one decimal place, each of the following.

   (a) $\displaystyle\int_0^1 x^2 dx$ (b) $\displaystyle\int_0^\pi x\cos(x)dx$ (c) $\displaystyle\int_0^3 e^{-x^2}dx$.

5. Divide $[0,1]$ into six equal parts and, using six decimal places, approximate each integral of Exercise 1 by Simpson's rule.

6. Complete the proof of Theorem 3.4.

7. Find both the exact error and, by Theorem 3.4, a bound for $|E|$ for each result of Exercise 5.

8. Using Simpson's rule, approximate, correct to four decimal places, each of the following

 (a) $\displaystyle\int_0^1 x^2 dx$ (b) $\displaystyle\int_0^\pi x\cos(x)dx$ (c) $\displaystyle\int_0^3 e^{-x^2}dx$.

9. Using Romberg integration, approximate each of the following with grid sizes h=0.1 and h=0.05. Then, evaluate each integral exactly and compare the results.

 (a) $\displaystyle\int_0^1 x^2 dx$ (b) $\displaystyle\int_0^\pi x\cos(x)dx$ (c) $\displaystyle\int_0^5 e^{-x}dx$.

10. Using Romberg integration with each of h=0.1 and h=0.05, approximate

$$\int_0^3 e^{-x^2}dx .$$

11. A linear function $f(x)=a_1+a_2x$ is known only approximately at x=0,1,2,3,4,5 by f(0)=4.1, f(1)=6.5, f(2)=9.0, f(3)=13.2, f(4)=17.1, f(5)=20.1. By least squares, fit the data and then approximate

$$\int_0^5 f(x)dx$$

by integrating the resulting least square fit.

12. Develop the three-point, forward difference formula:

$$f'(x_i) \approx \frac{-3f(x_i)+4f(x_{i+1})-f(x_{i+2})}{2h} ,$$

and show that the error in the approximation is

$$E = \frac{h^2}{3} f'''(\xi_1) - \frac{2}{3} h^2 f'''(\xi_2) \ ,$$

where, $x_i < \xi_1 < x_{i+1}$, $x_i < \xi_2 < x_{i+2}$. Then develop a three-point, backward formula for $f'(x_i)$ and a corresponding error bound.

13. Develop a five-point, central difference formula for $f''(x_i)$.

Supplementary Exercises

14. Using Simpson's rule, approximate π to four decimal places using the formula

$$\frac{\pi}{4} = \int_0^1 \frac{dx}{1+x^2} \ .$$

15. Approximate each of the following correct to four decimal places:

(a) $\displaystyle\int_{-1}^{1} \sqrt{(1-x^2)(2-x)} \ dx$ (b) $\displaystyle\int_0^1 \frac{\sin(x)}{x} dx$

(c) $\displaystyle\int_0^{\pi/2} \frac{dx}{\sin^2(x)+\frac{1}{4}\cos^2(x)}$ (d) $\displaystyle\int_0^{\pi/2} \sqrt{1-\frac{1}{2}\sin^2(x)} \ dx$

(e) $\displaystyle\int_0^1 e^{-x^2} dx$.

16. Approximate correct to two decimal places:

$$\int_0^1 \cos(x^2)dx .$$

17. Approximate correct to three decimal places:

$$\int_0^{0.5} \frac{x\,dx}{\cos(x)} .$$

18. Approximate correct to three decimal places:

$$\int_{-\infty}^{\infty} e^{-x^2}dx .$$

19. Using Romberg integration, approximate each of the following:

(a) $\int_0^1 \sqrt{x}\,\sin(x)\,dx$ (b) $\int_0^1 \frac{dx}{\sqrt{1+x^4}}$

(c) $\int_1^{\infty} e^{-x^2}dx .$

20. Approximate to four decimal places the total arc length of the ellipse whose equation is $x^2+2y^2=1$.

21. Consider the integral equation

$$\phi(x) = \frac{5}{6}x - \frac{1}{9} + \frac{1}{3}\int_0^1 (t+x)\phi(t)dt$$

for the unknown function ϕ on $0 \le x \le 1$. Approximate ϕ on the R_3 set $x_0=0$, $x_1=0.5$, $x_2=1.0$ by applying Simpson's rule to approximate the integral in the equation.

Compare your result with the exact solution $\phi=x$.

22. Generalize the trapezoidal rule to the case when [a,b] is subdivided into unequal parts.

23. Generalize the Simpson's rule to the case when [a,b] is subdivided into unequal parts.

24. Generalize the trapezoidal rule to double integrals.

25. Generalize Simpson's rule to double integrals.

4

Approximate Solution of Initial Value Problems for Ordinary Differential Equations

4.1 INTRODUCTION

One of the broad principles of science is that all things in Nature change with time. Some things change relatively slowly, like the erosion of rocks by wind and the movement of continents. Other things change relatively quickly, like the shape of a windblown cloud and the color of a chamelion when it moves from brown earth into green grass.

Since the fundamental mathematical concept which describes instantaneous change with respect to time is the derivative, it follows that equations which contain functions and their derivatives, that is differential equations, are the basic equations of all science.

Since velocity is the derivative of position with respect to time and acceleration is the derivative of velocity with respect to time, it is natural that equations which describe changes in position are second-order differential equations in position. Such equations are called dynamical equations and have been studied extensively since the development of calculus. The general second-order differential equation in one dependent variable and one independent variable is

$$(4.1) \qquad y'' = f(x,y,y') \, ,$$

in which, physically, y represents position and x represents time. We will be concerned with various problems connected with equation (4.1).

There are three basic problems of broad interest with respect to (4.1). The first is called the *initial value* problem. In it one is given three constants a, α, β, and is asked to find a continuous function y(x) on x≥a which satisfies (4.1) for x>a and also satisfies the beginning, or initial, conditions

(4.2) $y(a) = \alpha$, $y'(a) = \beta$.

Physically, α is an initial position and β is an initial velocity. The second problem is one in which no supplementary conditions are given and one has to find a *periodic* solution of (4.1). The third problem is called the ***boundary value*** problem. In it one is given four constants a, b, α, β, with a<b, and is asked to find a continuous function y(x) on [a,b] which satisfies (4.1) on a<x<b and also satisfies the boundary conditions y(a)=α, y(b)=β.

We will develop numerical techniques for each of the problems given above. In this chapter, we begin with the study of initial value problems, and, without loss of generality, we assume that a=0, so that all problems will begin at *zero time*.

Historically, problem (4.1)-(4.2) has been studied, primarily, by transforming it into a system of two first-order equations in two unknown functions and then solving the resulting system. This can always be done as follows. Let

(4.3) $y' = v$.

Then, (4.1) can be rewritten as

(4.4) $v' = f(x,y,v)$.

The initial data (4.2) then become

(4.5) $y(0) = \alpha$, $v(0) = \beta$.

Thus, (4.1)-(4.2) can be rewritten in the equivalent form

(4.6) $y' = v$, $y(0) = \alpha$

(4.7) $v' = f(x,y,v)$, $v(0) = \beta$,

which is a system of two first-order equations in two unknown functions y and v, with an initial condition for each. The mathematical value of transforming to a system is that one can then introduce the power of vector methodology and theory. From a general computer point of view, neither formulation (4.1)-(4.2) nor (4.6)-(4.7) is the more desirable. So, we will develop numerical techniques for both and we will begin with the system formulation.

4.2 EULER'S METHOD

We wish now to develop a numerical method for the system (4.6)-(4.7). For simplicity, let us begin with a study of *one* first-order equation in *one* unknown function with *one* initial value, and then show how all the resulting methodology and theory extend readily to system (4.6)-(4.7).

Consider, then, the problem of approximating a continuous function y(x) on x≥0 which satisfies the differential equation

(4.8) $y' = f(x,y)$, $x>0$,

and the initial condition

(4.9) $y(0) = \alpha$.

Using the simple forward difference approximation (3.30), L.Euler in 1768 developed an approximation for (4.8) which can be used as a numerical method, called *Euler's method*, in the following way. Since the computer cannot calculate forever, in (4.8), let us replace x>0 by 0<x≤L. The constant value L is usually determined by the physics of a dynamical phenomenon under study. If one is studying a phenomenon which is over after a short time period, then L can be chosen to be relatively small. If the phenomenon is over only after a lengthy time period, then L must be chosen to be relatively large. If the phenomenon seems to be endless, then one can choose L as large as his or her computer and available run time will allow. In any case, L is some constant, positive value. The interval [0,L] is then divided into n equal parts, each of length h, by the R_{n+1} set $x_i = ih$ i=0,1,2,...,n. Let $y(x_i) = y_i$ so that, by (4.9), y_0 is known and, in fact, $y_0 = \alpha$. Next, at each of the points $x_0, x_1, x_2, ..., x_{n-1}$, approximate the differential equation (4.8) by

(4.10) $\dfrac{y_{i+1} - y_i}{h} = f(x_i, y_i)$,   i=0,1,2,...,n–1,

or, in explicit recursion form,

(4.11) $y_{i+1} = y_i + hf(x_i, y_i)$,   i=0,1,2,...,n–1.

Then, beginning with

(4.12) $y_0 = \alpha$,

set i=0 in (4.11) and generate y_1. Knowing y_1, next set i=1 in (4.11) and generate y_2. Knowing y_2, set i=2 in (4.11) and generate y_3, and so forth, until, finally, y_n is generated. The resulting discrete function $y_0, y_1, y_2, \ldots, y_n$ is called the numerical solution.

EXAMPLE 1. Consider the initial value problem

(4.13) $y' = -y + x$, $y(0) = 1$,

and let us find a numerical solution on [0,1]. If we set n=5, then h=0.2, $x_0=0$, $x_1=0.2$, $x_2=0.4$, $x_3=0.6$, $x_4=0.8$, $x_5=1$, and the differential equation (4.13) is approximated by the difference equation

(4.14) $\dfrac{y_{i+1} - y_i}{0.2} = -y_i + x_i$,   i=0,1,2,3,4,

or, equivalently, by

(4.15) $y_{i+1} = (0.8)y_i + (0.2)x_i$,   i=0,1,2,3,4.

Since $y_0=1$, (4.15) yields, to three decimal places,

$$y_1 = (0.8)y_0 + (0.2)x_0 = (0.8)(1.000) + (0.2)(0.0) = 0.800$$
$$y_2 = (0.8)y_1 + (0.2)x_1 = (0.8)(0.800) + (0.2)(0.2) = 0.680$$
$$y_3 = (0.8)y_2 + (0.2)x_2 = (0.8)(0.680) + (0.2)(0.4) = 0.624$$
$$y_4 = (0.8)y_3 + (0.2)x_3 = (0.8)(0.624) + (0.2)(0.6) = 0.619$$
$$y_5 = (0.8)y_4 + (0.2)x_4 = (0.8)(0.619) + (0.2)(0.8) = 0.655 .$$

Thus, the numerical approximation with h=0.2 is

$$y(0) = 1$$
$$y(0.2) = 0.800$$
$$y(0.4) = 0.680$$

$$y(0.6) = 0.624$$
$$y(0.8) = 0.619$$
$$y(1) = 0.655 \ .$$

However, problem (4.13) is so simple that it can be solved exactly to yield

(4.16) $Y(x) = x-1+2e^{-x}$,

from which, to three decimal places, one has

$$Y(0) = 1$$
$$Y(0.2) = 0.837$$
$$Y(0.4) = 0.741$$
$$Y(0.6) = 0.698$$
$$Y(0.8) = 0.699$$
$$Y(1) = 0.736$$

Thus, even though h is relatively large, the numerical solution compares reasonably well on the grid points with the exact solution.

EXAMPLE 2. Let us reconsider (4.13) and see what happens if h is decreased to 0.1. This time n=10, the grid points are $x_0=0$, $x_1=0.1$, $x_2=0.2$, $x_3=0.3$, $x_4=0.4$, $x_5=0.5$, $x_6=0.6$, $x_7=0.7$, $x_8=0.8$, $x_9=0.9$, $x_{10}=1.0$, and the approximating difference equation is

(4.17) $\dfrac{y_{i+1}-y_i}{0.1} = -y_i+x_i$,   i=0,1,2,...,9,

or, equivalently and more conveniently,

$$y_{i+1} = (0.9)y_i+(0.1)x_i, \qquad\qquad i=0,1,2,...,9.$$

Calculating to three decimal places yields

$$y_0 = 1 \qquad\qquad y_1 = 0.900$$
$$y_2 = 0.820 \qquad\qquad y_3 = 0.758$$

$$y_4 = 0.712 \qquad y_5 = 0.681$$

$$y_6 = 0.663 \qquad y_7 = 0.657$$

$$y_8 = 0.661 \qquad y_9 = 0.675$$

$$y_{10} = 0.697 \ .$$

Thus, we have a more extensive approximation. Notice, also, in particular, that at the points whose abscissas are 0, 0.2, 0.4, 0.6 0.8, 1, the present numerical solution is a **better** approximation to the exact solution than was the one obtained in Example 1 with h=0.2.

To be of practical value, Euler's method and, indeed, all the methods to be developed later, must have a property which is implied heuristically in Example 2, above, that is, as h converges to zero, the numerical result should converge to the exact solution. Loosely speaking, this would imply that if one wants a highly accurate numerical result, then one need only choose h sufficiently small. It is necessary then, at the outset, that we prove such a theorem before continuing to more complex numerical techniques.

*4.3 CONVERGENCE OF EULER'S METHOD

In developing a convergence theorem, we will use the following results.

LEMMA 4.1. If the numbers $|E_i|$, i=0,1,2,...,n satisfy

$$(4.18) \qquad |E_{i+1}| \le A|E_i| + B \ , \qquad i=0,1,2,...,n-1,$$

where A and B are nonnegative constants and A≠1, then

$$(4.19) \qquad |E_i| \le A^i |E_0| + \frac{A^i - 1}{A - 1} B \ , \qquad i=1,2,...,n.$$

PROOF. For i=0, (4.18) yields

$$|E_1| \le A |E_0| + B = A^1 |E_0| + \frac{A^1 - 1}{A - 1} B \ .$$

Thus, (4.19) is certainly valid for i=1. The proof is now completed by mathematical induction. Assume (4.19) is valid for any i≥1, that is,

(4.20) $|E_i| \leq A^i |E_0| + \dfrac{A^i-1}{A-1} B$.

Then we must prove that

$$|E_{i+1}| \leq A^{i+1} |E_0| + \frac{A^{i+1}-1}{A-1} B .$$

From (4.18),

(4.21) $|E_{i+1}| \leq A|E_i| + B$,

so that (4.20) and (4.21) imply

$$|E_{i+1}| \leq A[A^i |E_0| + \frac{A^i-1}{A-1} B] + B = A^{i+1} |E_0| + \frac{A^{i+1}-1}{A-1} B ,$$

and the proof is complete.

Note that the value of Lemma 4.1 is as follows. If each term of a sequence $|E_0|$, $|E_1|$, $|E_2|$, ..., $|E_n|$ is related to the *previous* term by (4.18), then Lemma 4.1 enables one to relate each term directly *to* $|E_0|$ *only*, that is, to the very first term of the sequence.

LEMMA 4.2. **For all numbers x such that** $1+x\geq0$, **one has**

(4.22) $0 \leq (1+x)^a \leq e^{ax}$, $a\geq0$.

PROOF. Since the function e^x has continuous derivatives of all orders, we have, by Taylor's theorem,

(4.23) $e^x = 1+x+\dfrac{x^2}{2}e^{wx}$, $0<w<1$.

But the last term on the right-hand side of (4.23) is always nonnegative, thus,

$1+x\leq e^x$,

which implies (4.22).

We now proceed to three basic theorems for Euler's method. In each of these we consider the initial value problem

(4.24) $y' = f(x,y)$, $y(0) = \alpha$

on the interval [0,L], where L can be any fixed, positive constant, large or small. Let the exact solution $Y(x)$ of (4.24) be approximated by Euler's method with nh=L as follows:

(4.25) $y_{i+1} = y_i + hf(x_i,y_i)$, $i=0,1,2,\ldots,n-1$

(4.26) $y_0 = \alpha$.

Denote by ε_i the error between the exact solution and the numerical solution obtained with (4.25), that is

(4.27) $\varepsilon_i = Y(x_i) - y_i$, $i=0,1,2,\ldots,n.$

Then the following theorem results.

THEOREM 4.1. Assume that (4.24) has a unique solution $Y(x)$ which has continuous derivatives through the second order on [0,L]. Assume also that $\partial f/\partial y$ exists, is continuous and is bounded for all y and for $x\in$ [0,L]. Then, there exist positive constants N and M, independent of grid size h, such that

(4.28) $|\varepsilon_i| \le \dfrac{(e^{LM}-1)Nh}{2M}$, $i=1,2,\ldots,n.$

PROOF. Choose the positive constants M and N such that

(4.29) $|Y''(x)| \le N$, $0 \le x \le L$,

(4.30) $|\dfrac{\partial f}{\partial y}| \le M$, $0 \le x \le L$, $-\infty < y < \infty$.

Then, from (4.24) and (4.25),

$$|\varepsilon_{i+1}| = |Y(x_{i+1}) - y_{i+1}| = |Y(x_i+h) - y_{i+1}|$$

$$= \left|Y(x_i) + hY'(x_i) + \frac{h^2}{2}Y''(\xi) - [y_i + hf(x_i, y_i)]\right|$$

$$= \left|[Y(x_i) - y_i] + h[f(x_i, Y(x_i)) - f(x_i, y_i)] + \frac{h^2}{2}Y''(\xi)\right|$$

$$= \left|[Y(x_i) - y_i] + h\frac{\partial f(x_i, \mu)}{\partial y}[Y(x_i) - y_i] + \frac{h^2}{2}Y''(\xi)\right|$$

$$= \left|[1 + h\frac{\partial f(x_i, \mu)}{\partial y}][Y(x_i) - y_i] + \frac{h^2}{2}Y''(\xi)\right|$$

$$\leq (1+hM)|Y(x_i) - y_i| + N\frac{h^2}{2} = (1+hM)|\varepsilon_i| + N\frac{h^2}{2}.$$

Hence,

$$|\varepsilon_{i+1}| \leq (1+hM)|\varepsilon_i| + N\frac{h^2}{2}.$$

Moreover, by Lemma 4.1, then,

$$|\varepsilon_i| \leq (1+hM)^i|\varepsilon_0| + \frac{(1+hM)^i - 1}{(1+hM) - 1}\frac{Nh^2}{2}.$$

But, $\varepsilon_0 = 0$, so that

$$|\varepsilon_i| \leq \frac{(1+hM)^i - 1}{hM}\frac{Nh^2}{2} \leq \frac{(1+hM)^n - 1}{2M}Nh.$$

Furthermore, by Lemma 4.2,

$$|\varepsilon_i| \leq \frac{(e^{nhM} - 1)Nh}{2M} = \frac{(e^{LM} - 1)Nh}{2M},$$

and the theorem is proved.

Note that (4.28) is, once again, an error bound formula. Such formulas are most convenient for convergence proofs. Indeed, convergence follows readily from it.

THEOREM 4.2. Under the assumptions of Theorem 4.1,

$$\lim_{h \to 0} |\varepsilon_i| = 0 , \qquad\qquad i=1,2,\dots,n.$$

The proof follows directly from (4.28).

It should be noted now that the above theorem, like previous ones, does not include the fact that there is *roundoff* error when one does the actual calculations. The analysis of roundoff error is, in general, extremely complex, and no methodology available at present works uniformly well. Let us then merely indicate for Euler's method what kind of strategies could be employed to avoid excessive roundoff accumulation by considering several interesting results.

The numbers actually obtained from a computer will not be the y_i but, say, some quantities u_i. These numbers satisfy an equation of the form

(4.31) $u_{i+1} = u_i + hf(x_i,u_i) + r_{i+1}(h) ,$ $i=0,1,2,\dots,n-1,$

where $r_{i+1}(h)$ represents the local roundoff error introduced by inexact evaluation of the quantity

$$u_i + hf(x_i,u_i) .$$

If r_0 denotes the initial roundoff error committed in evaluating y_0, then the initial condition for (4.31) becomes

$$u_0 = \alpha + r_0 .$$

Now, let us consider the total errors E_i which are defined by

(4.32) $E_i = Y(x_i) - u_i ,$ $i=0,1,2,\dots,n,$

between the exact solution, $Y(x_i)$, and the actual numerical solution u_i. Thus, $E_0 = r_0$. Assume also, for simplicity, that the computation is being done in such a fashion that there is a positive number R which is equal to the absolute value of the maximum possible roundoff error. For example, if one is constrained, for one reason or another, to round y_i to four decimal places, then $R=5\cdot10^{-5}$. We then have the following interesting result.

THEOREM 4.3. Under the assumptions of Theorem 4.1, the total error $|E_i|$ satisfies the inequality:

$$(4.33) \qquad |E_i| \le e^{LM}\Big[|r_0| + \frac{1}{M}\Big(\frac{Nh}{2} + \frac{R}{h}\Big)\Big]\,, \qquad\qquad i=1,2,\dots,n.$$

PROOF. From (4.24) and (4.31), we have

$$|E_{i+1}| = |Y(x_{i+1})-u_{i+1}| = |Y(x_i+h)-u_{i+1}|$$

$$= \Big|Y(x_i)+hY'(x_i)+\frac{h^2}{2}Y''(\xi)-[u_i+hf(x_i,u_i)+r_{i+1}(h)]\Big|$$

$$= \Big|E_i+h\frac{\partial f(x_i,\mu)}{\partial y}E_i+\frac{h^2}{2}Y''(\xi)-r_{i+1}(h)\Big|$$

$$\le (1+hM)|E_i|+N\frac{h^2}{2}+R\,.$$

Thus, by Lemma 4.1,

$$|E_i| \le (1+hM)^i|E_0|+\frac{(1+hM)^i-1}{hM}\Big(\frac{Nh^2}{2}+R\Big)$$

$$\le (1+hM)^i|r_0|+\frac{(1+hM)^i}{hM}\Big(\frac{Nh^2}{2}+R\Big)\,,$$

so that, by Lemma 4.2,

$$|E_i| \le e^{ihM}\Big[|r_0|+\frac{1}{M}\Big(\frac{Nh}{2}+\frac{R}{h}\Big)\Big] \le e^{LM}\Big[|r_0|+\frac{1}{M}\Big(\frac{Nh}{2}+\frac{R}{h}\Big)\Big]\,,$$

and the theorem is proved.

The dependence of the total error bound given in Theorem 4.3 on the mesh size h is illustrated in Figure 4.1. The choice of h for which the error bound in (4.33) is a *minimum* is obtained by differentiation of the right-hand side with respect to h and yields

$$h_{opt} = \sqrt{\frac{2R}{N}}\,.$$

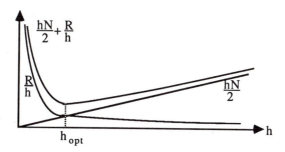

Figure 4.1

In most fixed point calculations performed on modern computers R<<N, and hence the "optimal" value for h will be unnecessarily small and need not be employed. Furthermore, Figure 4.1 indicates that for fixed $h>h_{opt}$ no significant increase in accuracy may result by reducing R. However, there are problems in which a reduction in h may result in a significant increase in roundoff error. In these cases the numerical results can be improved by using the "double precision" arithmetic available in most computers.

EXAMPLE. Consider, again, the initial value problem (4.13), and let us analyze the behavior of the numerical solution at x=1 obtained by using different grid sizes and different values for R. The exact solution, rounded to four decimal places, is $Y(1)=2e^{-1}=0.7358$. Table 4.1 shows the numerical solution that one obtains for grid sizes h=0.2, 0.1, 0.02, 0.01 and 0.002, and with one, two, three and four decimal places accuracy so that, R=0.05, 0.005, 0.0005 and 0.00005, respectively. It is clearly shown that best results are obtained with small h provided we have enough decimal places to perform our computations, which always results by use of double precision.

Table 4.1

	R=0.05	R=0.005	R=0.0005	R=0.00005
h=0.2	0.6	0.66	0.655	0.6554
h=0.1	0.7	0.69	0.698	0.6973
h=0.02	1.0	0.72	0.729	0.7285
h=0.01	1.0	0.75	0.732	0.7321
h=0.002	1.0	1.00	0.729	0.7351

4.4 A RUNGE-KUTTA METHOD

We have developed one of the simplest numerical methods, Euler's method, for initial value problems and have provided, in Theorems 4.1 - 4.3, mathematical support for its value. We proceed next to show how to obtain improved accuracy *without changing* h, but by "improving" on the recursion formula (4.10). In the present development, however, we will assume that we can use enough decimal places so that roundoff error can be neglected. Consider, then, the more general formula

$$(4.34) \qquad \frac{y_{i+1} - y_i}{h} = \eta f(x_i, y_i) + \beta f(x_i + \gamma h, y_i + \delta h) ,$$

where η, β, γ and δ are parameters to be determined. Of course, if one sets $\eta = 1$ and $\beta = 0$, then Euler's formula results. The motivation for considering (4.34) lies in the fact that $f(x,y)$ is known for *all* x in $[x_i, x_{i+1}]$ and for *all* y, and we would like to incorporate this available knowledge into a formula which will yield greater accuracy.

First, let us rewrite (4.34) in the equivalent form

$$(4.35) \qquad y_{i+1} = y_i + \eta h f(x_i, y_i) + \beta h f(x_i + \gamma h, y_i + \delta h) ,$$

Recalling the general Taylor expansion for a function of two variables, that is,

$$f(x+h, y+k) = f(x,y) + \left[h \frac{\partial f}{\partial x} + k \frac{\partial f}{\partial y} \right] + \frac{1}{2} \left[h^2 \frac{\partial^2 f}{\partial x^2} + 2hk \frac{\partial^2 f}{\partial x \partial y} + k^2 \frac{\partial^2 f}{\partial y^2} \right] + \ldots + R_n ,$$

and, using the notation

$$\frac{\partial f}{\partial x} = f_x , \qquad \frac{\partial f}{\partial y} = f_y , \qquad \frac{\partial^2 f}{\partial x^2} = f_{xx} , \qquad \frac{\partial^2 f}{\partial x \partial y} = f_{xy} , \qquad \frac{\partial^2 f}{\partial y^2} = f_{yy} , \qquad \ldots ,$$

we have

$$(4.36) \qquad f(x_i + \gamma h, y_i + \delta h) = f(x_i, y_i) + \gamma h f_x(x_i, y_i) + \delta h f_y(x_i, y_i)$$

$$+ \frac{1}{2} [\gamma^2 h^2 f_{xx}(x_i, y_i) + 2\gamma\delta h^2 f_{xy}(x_i, y_i) + \delta^2 h^2 f_{yy}(x_i, y_i)] + O(h^3) .$$

Substitution of (4.36) into (4.35) and recombination yields

(4.37) $\quad y_{i+1} = y_i + h(\eta + \beta)f(x_i, y_i) + \dfrac{h^2}{2}[2\beta\gamma f_x(x_i, y_i) + 2\beta\delta f_y(x_i, y_i)]$

$\quad\quad\quad + \dfrac{h^3}{6}[3\beta\gamma^2 f_{xx}(x_i, y_i) + 6\beta\gamma\delta f_{xy}(x_i, y_i) + 3\beta\delta^2 f_{yy}(x_i, y_i)] + \beta\ O(h^4)\ .$

Note that the right-hand side of (4.37) is of the particular form

(4.38) $\quad y_{i+1} = y_i + hF_1 + \dfrac{h^2}{2}F_2 + \dfrac{h^3}{6}F_3 + O(h^4)\ ,$

where F_1, F_2, F_3 are the respective coefficients in (4.37). Note also that (4.38) is somewhat suggestive of a Taylor expansion for a function of one variable. So, suppose next that the exact solution of (4.8)-(4.9) is $Y(x)$ and that $Y(x)$ can be written as the Taylor series:

(4.39) $\quad Y_{i+1} = Y_i + hY_i' + \dfrac{h^2}{2}Y_i'' + \dfrac{h^3}{6}Y_i'''\ldots\ .$

From

$$Y' = f(x, Y)\ ,$$

it follows by the chain rule that

$$Y'' = f_x + f_y f$$

$$Y''' = f_{xx} + 2ff_{xy} + f^2 f_{yy} + f_x f_y + ff_y^2\ ,$$

which, upon substitution in (4.39), yields

(4.40) $\quad Y_{i+1} = Y_i + hf(x_i, Y_i) + \dfrac{h^2}{2}[f_x(x_i, Y_i) + f_y(x_i, Y_i)f(x_i, Y_i)]$

$\quad\quad\quad + \dfrac{h^3}{6}\big\{ f_{xx}(x_i, Y_i) + 2f(x_i, Y_i)f_{xy}(x_i, Y_i) + f_{yy}(x_i, Y_i)[f(x_i, Y_i)]^2$

$\quad\quad\quad + f_x(x_i, Y_i)f_y(x_i, Y_i) + f(x_i, Y_i)[f_y(x_i, Y_i)]^2 \big\} + O(h^4)\ .$

Now, (4.37) is an expansion for the *numerical* solution, while (4.40) is an expansion for the *exact* solution. Let us examine the ways in which η, β, γ and δ can be chosen so that if $y_i=Y_i$ then (4.37) for y_{i+1} agrees, in some sense, with (4.40) for Y_{i+1}.

For the special choices $\eta=1$ and $\beta=0$, (4.37) then agrees with (4.40) up to, but *not* including the h^2 terms. This choice, as noted previously, yields Euler's formula. If we wish to construct a formula for which (4.37) agrees with (4.40) through, at least, the h^2 terms, and in this sense is a better approximation than Euler's formula, we need only set the corresponding coefficients of the h *and* the h^2 terms equal, so that

$$\eta+\beta = 1$$
$$2\beta\gamma f_x+2\beta\delta f_y = f_x+ff_y ,$$

which is a system of two equations in the four parameters η,β,γ,δ. Since four parameters are involved, we can, without loss, expand the above system of two equations to the following three equations:

$$\eta+\beta = 1$$
$$2\beta\gamma = 1$$
$$2\beta\delta = f ,$$

by requiring that the corresponding coefficients of f_x and f_y also be equal. Now we have a system of three equations for the four parameters. This system has an infinite number of solutions, a very convenient one being

(4.41) $\eta = \beta = \dfrac{1}{2} ,$ $\gamma = 1 ,$ $\delta = f(x_i,y_i) .$

Substitution of these parameters into (4.35) yields

(4.42) $y_{i+1} = y_i+ \dfrac{h}{2} f(x_i,y_i)+ \dfrac{h}{2} f(x_{i+1}, y_i+hf(x_i,y_i)) ,$

which is called a second-order, *Runge-Kutta* formula. The name "Runge-Kutta" is derived from the two mathematicians who first developed the methods and formulas under discussion. The particular formula (4.42) is called a second-order formula because the Taylor expansion (4.40) for the exact solution Y_{i+1} and the Taylor expansion (4.37) for the numerical solution y_{i+1} agree, or are identical, through the terms of order h^2 under the assumption that $y_i = Y_i$. The formula (4.42) is a "better" approximation than Euler's formula because the Taylor expansion for y_{i+1} in Euler's formula and that for the exact solution agree only through the terms of order h. The terms of the exact Taylor expansion with which the numerical solution does **not** agree are called the truncation error terms, or, simply, the **truncation error**.

Calculation with (4.42) can be done efficiently in three steps, which can be derived as follows. First, let

$$K_0 = hf(x_i, y_i) .$$

Then, (4.42) becomes

(4.43) $$y_{i+1} = y_i + \frac{1}{2} K_0 + \frac{h}{2} f(x_{i+1}, y_i + K_0) .$$

Next, let

$$K_1 = hf(x_{i+1}, y_i + K_0) ,$$

so that (4.43) becomes

$$y_{i+1} = y_i + \frac{1}{2} (K_0 + K_1) .$$

In summary, then, (4.42) is equivalent to the three-step calculation

(4.44) $$K_0 = hf(x_i, y_i) ,$$

(4.45) $$K_1 = hf(x_{i+1}, y_i + K_0) ,$$

(4.46) $$y_{i+1} = y_i + \frac{1}{2} (K_0 + K_1) .$$

EXAMPLE. Consider again the initial value problem

$$y' = -y + x , \qquad y(0) = 1 .$$

Using h=0.1, let us now solve numerically on [0,1] with (4.44)-(4.46). From the given differential equation, we have

$$f(x,y) = -y + x .$$

Since h=0.1, (4.44)-(4.46) become

$$K_0 = 0.1(-y_i + x_i)$$

$$K_1 = 0.1(-y_i - K_0 + x_{i+1})$$

$$y_{i+1} = y_i + 0.5(K_0 + K_1) .$$

Thus, for i=0, we find

$$K_0 = 0.1(-y_0 + x_0) = 0.1(-1) = -0.100$$

$$K_1 = 0.1(-y_0 - K_0 + x_1) = 0.1(-1 + 0.1 + 0.1) = -0.080$$

$$y_1 = y_0 + 0.5(K_0 + K_1) = 1 + 0.5(-0.1 - 0.08) = 0.910 .$$

Next, for i=1, we have

$$K_0 = 0.1(-y_1 + x_1) = 0.1(-0.910 + 0.1) = -0.081$$

$$K_1 = 0.1(-y_1 - K_0 + x_2) = 0.1(-0.910 + 0.081 + 0.2) = -0.0629$$

$$y_2 = 0.910 + 0.5(-0.081 - 0.0629) = 0.838 .$$

Continuing in the indicated fashion, we find, to three decimal places,

$$y_3 = 0.782 , \qquad y_4 = 0.742 ,$$
$$y_5 = 0.714 , \qquad y_6 = 0.699 ,$$
$$y_7 = 0.694 , \qquad y_8 = 0.700 ,$$
$$y_9 = 0.714 , \qquad y_{10} = 0.737 ,$$

Table 4.2

	A	B	C
$x_0=0.0$	1.000	1.000	1.000
$x_1=0.1$	0.910	0.900	0.910
$x_2=0.2$	0.837	0.820	0.838
$x_3=0.3$	0.782	0.758	0.782
$x_4=0.4$	0.741	0.712	0.742
$x_5=0.5$	0.713	0.681	0.714
$x_6=0.6$	0.698	0.663	0.699
$x_7=0.7$	0.693	0.657	0.694
$x_8=0.8$	0.699	0.661	0.700
$x_9=0.9$	0.713	0.675	0.714
$x_{10}=1.0$	0.736	0.697	0.737

which is a much better approximation to the exact solution $Y=x-1+2e^{-x}$ than the Euler solution generated in Example 2 of Section 4.2. For clarity in this comparison, we have listed in column A of Table 4.2 the values of the exact solution, in column B the results by Euler's method from Example 2 in section 4.2, and in column C the results just generated.

4.5 HIGHER ORDER RUNGE-KUTTA FORMULAS

In general, a Runge-Kutta difference approximation of

$$(4.47) \qquad y' = f(x,y)$$

on $[x_i, x_{i+1}]$ is a difference equation of the form

$$(4.48) \qquad y_{i+1} = y_i + h[\eta_1 f(\xi_1, \mu_1) + \eta_2 f(\xi_2, \mu_2) + \ldots + \eta_k f(\xi_k, \mu_k)] ,$$

where $x_i \leq \xi_j \leq x_{i+1}$, $j=1,2,\ldots,k$. Of course (4.34) is a special case of (4.48). A large number of Runge-Kutta formulas have been developed, studied, and applied. For example, just as (4.42) agrees with (4.40) through the h^2 terms, so it was shown by Heun [K.Heun (1900)] that

(4.49)

$$K_0 = hf(x_i,y_i)$$

$$K_1 = hf(x_i+h/3, y_i+K_0/3)$$

$$K_2 = hf(x_i+2h/3, y_i+2K_1/3)$$

$$y_{i+1} = y_i+(1/4)(K_0+3K_2)$$

agrees with the Taylor expansion (4.40) through the h^3 terms.

But, perhaps, the most practical formula of all, from the point of view of accuracy and economy, is the Kutta fourth-order formula [W.Kutta (1901)]. It is often the first formula applied by a professional numerical analyst in approximating the solution of an initial value problem. It is given in the way it is used for calculation as follows:

(4.50)

$$K_0 = hf(x_i,y_i)$$

$$K_1 = hf(x_i+h/2, y_i+K_0/2)$$

$$K_2 = hf(x_i+h/2, y_i+K_1/2)$$

$$K_3 = hf(x_{i+1}, y_i+K_2)$$

$$y_{i+1} = y_i+(1/6)(K_0+2K_1+2K_2+K_3) \ .$$

It agrees with the Taylor expansion of the exact solution through the h^4 terms. Because of its significance, we will illustrate how to implement it, even though the ideas are entirely analogous to those of the example in Section 4.4.

EXAMPLE. Consider the initial value problem

(4.51) $\qquad$ $y' = -y+x , \qquad y(0) = 1 \ .$

Using h=0.1 and rounding to three decimal places, let us solve (4.51) numerically up to x=1 with (4.50). From (4.51), the iteration formulas are

$$K_0 = (0.1)f(x_i, y_i) = (0.1)[-y_i+x_i]$$

$$K_1 = (0.1)f(x_i+0.05, y_i+K_0/2) = (0.1)[-y_i-(K_0/2)+x_i+0.05]$$

$$K_2 = (0.1)f(x_i+0.05, y_i+K_1/2) = (0.1)[-y_i-(K_1/2)+x_i+0.05]$$

$$K_3 = (0.1)f(x_i+0.1, y_i+K_2) = (0.1)[-y_i-K_2+x_i+0.1]$$

$$y_{i+1} = y_i+(1/6)[K_0+2K_1+2K_2+K_3] \ .$$

Hence, for i=0, we have

$$K_0 = (0.1)[-y_0+x_0] = -0.100$$
$$K_1 = (0.1)[-y_0+0.05+x_0+0.05] = -0.090$$
$$K_2 = (0.1)[-y_0+0.045+x_0+0.05] = -0.091$$
$$K_3 = (0.1)[-y_0+0.091+x_0+0.1] = -0.081$$
$$y_1 = 1+(1/6)[-0.100+2(-0.090)+2(-0.091)-0.081] = 0.910 .$$

Next, for i=1, we have

$$K_0 = (0.1)[-y_1+x_1] = -0.081$$
$$K_1 = (0.1)[-y_1+0.0405+x_1+0.05] = -0.072$$
$$K_2 = (0.1)[-y_1+0.036+x_1+0.05] = -0.072$$
$$K_3 = (0.1)[-y_1+0.0724+x_1+0.1] = -0.064$$
$$y_2 = 0.910+(1/6)[-0.081+2(-0.072)+2(-0.072)-0.064] = 0.838 .$$

The process continues next with i=2, and so on, until i=9. The results, given to three decimal places, are identical with those of the exact solution $y=x-1+2e^{-x}$. As a matter of fact, even with h=0.1, calculation with more decimal places reveals that the results agree with the exact solution to at least *five* decimal places. *In a rough way*, this is consistent with the way in which the Kutta formula was constructed, for the numerical and the analytical solutions agree through the h^4 terms in the Taylor series. Thus, the error term behaves like h^5. In this example, h=0.1, so that one would expect an error, at each calculation step, like $(0.1)^5=10^{-5}$. By changing the grid to h=0.01, one would expect an error at each calculation step of about $(0.01)^5=10^{-10}$.

4.6 KUTTA'S FOURTH-ORDER METHOD FOR A SYSTEM OF TWO FIRST-ORDER EQUATIONS

For a system of two first-order equations with prescribed initial values, that is, for

(4.52) $y' = F(x,y,v) ,$ $y(0) = \alpha$

(4.53) $v' = G(x,y,v) ,$ $v(0) = \beta ,$

the Kutta fourth-order technique can be extended directly. This can be done by extensive Taylor series considerations, similar to those in Sections 4.4 and 4.5, or can be deduced easily by analogy with (4.50) as follows. First, since there are two equations, we must have two sets of K's. For notational simplicity, we take, instead, a set of K's and a set of M's, as follows:

$$K_0 = hF(x_i, y_i, v_i) \qquad\qquad M_0 = hG(x_i, y_i, v_i)$$
$$K_1 = hF(x_i+h/2, y_i+K_0/2, v_i+M_0/2) \qquad M_1 = hG(x_i+h/2, y_i+K_0/2, v_i+M_0/2)$$
$$K_2 = hF(x_i+h/2, y_i+K_1/2, v_i+M_1/2) \qquad M_2 = hG(x_i+h/2, y_i+K_1/2, v_i+M_1/2)$$
$$K_3 = hF(x_{i+1}, y_i+K_2, v_i+M_2) \qquad M_3 = hG(x_{i+1}, y_i+K_2, v_i+M_2) .$$

Note that the formulas for the K's are related directly to equation (4.52), while the formulas for the M's are related directly to equation (4.53). The formulas are direct analogues of (4.50), with those for the K's using the function $F(x,y,v)$ and those for the M's using the function $G(x,y,v)$. The calculation of y_{i+1} and v_{i+1} is given, finally, by

(4.54) $\qquad y_{i+1} = y_i + (1/6)(K_0 + 2K_1 + 2K_2 + K_3)$

(4.55) $\qquad v_{i+1} = v_i + (1/6)(M_0 + 2M_1 + 2M_2 + M_3) .$

4.7 KUTTA'S FOURTH-ORDER FORMULAS FOR SECOND-ORDER DIFFERENTIAL EQUATIONS

At long last, we return to initial value problem (4.1)-(4.2), that is, to

(4.56) $\qquad y'' = f(x,y,y') ,$

(4.57) $\qquad y(0) = \alpha , \qquad y'(0) = \beta .$

Given any such problem, Kutta's formulas can be applied if one first converts it into an equivalent problem for systems. Thus, let $y'=v$, so that (4.56)-(4.57) is equivalent to the system

(4.58) $\qquad y' = v , \qquad\qquad y(0) = \alpha$

(4.59) $\qquad v' = f(x,y,v) , \qquad\quad v(0) = \beta .$

System (4.58)-(4.59) is, however, simpler than the general system (4.52)-(4.53), for $F(x,y,v)=v$ and $G(x,y,v)=f(x,y,v)$. Indeed, the elementary structure of F allows us to simplify the general formulas of the last section as follows. First, K_0, K_1, K_2, K_3 reduce to

$$K_0 = hv_i$$
$$K_1 = h(v_i+M_0/2)$$
$$K_2 = h(v_i+M_1/2)$$
$$K_3 = h(v_i+M_2) .$$

Substitution of these into the formulas of section 4.6 allows us to *eliminate* all the K's from the formulas. The following simple algorithm then results. Calculate

(4.60)
$$M_0 = hf(x_i,y_i,v_i)$$
$$M_1 = hf(x_i+h/2, y_i+hv_i/2, v_i+M_0/2)$$
$$M_2 = hf(x_i+h/2, y_i+hv_i/2+hM_0/4, v_i+M_1/2)$$
$$M_3 = hf(x_{i+1}, y_i+hv_i+hM_1/2, v_i+M_2) .$$

Then,

(4.61)
$$y_{i+1} = y_i+hv_i+(h/6)(M_0+M_1+M_2)$$
$$v_{i+1} = v_i+(1/6)(M_0+2M_1+2M_2+M_3) .$$

EXAMPLE. With h=0.1, let us show how to use the Kutta's fourth-order formulas (4.60)-(4.61) to approximate y and y' on [0,1] for the initial value problem

(4.62)
$$y''+xy'+y = 3+5x^2$$
$$y(0) = 1 , \qquad y'(0) = 0 .$$

Converting (4.62) to a system yields

$$y' = v, \qquad\qquad y(0) = 1$$
$$v' = 3+5x^2-xv-y, \qquad v(0) = 0.$$

Thus,

$$f(x,y,v) = 3+5x^2-xv-y.$$

Since $y_0=1$ and $v_0=0$, let $i=0$ and consider $x_1=0.1$. From (4.60) with $i=0$, we have, to four decimal places,

$$M_0=(0.1)f(x_0,y_0,v_0) = (0.1)f(0,1,0) = 0.2000$$
$$M_1=(0.1)f(x_0+0.05,y_0+0.05v_0,v_0+0.1) = (0.1)f(0.05,1,0.1) = 0.2008$$
$$M_2=(0.1)f(x_0+0.05,y_0+0.05v_0+0.025M_0,v_0+0.5M_1)=(0.1)f(0.05,1.005,0.1004)=0.2002$$
$$M_3=(0.1)f(x_1,y_0+0.1v_0+0.05M_1,v_0+M_2) = (0.1)f(0.1,1.01,0.2002) = 0.2020.$$

Thus, from (4.61), we have at x_1, to four decimal places,

$$y_1 = y_0+(0.1)v_0+(0.1)(M_0+M_1+M_2)/6 = 1.0100$$
$$v_1 = v_0+(1/6)(M_0+2M_1+2M_2+M_3) = 0.2007.$$

To proceed to the next grid point, that is, $x_2=0.2$, we set $i=1$. One then calculates a new set of M's as follows:

$$M_0 = (0.1)f(x_1, y_1, v_1) = 0.2020$$
$$M_1 = (0.1)f(x_1+0.05, y_1+0.05v_1, v_1+0.5M_0) = 0.2047$$
$$M_2 = (0.1)f(x_1+0.05, y_1+0.05v_1+0.025M_0, v_1+0.5M_1) = 0.2042$$
$$M_3 = (0.1)f(x_2, y_1+0.1v_1+0.05M_1, v_1+M_2) = 0.2079,$$

so that

$$y_2 = 1.0403, \qquad v_2 = 0.4053.$$

Proceeding in the indicated fashion, but with six decimal places, one finds

$$
\begin{array}{lll}
x_0 = 0.0 & y_0 = 1.000000 & v_0 = 0.000000 \\
x_1 = 0.1 & y_1 = 1.010017 & v_1 = 0.200665 \\
x_2 = 0.2 & y_2 = 1.040265 & v_2 = 0.405280 \\
x_3 = 0.3 & y_3 = 1.091330 & v_3 = 0.617601 \\
x_4 = 0.4 & y_4 = 1.164155 & v_4 = 0.841005 \\
x_5 = 0.5 & y_5 = 1.259996 & v_5 = 1.078335 \\
x_6 = 0.6 & y_6 = 1.380360 & v_6 = 1.331784 \\
x_7 = 0.7 & y_7 = 1.526939 & v_7 = 1.602809 \\
x_8 = 0.8 & y_8 = 1.701532 & v_8 = 1.892108 \\
x_9 = 0.9 & y_9 = 1.905969 & v_9 = 2.199628 \\
x_{10} = 1.0 & y_{10} = 2.142040 & v_{10} = 2.524626 \;.
\end{array}
$$

4.8 THE METHOD OF TAYLOR EXPANSIONS

When calculating on a computer, it is always of value to be able to check one's work. This is especially true if a scientific principle or phenomenon is being modeled or analyzed. For this reason, we will develop a second numerical method, called the ***method of Taylor expansions***, for the initial value problem

(4.63) $y'' = f(x,y,y')$, $y(0) = \alpha$, $y'(0) = \beta$.

The method of Taylor expansions, which is both old and well studied, was considered to be relatively cumbersome at the time when high speed computers first became available, and even for some time thereafter. The reason is that often the method requires extensive differentiation. However, with the very recent advent of computer techniques for symbol manipulation, by means of which the computer also does this differentiation, the method of Taylor expansions has returned to the class of highly valuable procedures [G.Corliss and Y.F.Chang (1982); L.B.Rall (1979); R.E.Moore (1979)].

The Taylor expansion method, unlike the Runge-Kutta method, interestingly enough, does ***not*** require that (4.63) be transformed into an equivalent first-order system, and proceeds as follows. Fix L>0 and divide the interval [0,L] into n equal parts by the R_{n+1} set $x_i = ih$, i=0,1,2,...,n, with nh=L. We wish to determine approximate values for y_{i+1} and y'_{i+1} from known values of y_0 and y'_0. Assuming the validity of finite Taylor expansions, we have

(4.64) $y_{i+1} = y_i + hy_i' + \dfrac{h^2}{2} y_i'' + \dfrac{h^3}{3!} y_i''' + \ldots + \dfrac{h^k}{k!} y_i^{(k)} + R_1$,

(4.65) $y_{i+1}' = y_i' + hy_i'' + \dfrac{h^2}{2} y_i''' + \dfrac{h^3}{3!} y_i^{iv} + \ldots + \dfrac{h^k}{k!} y_i^{(k+1)} + R_2$,

where

$$R_1 = \frac{h^{k+1}}{(k+1)!} y^{(k+1)}(\xi_1) , \quad x_i < \xi_1 < x_{i+1} ; \qquad R_2 = \frac{h^{k+1}}{(k+1)!} y^{(k+2)}(\xi_2) , \quad x_i < \xi_2 < x_{i+1} .$$

The method of Taylor expansions, then, simply drops the remainder terms R_1 and R_2 from (4.64)-(4.65) and uses the resulting formulas for computation. Thus, the Taylor expansion method uses the formulas

(4.66) $y_{i+1} = y_i + hy_i' + \dfrac{h^2}{2} y_i'' + \dfrac{h^3}{3!} y_i''' + \ldots + \dfrac{h^k}{k!} y_i^{(k)}$

(4.67) $y_{i+1}' = y_i' + hy_i'' + \dfrac{h^2}{2} y_i''' + \dfrac{h^3}{3!} y_i^{iv} + \ldots + \dfrac{h^k}{k!} y_i^{(k+1)}.$

Formulas (4.66) and (4.67) are called k-th order approximations because y_{i+1} and y_{i+1}' agree with their Taylor expansions through the terms of order h^k. The basic difficulty in applying (4.66)-(4.67) is that we will know y_i and y_i' but will have to determine all the higher order derivatives required by the formulas. These are usually derived from the given differential equation itself by successive differentiations, which require repeated use of the chain rule.

EXAMPLE. Consider again the initial value problem (4.62) which was solved numerically in the previous section with Kutta's fourth-order method. In order to check both the method and the corresponding numerical results, let us solve the same problem with the method of Taylor expansions of order four. Thus, the initial value problem to be solved can be rewritten as

(4.68) $y'' = 3 + 5x^2 - y - xy'$

(4.69) $y(0) = 1 , \qquad y'(0) = 0 .$

By using again L=1 and n=10, we get h=0.1, and formulas (4.66)-(4.67) become

$$(4.70) \qquad y_{i+1} = y_i + (0.1)y_i' + \frac{(0.1)^2}{2} y_i'' + \frac{(0.1)^3}{6} y_i''' + \frac{(0.1)^4}{24} y_i^{iv}$$

$$(4.71) \qquad y_{i+1}' = y_i' + (0.1)y_i'' + \frac{(0.1)^2}{2} y_i''' + \frac{(0.1)^3}{6} y_i^{iv} + \frac{(0.1)^4}{24} y_i^{v} \ .$$

Note, immediately, that the five coefficients in each of the expansions (4.70) and (4.71) are the same, so that these need to be computed only once. Also, from (4.68), it follows that

$$(4.72) \qquad y'' = 3 + 5x^2 - y - xy'$$

$$(4.73) \qquad y''' = 10x - 2y' - xy''$$

$$(4.74) \qquad y^{iv} = 10 - 3y'' - xy'''$$

$$(4.75) \qquad y^{v} = -4y''' - xy^{iv} \ .$$

Now, for i=0, one has from (4.69) and (4.72)-(4.75)

$$y_0 = 1$$
$$y_0' = 0$$
$$y_0'' = 3 + 5x_0^2 - y_0 - x_0 y_0' = 2$$
$$y_0''' = 10x_0 - 2y_0' - x_0 y_0'' = 0$$
$$y_0^{iv} = 10 - 3y_0'' - x_0 y_0''' = 4$$
$$y_0^{v} = -4y_0''' - x_0 y_0^{iv} = 0 \ ,$$

which, upon substitution into (4.70) and (4.71), yields, to six decimal places,

$$y_1 = 1.010017 \ , \qquad y_1' = 0.200667 \ .$$

If one then continues for i=1,2,...,9, one finds, to six decimal places, the following numerical solution:

$$x_0 = 0.0 \qquad y_0 = 1.000000 \qquad y_0' = 0.000000$$
$$x_1 = 0.1 \qquad y_1 = 1.010017 \qquad y_1' = 0.200667$$
$$x_2 = 0.2 \qquad y_2 = 1.040265 \qquad y_2' = 0.405284$$
$$x_3 = 0.3 \qquad y_3 = 1.091331 \qquad y_3' = 0.617605$$
$$x_4 = 0.4 \qquad y_4 = 1.164157 \qquad y_4' = 0.841010$$
$$x_5 = 0.5 \qquad y_5 = 1.259999 \qquad y_5' = 1.078341$$
$$x_6 = 0.6 \qquad y_6 = 1.380364 \qquad y_6' = 1.331789$$
$$x_7 = 0.7 \qquad y_7 = 1.526945 \qquad y_7' = 1.602814$$
$$x_8 = 0.8 \qquad y_8 = 1.701538 \qquad y_8' = 1.892111$$
$$x_9 = 0.9 \qquad y_9 = 1.905976 \qquad y_9' = 2.199630$$
$$x_{10} = 1.0 \qquad y_{10} = 2.142049 \qquad y_{10}' = 2.524626 \ .$$

By comparing now the results thus obtained, with those of the example in the previous section, we observe that both results agree to five decimal places.

4.9 INSTABILITY

In solving problems with initial conditions on a computer, the occurrence of overflow is a relatively common and disturbing event. Any such computation will be called *unstable*. This definition is the computer analogue of a classical mathematical definition in which a solution of a differential equation is called unstable if it is *unbounded* on $0 < x < \infty$. Our definition simply considers "unbounded" as being greater than the largest number in one's computer. A computation which is *not unstable* is called *stable*.

There are several steps one can take to remedy an unstable calculation, although none may work. These are as follows. First, *check the program*. This can be done by explaining it to another person or by running it for a nontrivial problem for which the exact solution is known. If the instability persists, one next *checks the computer*. Overheating can cause erratic computer performance and miswiring has been known to occur. The most practical check is to run the same program on a second computer. If the instability still persists, one can attempt a mathematical analysis, when possible. Such analyses are given in the next two examples.

EXAMPLE 1. Consider the initial value problem

$$(4.76) \qquad y' = -100y \ ,$$

$$(4.77) \qquad y(0) = 1 \ .$$

The exact solution is $Y(x)=e^{-100x}$, which converges to zero as x goes to infinity. Now, suppose one solves the initial value problem (4.76)-(4.77) using Euler's method with h=0.1. Then,

$$\frac{y_{i+1}-y_i}{0.1} = -100y_i, \qquad y_0 = 1, \qquad i=0,1,2,3,\ldots,$$

or,

$$y_{i+1} = -9y_i, \qquad y_0 = 1, \qquad i=0,1,2,3,\ldots.$$

Then,

$$y_1 = -9$$

$$y_2 = (-9)^2$$

$$y_3 = (-9)^3$$

$$y_4 = (-9)^4$$

$$\cdot$$
$$\cdot$$
$$\cdot$$

and so forth, which yields overflow quickly. To analyze this instability, we redo the calculation, but **do not** fix h. Hence,

$$\frac{y_{i+1}-y_i}{h} = -100y_i, \qquad y_0 = 1, \qquad i=0,1,2,3,\ldots,$$

so that

$$y_{i+1} = (1-100h)y_i, \qquad y_0 = 1, \qquad i=0,1,2,3,\ldots.$$

Thus,

$$y_1 = (1-100h)y_0 = (1-100h) \, ,$$
$$y_2 = (1-100h)y_1 = (1-100h)^2,$$
$$y_3 = (1-100h)^3,$$
$$y_4 = (1-100h)^4,$$

$$\cdot$$
$$\cdot$$
$$\cdot$$

from which it follows that

$$y_i = (1-100h)^i, \qquad\qquad i=0,1,2,3,\dots \ .$$

For stability, one must have

$$|1-100h| \le 1 \, ,$$

or,

$$-1 \le 1-100h \le 1 \, ,$$

or,

$$-2 \le -100h \le 0 \, .$$

Thus, one requires

$$100h \le 2 \, ,$$

or

$$(4.78) \qquad h \le 0.02 \, .$$

The constraint (4.78) is called a stability condition on the grid size. The choice h=0.1 violated this condition. Moreover, when the strict inequality in (4.78) is satisfied, the numerical solution converges to zero as x_i goes to infinity.

In the above example, the equation (4.76) was linear. If the equation were nonlinear, one usually linearizes and develops an approximate stability condition.

Sometimes one can develop energy inequalities for the stability analysis of a nonlinear problem, but these may be very difficult to obtain [D.Greenspan (1973); R.D.Richtmyer and K.W.Morton (1967)]. In any case, the first rule of thumb to be derived is that if instability occurs, *decrease the grid size* and repeat the calculation. This rule works so often that it will probably solve the problem. In some cases, however, it may not, and we consider such an example next.

EXAMPLE 2. Suppose one is given the following recursion formula and additional conditions:

$$(4.79) \qquad y_{i+2} = -\frac{3}{2}y_{i+1} + y_i, \qquad y_0 = \frac{1}{2}, \qquad y_1 = \frac{1}{4}, \qquad i=0,1,2,3,....$$

Then

$$y_2 = -\frac{3}{2}y_1 + y_0 = \frac{1}{8}$$

$$y_3 = -\frac{3}{2}y_2 + y_1 = \frac{1}{16}$$

$$y_4 = -\frac{3}{2}y_3 + y_2 = \frac{1}{32}$$

$$\vdots$$

from which it follows that

$$y_i = \left(\frac{1}{2}\right)^{i+1}, \qquad i=0,1,2,3,...,$$

which converges to zero as i goes to infinity. Suppose, however, as in all computers, one has to round. For simplicity, we will round all y values to one decimal place. Rounding to a higher number of decimal places merely delays the onset of the phenomenon to be shown next. Hence, recalculating with rounding, we find first that (4.79) becomes

$$(4.80) \qquad y_{i+2} = -(1.5)y_{i+1} + y_i, \qquad y_0 = 0.5, \qquad y_1 = 0.3.$$

Then,

$$y_2 = -(1.5)y_1 + y_0 = 0.1$$
$$y_3 = -(1.5)y_2 + y_1 = 0.2$$
$$y_4 = -(1.5)y_3 + y_2 = -0.2$$
$$y_5 = -(1.5)y_4 + y_3 = 0.5$$
$$y_6 = -(1.5)y_5 + y_4 = -1.0$$
$$y_7 = -(1.5)y_6 + y_5 = 2.0$$
$$y_8 = -(1.5)y_7 + y_6 = -4.0$$
$$y_9 = 8.0$$
$$y_{10} = -16.0$$
$$\cdot$$
$$\cdot$$
$$\cdot$$
$$y_k = -(-2)^{k-6}, \qquad\qquad k \geq 6 ,$$

so that y_i becomes unbounded with increasing i. Let us show how to analyze this instability.

First, let us rewrite (4.79) as

(4.81) $\qquad$ $2y_{i+2} + 3y_{i+1} - 2y_i = 0 ,$

Equation (4.81) is called a second-order linear difference equation with constant coefficients. In analogy with second-order linear differential equations with constant coefficients, it can be solved in general as follows. Set

$$y_i = \lambda^i, \qquad \lambda \text{ a nonzero constant.}$$

Then,

$$2\lambda^{i+2} + 3\lambda^{i+1} - 2\lambda^i = 0 ,$$

or,

$$2\lambda^2 + 3\lambda - 2 = 0 .$$

Thus,

$$(2\lambda - 1)(\lambda + 2) = 0$$

and

$$\lambda = \tfrac{1}{2}, \qquad \lambda = -2 .$$

Each of

$$y_i = \left(\tfrac{1}{2}\right)^i, \qquad y_i = (-2)^i$$

is a solution of (4.81) and the general solution is

$$(4.82) \qquad y_i = c_1 \left(\tfrac{1}{2}\right)^i + c_2 (-2)^i, \qquad i = 0,1,2,3,\dots,$$

where c_1 and c_2 are arbitrary constants. For our first calculation above we had $y_0 = \tfrac{1}{2}$, $y_1 = \tfrac{1}{4}$. Thus, from (4.82), for $i=0$ and $i=1$,

$$c_1 + c_2 = \tfrac{1}{2}$$

$$c_1 \left(\tfrac{1}{2}\right) + c_2(-2) = \tfrac{1}{4},$$

which imply $c_1 = \tfrac{1}{2}$, $c_2 = 0$. Thus,

$$y_i = \left(\tfrac{1}{2}\right)^{i+1},$$

which is the same as the result we got when the calculations were exact. But, when including rounding, we had $y_0 = 0.5$, $y_1 = 0.3$. For $i=0$ and $i=1$, then, from (4.82),

$$0.5 = c_1 + c_2$$

$$0.3 = c_1 \left(\tfrac{1}{2}\right) + c_2(-2) .$$

The solution of this system is $c_1 = \tfrac{13}{25}$, $c_2 = -\tfrac{1}{50}$, and the general solution is now

(4.83) $y_i = \frac{13}{25} \left(\frac{1}{2}\right)^i - \frac{1}{50}(-2)^i$,   i=0,1,2,3,... .

However, in (4.83), the term $(-2)^i$ becomes unbounded as i becomes large, which is why the instability results.

Note, now, that the difference equation (4.81) can be regarded as an approximation for a first-order differential equation. Consider, in fact, equation (4.76):

(4.84) $y' = -100y$,

Fix h>0 and consider the grid points $x_i=ih$, i=0,1,2,3,.... At x_{i+1}, i=0,1,2,..., substitution of y' with the corresponding central difference approximation, equation (4.84) yields

$$\frac{y_{i+2}-y_i}{2h} = -100y_{i+1} ,$$

or, equivalently,

(4.85) $y_{i+2}+2(100)hy_{i+1}-y_i = 0$.

For h=3/400, equation (4.85) is equivalent to (4.81).

Note, finally, that the difference equation (4.85) leads to unstable computation for *any* positive h. In fact, the general solution of equation (4.85) is

$$y_i = c_1(\lambda_1)^i+c_2(\lambda_2)^i ,$$

where λ_1 and λ_2 are given by

$$\lambda_{1,2} = -100h\pm\sqrt{(100h)^2+1} .$$

Thus, no matter how small we chose h, since

$$| -100h - \sqrt{(100h)^2 + 1} \, | > 1 \, ,$$

one of $|\lambda_1|$, $|\lambda_2|$ is always greater than unity and the resulting calculations will be unstable for any nonzero choice of c_1 and c_2.

From the above example follows the second rule of thumb. If decreasing the grid size always continues to yield instability, then *change the difference equation* being used. The way to do this is always to choose a formula of the form

$$ay_{i+2} + by_{i+1} + cy_i = 0 \, , \qquad\qquad a \neq 0 \, ,$$

such that the two roots λ_1, λ_2 of

$$a\lambda^2 + b\lambda + c = 0$$

satisfy $|\lambda| \leq 1$. The reason is that if this condition is valid then all particular solutions of the general solution

$$y_i = c_1(\lambda_1)^i + c_2(\lambda_2)^i$$

are bounded for all i.

*4.10 APPROXIMATION OF PERIODIC SOLUTIONS OF DIFFERENTIAL EQUATIONS

Thus far, we have discussed the initial value problem for the second-order differential equation

(4.86) $y'' = f(x,y,y')$.

A second problem of wide interest for (4.86) asks for the periodic solutions of the equation. Such solutions correspond physically to stable, periodic motions of an oscillator. However, no additional data, like initial conditions are provided.

To show how to approach such problems numerically, we will select a particular equation of interest and, guided by known analytical results, develop an efficient and practical method for approximating the periodic solutions. Consider, then, the van der Pol equation

$$(4.87) \qquad y'' - \lambda(1-y^2)y' + y = 0 , \qquad\qquad \lambda > 0 ,$$

in which λ is constant. It is known that for each such λ the equation has a periodic solution. To be precise, we quote next a relevant theorem [C.W.Clenshaw (1966)].

THEOREM 4.4. **For each $\lambda > 0$, there exists exactly one periodic solution $y(x)$ of (4.87). If $y'(0)=0$, $y(0)=\alpha > 0$, and $T=T(\lambda)$ is the period of $y(x)$, then $y'(T/2)=0$, $y(T/2)=-\alpha$, and $y(x)$ is monotonic decreasing on $[0,T/2]$, while it is monotonic increasing on $[T/2,T]$.**

The essential content of Theorem 4.4 for the half-period $[0,T/2]$ is shown in Figure 4.2.

Numerically, Theorem 4.4 assures us that what we plan to approximate does exist and is unique. In addition, it provides us with sufficient information to formulate an appropriate algorithm, which is described as follows.

Set $y'(0)=0$. Next, choose the following initial approximations for α: $\alpha_1=1$, $\alpha_2=2$, $\alpha_3=3$, ..., $\alpha_{10}=10$. Beginning with the smallest α, that is $\alpha_1=1$, we solve numerically, and, in sequence, the set of initial value problems

$$(4.88) \qquad y'' - \lambda(1-y^2)y' + y = 0 , \qquad y(0) = \alpha_i , \qquad y'(0) = 0 , \qquad i=1,2,3,...,10.$$

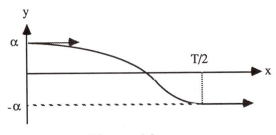

Figure 4.2

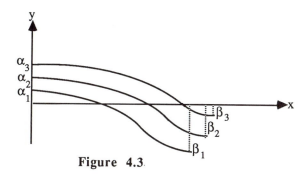

Figure 4.3

For each α_i, let β_i be the *first* minimum value which the numerical solution yields. Figure 4.3 shows the β_1, β_2, β_3 which correspond to α_1, α_2, α_3, respectively. Let $(T_i/2)$ be the x value which corresponds to β_i.

Now, in Figure 4.3, note that $\alpha_1<-\beta_1$, $\alpha_2<-\beta_2$, $\alpha_3>-\beta_3$. Since we want $\alpha_i=-\beta_i$, this means that the desired α is in the range $2<\alpha<3$. We next consider a new set of α's, namely $\alpha_0=2.0$, $\alpha_1=2.1$, $\alpha_2=2.2$, ..., $\alpha_{10}=3.0$, and solve numerically, and in sequence, the initial value problems (4.88). For each α_i, let β_i be the first minimum value which the numerical solution yields. As above, comparison of each α_i with the corresponding $-\beta_i$ leads to new bounds on the exact α sought. For example, we may find that $\alpha_6<\alpha<\alpha_7$, that is, $2.6<\alpha<2.7$. We would then continue by considering the new set $\alpha_0=2.60$, $\alpha_1=2.61$, $\alpha_2=2.62$, ..., $\alpha_{10}=2.70$, and repeat the process in the indicated fashion. In this way we can find an approximation α_i to α to any fixed number of decimal places. The corresponding $x=T_i/2$ at which $y(T_i/2)=\beta_i$ yields the half period.

EXAMPLE. Consider the van der Pol equation for each of the three values $\lambda=0.1$, $\lambda=1$, $\lambda=10$. We wish to approximate α and $T/2$ for each. In implementing the method described above, we applied Kutta's fourth-order formulas with $h=0.001$. The results are summarized as follows:

$$\lambda = 0.1 \quad \rightarrow \quad \alpha = 2.000 , \quad T/2 = 3.148$$

$$\lambda = 1 \quad \rightarrow \quad \alpha = 2.009 , \quad T/2 = 3.335$$

$$\lambda = 10 \quad \rightarrow \quad \alpha = 2.014 , \quad T/2 = 9.538 .$$

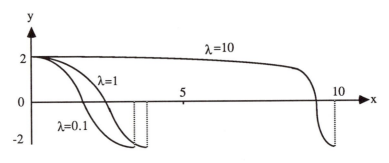

Figure 4.4

The graphs of the approximate periodic solutions are shown in Figure 4.4. As a check, the calculations were redone using fourth-order Taylor expansions. For this purpose, note that

$$y'' = \lambda y' - \lambda y^2 y' - y$$

$$y''' = \lambda y'' - 2\lambda y(y')^2 - \lambda y^2 y'' - y'$$

$$y^{iv} = \lambda y''' - 2\lambda(y')^3 - 6\lambda yy'y'' - \lambda y^2 y''' - y''$$

$$y^v = \lambda y^{iv} - 12\lambda(y')^2 y'' - 6\lambda y(y'')^2 - 8\lambda yy'y''' - \lambda y^2 y^{iv} - y'''.$$

The results, to three decimal places, were the same as those using the Kutta formulas.

4.11 HINTS AND REMARKS

A large number of methods are available for the numerical solution of initial value problems for ordinary differential equations. These can be categorized reasonably by their utilization of predictor-corrector iteration (see Exercise 19), multisteps, numerical integration, isoclines, interpolation (high-order and trigonometric), series (trigonometric, Lie, Chebyshev), Monte Carlo simulation, energy conservation, leap frogging, and boundary value techniques. Nevertheless, the methods we have developed are entirely sufficient for the study of most problems in science and engineering. Indeed, even for such sensitive problems as "stiff" problems, in which one of the variables x or y varies relatively slowly while the other varies relatively rapidly, high-order Taylor approximations usually suffice.

Thus, for example, for $\lambda=25$ in (4.87), a tenth-order Taylor approximation yields excellent results.

With regard to accuracy in extended calculations, it is often useful to change the grid size h at appropriate times. Though no precise, general rule exists for any related adaptive grid procedure, two rules in common use are as follows. First, if both h and h/2 yield results which agree to a desired number of decimal places at any time step, then h is considered to be adequate at that time step. Second, if the physics of a problem includes a force of magnitude F, then at each time step one can choose h to satisfy $hF<\varepsilon$, where ε is a prescribed, positive constant. The latter rule is especially useful in n-body problem.

EXERCISES

Basic Exercises

1. Transform each of the following initial value problems into an equivalent problem for a first order system.

   (a) $y''-3y'-y = 7x+25$, $y(0) = 1$, $y'(0) = 2$

   (b) $y''-3xy'-x^2y = 7x+25$, $y(0) = 1$, $y'(0) = 2$

   (c) $y''-3xyy' = 7x+25$, $y(0) = 1$, $y'(0) = 2$

   (d) $y''-e^{3x}yy' = 7xe^x+25$, $y(0) = 1$, $y'(0) = 2$.

2. For the initial value problem

   $$y' = y^2+2x-x^4, \qquad\qquad y(0) = 0 ,$$

   use Euler's method with h=0.1 to generate a numerical solution on $0 \le x \le 1$. Give the results to three decimal places and compare with the exact solution $y=x^2$.

3. Using Euler's method with h=0.2, find to three decimal places the numerical solution on $0 \le x \le 1$ for each of the following.

   (a) $y' = xy-2x$, $y(0) = 4$

   (b) $y' = y^2$, $y(0) = 1$

   (c) $y' = xy^2-2y$, $y(0) = 1$.

4. Each of the initial value problems in Exercise 3 is elementary in the sense that it can be

solved exactly. Find each such solution and compare it with the corresponding numerical solution.

5. Using Euler's method with h=0.01, find the numerical solution on $0 \le x \le 2$ of the initial value problem

$$y' = -2xy^2, \qquad y(0) = 1 .$$

When doing the calculations, carry as many decimal places as possible, but give the answers to six decimal places. Compare your results with the exact solution

$$y = (1+x^2)^{-1} .$$

6. Prove that if a set of constants $a_0, a_1, a_2, \ldots, a_n$ and B satisfy

$$|a_{k+1}| \le |a_k| + B , \qquad k=0,1,2,\ldots,n-1,$$

then

$$|a_k| \le |a_0| + kB , \qquad k=1,2,\ldots,n.$$

7. For each of the following, estimate the M of Theorem 4.1.

   (a) $y' = x+y$, $0<x<1$
   (b) $y' = x+\sin(y)$, $0<x<2$
   (c) $y' = x^2y$, $0<x<2$
   (d) $y' = x^2\cos(y)$, $0<x<2$
   (e) $y' = x^2-\cos(y)$, $0<x<2$.

8. Recalculate the numerical solution of the initial value problem in Exercise 2, but use the Runge-Kutta formula (4.42). Compare the results.

9. Recalculate the numerical solution of the initial value problem in Exercise 5, but use Kutta's formula (4.50). Compare the results.

10. Using Kutta's formulas (4.60) and (4.61) with h=0.001, find y_{i+1} and v_{i+1} up to

x_{1000} for the initial value problem

$$y'' + (y')^3 - 8xy = 2 , \qquad y(0) = y'(0) = 0 .$$

Compare your results with the exact solution $y=x^2$.

11. Using Kutta's formulas (4.60) and (4.61) with h=0.01, find y_{i+1} and v_{i+1} on $0 \le x \le 25$ for the initial value problem

$$y'' + 0.3 y' + \sin(y) = 0 , \qquad y(0) = \pi/4 , \qquad y'(0) = 0$$

and graph the solution.

12. Using third-order Taylor expansions and h=0.2, find the numerical solution on $0 \le x \le 0.6$ of the initial value problem

$$y'' = y^2 - x^2, \qquad y(0) = 0 , \qquad y'(0) = 1 .$$

13. Using fourth-order Taylor expansions with h=0.001, find the numerical solution up to x_{1000} for the initial value problem in Exercise 10 and compare your results with those of Exercise 10.

14. Using fourth-order Taylor expansions with h=0.01, find the numerical solution on $0 \le x \le 25$ of the initial value problem in Exercise 11 and compare your results with those of Exercise 11.

15. Determine the values of h for which Euler's method is stable when applied to the following initial value problems:

(a) $y' = -1000y ,$ $y(0) = 1$
(b) $y' = -0.01y ,$ $y(0) = 1$
(c) $y' = -0.01y ,$ $y(0) = -1$
(d) $y' = 0.01y ,$ $y(0) = 1 .$

16. Determine which of the following recursion formulas are stable and which are unstable:

(a)	$y_{i+2} = y_{i+1} - 6y_i$,	$i=0,1,2,\ldots$
(b)	$y_{i+2} = (1/6)(y_{i+1} + y_i)$,	$i=0,1,2,\ldots$
(c)	$y_{i+2} = 5y_{i+1} - 6y_i$,	$i=0,1,2,\ldots$
(d)	$y_{i+2} = 4y_i$,	$i=0,1,2,\ldots$
(e)	$y_{i+2} = -4y_i$,	$i=0,1,2,\ldots$
(f)	$y_{i+2} = y_i$,	$i=0,1,2,\ldots$
(g)	$y_{i+2} = -(1/4)y_i$,	$i=0,1,2,\ldots$.

17. Approximate the periodic solution of the van der Pol equation for each of $\lambda=0.5$ and $\lambda=5.0$. Compare your solutions with those given in Figure 4.4.

18. Show that use of the "backward" Euler formula

$$\frac{y_{i+1} - y_i}{h} = f(x_{i+1}, y_{i+1})$$

for (4.76) is stable for all h, provided the exact solution is bounded.

Supplementary Exercises

19. Consider the Runge-Kutta formula (4.42) in the form

$$y_{i+1}^{(1)} = y_i + hf(x_i, y_i)$$
$$y_{i+1} = y_i + \frac{h}{2}[f(x_i, y_i) + f(x_{i+1}, y_{i+1}^{(1)})] .$$

The first equation can be interpreted as an initial approximation, or *predictor*, by Euler's formula for y_{i+1}. If the second formula is rewritten as

$$y_{i+1}^{(k+1)} = y_i + \frac{h}{2} [f(x_i, y_i) + f(x_{i+1}, y_{i+1}^{(k)})] , \qquad k=1,2,3,\ldots,$$

then it is called a ***corrector*** formula. Iteration with the corrector formula continues until $y_{i+1}^{(K+1)} = y_{i+1}^{(K)}$, for some value K, and one sets $y_{i+1} = y_{i+1}^{(K)}$. The resulting method, performed for each value of i, is called a ***predictor-corrector*** method. Apply the above predictor-corrector method to the initial value problem of Exercise 2 and compare your results with the exact solution, with the numerical solution obtained in Exercise 2, and with the numerical solution obtained in Exercise 8.

20. Using Kutta's fourth-order formulas with h=0.2, approximate y at x=0.8 for the initial value problem

$$y' = \frac{1}{x+y} , \qquad y(0) = 2 .$$

21. Derive Heun's formula.

22. Using a fourth-order Taylor expansion with h=0.1, approximate the solution of

$$y' = xy^{1/3}, \qquad y(1) = 1$$

at x=5.0. Compare your result with the exact solution.

23. Using a third-order Taylor expansion with h=0.1, approximate the solution of

$$y' = \frac{2xy + e^y}{x^2 + xe^y} , \qquad y(1) = 0$$

at x=2.0.

24. Repeat Exercise 22 but use Kutta's fourth-order formulas and compare results.

25. Using formulas (4.54) and (4.55) with h=0.1, find a numerical solution of each of the following systems on $0 \le x \le 4$:

(a) $y' = y+v+1$, $y(0) = 0$
       $v' = y+v-1$, $v(0) = 0$

(b) $y' = y^2-v^2+1$, $y(0) = 0$
       $v' = v^2-y^2-1$, $v(0) = 0$

(c) $y' = 1+v-y^2-v^2$, $y(0) = 0$
       $v' = 1-y-y^2-v^2$, $v(0) = 1$.

26. Using h=0.01, find a numerical solution on $0 \leq x \leq 10$ for the system

$$\frac{d^2y}{dx^2} = -\frac{y}{(y^2+v^2)^{3/2}}, \qquad y(0) = 1, \qquad \frac{dy}{dx}\Big|_{x=0} = 0$$

$$\frac{d^2v}{dx^2} = -\frac{v}{(y^2+v^2)^{3/2}}, \qquad v(0) = 0, \qquad \frac{dv}{dx}\Big|_{x=0} = 1.$$

27. Using either Kutta's fourth-order formulas or fourth-order Taylor expansions, approximate the solution of

$$y''-(x^2-1)y = 0, \qquad y(0) = 1, \qquad y'(0) = 0$$

on $0 \leq x \leq 20$. Compare your answers with the exact solution

$$y = e^{-x^2/2}.$$

Analyze the difficulties when x>8.

5

Approximate Solution of Boundary Value Problems for Ordinary Differential Equations

5.1 INTRODUCTION

Physically, a boundary value problem for a second-order differential equation is one in which one knows where he is located at the initial time x=a, knows where he wants to be at a later time x=b, and has to determine his motion from the differential equation in the time interval between x=a and x=b.

Formally, a boundary value problem is defined as follows. Given four constants a, b, α, β, with a<b, and a second-order differential equation

$$(5.1) \qquad y'' = f(x,y,y') ,$$

find a function y(x) which is continuous on [a,b], which satisfies (5.1) on a<x<b, and which satisfies the two conditions

$$(5.2) \qquad y(a) = \alpha , \qquad y(b) = \beta .$$

Conditions (5.2) are called boundary conditions, or boundary data. Physically, α is an initial position and β is a final position.

Because boundary data are distinctly different from initial data, and because boundary value problems are defined on finite intervals, not unbounded intervals, most of the considerations in this chapter will be distinctly different from those of the last.

5.2 CENTRAL DIFFERENCE METHOD FOR LINEAR BOUNDARY VALUE PROBLEMS

Let us begin with a simple, illustrative example which will reveal the basic ideas. Consider the boundary value problem

(5.3) $y'' = 0$

(5.4) $y(0) = -1$, $y(6) = 5$.

The exact solution is, of course,

(5.5) $y = x - 1$.

Proceeding numerically, however, let us first divide [0,6] into n equal parts. For example, let n=6. Thus, h=1 and $x_0=0$, $x_1=1$, $x_2=2$, $x_3=3$, $x_4=4$, $x_5=5$, $x_6=6$ is the R_7 set of grid points. On this R_7 set, let $y_i=y(x_i)$ be a discrete function for which $y_0=-1$, $y_6=5$, while y_1, y_2, y_3, y_4, y_5 are unknown. It is these unknown values that we have to determine. This will be accomplished by approximating the differential equation (5.3) at **each** of the interior grid points x_1, x_2, x_3, x_4, x_5 as follows.

Using the approximation (3.40), that is,

$$y''(x_i) \approx \frac{y_{i+1} - 2y_i + y_{i-1}}{h^2} ,$$

in (5.3) with h=1 implies

(5.6) $y_{i+1} - 2y_i + y_{i-1} = 0$,

which at x_1, x_2, x_3, x_4, x_5, yields

(5.7) $y_2 - 2y_1 + y_0 = 0$

(5.8) $y_3 - 2y_2 + y_1 = 0$

(5.9) $y_4 - 2y_3 + y_2 = 0$

(5.10) $y_5 - 2y_4 + y_3 = 0$

(5.11) $y_6 - 2y_5 + y_4 = 0$.

Substituting (5.4) into (5.7)-(5.11) then implies

$$-2y_1 +y_2 \qquad\qquad = 1$$
$$y_1 -2y_2 +y_3 \qquad = 0$$
$$y_2 -2y_3 +y_4 \quad = 0$$
$$y_3 -2y_4 +y_5 = 0$$
$$y_4 -2y_5 = -5 \,,$$

which is a tridiagonal system which satisfies all the assumptions of Theorem 1.2. Thus, the solution exists and is unique. Moreover, the solution can be found either by the algorithm of Section 1.4 or by the generalized Newton's method which, in this case, converges for all initial guesses and for all ω in the range $0<\omega<2$. One finds, then, that the solution of the system is

(5.12) $\qquad y_1 = 0 \,, \qquad y_2 = 1 \,, \qquad y_3 = 2 \,, \qquad y_4 = 3 \,, \qquad y_5 = 4 \,,$

which, in addition to $y_0=-1$ and $y_6=5$, is called the *numerical* solution. Note, incidentally, that in this simple example the *numerical* solution and the exact solution are *identical* on the R_7 set because of the simplicity of the problem.

The ideas developed in the above example will be extended next to *general* linear boundary value problems as follows.

Let I be an open interval $a<x<b$ and let α and β be constants. Let $P(x)$, $Q(x)$, $R(x)$ be continuous on $[a,b]$. Then consider the boundary value problem defined by

(5.13) $\qquad y''+P(x)y'+Q(x)y = R(x) \,, \qquad x\in I$

(5.14) $\qquad y(a) = \alpha \,, \qquad y(b) = \beta \,.$

Problem (5.13)-(5.14) is called a linear boundary value problem because (5.13) is the general linear differential equation of second order in one independent and one dependent variable. Rarely can we find an exact solution of this problem.

For numerical purposes, it is desirable to know, *a priori*, that problem (5.13)-(5.14) has a unique solution. For this reason, we shall assume, in addition, that

(5.15) $\qquad Q(x) \le 0 \,, \qquad\qquad a\le x\le b \,.$

Condition (5.15) is *sufficient* to insure uniqueness, but is not necessary. This result is a particular case of a very comprehensive theory which exists for boundary value problems [R.Courant and D.Hilbert (1962), pp.336-341]. Moreover, for $R(x) \equiv 0$ and $Q(x) \leq 0$, any nonconstant solution $y(x)$ of (5.13) has the following (weak) *max-min* property: $y(x)$ cannot assume a positive maximum or a negative minimum on I. Also, if $R(x) \equiv 0$ and $Q(x) \equiv 0$, then any nonconstant solution $y(x)$ of equation (5.13) has the following (strong) max-min property: $y(x)$ cannot assume its maximum or minimum on I [M.H.Protter and H.F.Weinberger (1984), pp. 1-6]. As an example, note that equation (5.3) has the general solution of the form $y(x)=a+bx$. Thus, each one of these solutions has its maximum and minimum at the end points of any interval.

Although (5.15) assures existence and uniqueness, it provides no insight as to how to find the solution, and in general, numerical methodology is required. Our numerical approach will be, as described in the example above, to approximate problem (5.13)-(5.14) by an appropriate problem in algebra. This will be done, first, by substituting approximations (3.37) and (3.40) into (5.13) to yield an algebraic, or *difference*, equation. This difference equation, when written consecutively at the interior grid points $x_1, x_2, \ldots, x_{n-1}$ of an R_{n+1} set will result in a tridiagonal, linear algebraic system, whose solution, together with the boundary values, is called the numerical solution.

Specifically, we proceed as follows. Divide [a,b] into n equal parts, each of length h, by the R_{n+1} set $a=x_0<x_1<x_2<\ldots<x_n=b$. Then, at each interior grid point x_i, substitution of

$$(5.16) \qquad y'' \approx \frac{y_{i+1}-2y_i+y_{i-1}}{h^2}$$

and

$$(5.17) \qquad y' \approx \frac{y_{i+1}-y_{i-1}}{2h}$$

into (5.13) yields the following difference equation approximation of the differential equation.

$$(5.18) \qquad \frac{y_{i+1}-2y_i+y_{i-1}}{h^2}+P(x_i)\frac{y_{i+1}-y_{i-1}}{2h}+Q(x_i)y_i = R(x_i) , \qquad i=1,2,\ldots,n-1,$$

or, equivalently,

(5.19) $[2-hP(x_i)]y_{i-1}+[-4+2h^2Q(x_i)]y_i+[2+hP(x_i)]y_{i+1} = 2h^2R(x_i)$, $i=1,2,...,n-1$.

If one then writes (5.19) consecutively for $i=1,2,...,n-1$ *and* inserts the known values (5.14), there results a tridiagonal linear algebraic system of $n-1$ equations in the $n-1$ unknowns $y_1, y_2, ..., y_{n-1}$. A solution of this system, with the values y_0 and y_n, is the numerical solution on the given R_{n+1} set. This method is called the ***central difference method*** because a central difference formula has been use to approximate $y'(x_i)$.

Let us now examine examples of the general method just described.

EXAMPLE 1. Consider the boundary value problem

(5.20) $$y''+ \frac{x-3}{2}y'-y = 0$$

(5.21) $y(0) = 11$, $y(6) = 11$.

The exact solution is $y=x^2-6x+11$. To generate the numerical solution, let $n=6$, so that $h=1$. Then, $x_0=0$, $x_1=1$, $x_2=2$, $x_3=3$, $x_4=4$, $x_5=5$, $x_6=6$. Setting $y(x_i)=y_i$, then $y_0=11$, $y_6=11$. At each point x_i, $i=1,2,3,4,5$, the differential equation (5.20) is approximated by the difference equation

(5.22) $$\frac{y_{i+1}-2y_i+y_{i-1}}{h^2} + \frac{x_i-3}{2}\frac{y_{i+1}-y_{i-1}}{2h} - y_i = 0$$

which simplifies to

(5.23) $(7-x_i)y_{i-1}-12y_i+(1+x_i)y_{i+1} = 0$.

Writing (5.23) for each of $i=1,2,3,4,5$, in order, yields

$$(7-x_1)y_0-12y_1+(1+x_1)y_2 = 0$$
$$(7-x_2)y_1-12y_2+(1+x_2)y_3 = 0$$
$$(7-x_3)y_2-12y_3+(1+x_3)y_4 = 0$$
$$(7-x_4)y_3-12y_4+(1+x_4)y_5 = 0$$
$$(7-x_5)y_4-12y_5+(1+x_5)y_6 = 0 .$$

Substitution of the known values for y_0, y_6, x_1, x_2, x_3, x_4 and x_5 into the system yields, then,

$$-12y_1 + 2y_2 \qquad\qquad = -66$$
$$5\,y_1 - 12y_2 + 3y_3 \qquad = 0$$
$$4y_2 - 12y_3 + 4y_4 \quad = 0$$
$$3y_3 - 12y_4 + 5y_5 = 0$$
$$2y_4 - 12y_5 = -66$$

the solution of which is

$$y_1 = 6, \qquad y_2 = 3, \qquad y_3 = 2, \qquad y_4 = 3, \qquad y_5 = 6.$$

These values, together with $y_0 = 11$ and $y_6 = 11$, constitute the numerical solution, which, again, coincides with the exact solution on the R_7 set. However, one can no longer say that this is because problem (5.20)-(5.21) is simple, since (5.20)-(5.21) is *not* simple. Indeed, what is simple is the *solution* $y = x^2 - 6x + 11$ of (5.20)-(5.21) and, as will be shown later, it is the simplicity of the analytical solution which has resulted now, *and* in the previous example, in the *unusual* situation where the numerical solution has been devoid of error.

Before continuing to the next example, it is important to make the following observation. With respect to the matrix of the tridiagonal system in the numerical method, the coefficient of y_{i-1} in the difference equation yields the *subdiagonal* elements, the coefficient of y_i yields the *main diagonal* elements, and the coefficient of y_{i+1} yields the *superdiagonal* elements. This can be verified easily from (5.23). It is essential to understand and appreciate this before proceeding.

EXAMPLE 2. Consider the boundary value problem

(5.24) $y'' + 2(2-x)y' = 2(2-x)$

(5.25) $y(0) = -1, \quad y(6) = 5,$

the exact solution of which is $y = x - 1$. To generate the numerical solution, let $n=6$ so that $h=1$. Then $x_i = ih$, $i = 0,1,2,3,4,5,6$. Thus, by using (5.16) and (5.17), the differential equation (5.24) is approximated by the difference equation

(5.26) $\dfrac{y_{i+1} - 2y_i + y_{i-1}}{h^2} + 2(2-x_i)\dfrac{y_{i+1} - y_{i-1}}{2h} = 2(2-x_i),$

or, equivalently, by

(5.27) $[2-2(2-x_i)]y_{i-1}-4y_i+[2+2(2-x_i)]y_{i+1} = 4(2-x_i)$.

For i=1,2,3,4,5, and with the insertion of y_0, y_6, x_1, x_2, x_3, x_4, x_5, one has from (5.27)

$$
\begin{aligned}
-4y_1+4y_2 &= 4 \\
2y_1-4y_2+2y_3 &= 0 \\
4y_2-4y_3 &= -4 \\
6y_3-4y_4-2y_5 &= -8 \\
8y_4-4y_5 &= 8 .
\end{aligned}
$$

Unfortunately, the resulting tridiagonal system is not diagonally dominant and even has a zero and a negative value on the superdiagonal. Thus, no theorem from Section 1.4 is applicable to the system to guarantee that its solution is unique. Indeed, the determinant of the associated matrix is *zero*, and consequently the system cannot be solved uniquely. We would like to avoid such a quandary whenever possible. The next example shows how this can be accomplished.

EXAMPLE 3. Consider, again, boundary value problem (5.24)-(5.25). This time, however, let us *not* fix h at first. Use of (5.16) and (5.17), yields, again, (5.26), but this time h is *not* necessarily unity. Equation (5.26) can be rewritten as

(5.28) $[2-2h(2-x_i)]y_{i-1}-4y_i+[2+2h(2-x_i)]y_{i+1} = 4h^2(2-x_i)$.

The subdiagonal elements of the resulting tridiagonal system will be $[2-2h(2-x_i)]$, the main diagonal elements will be -4, and the superdiagonal elements will be $[2+2h(2-x_i)]$. Hence, first we would like to have the conditions (1.20)-(1.22) valid, that is,

(5.29) $2-2h(2-x_i) > 0$

(5.30) $-4 < 0$

(5.31) $2+2h(2-x_i) > 0$.

Inequality (5.30) is always valid. Hence, h must be chosen in such a fashion that (5.29) and (5.31) are valid. Note now that, since h>0, (5.29) and (5.31) combine as follows

$$2 > h|2(2-x_i)| , \qquad 0 < x_i < 6 .$$

Thus, the required condition on h is

(5.32) $\qquad h < \dfrac{2}{\underset{[0,6]}{\max} |2(2-x)|} = \dfrac{1}{4} .$

In Example 2, then, h was chosen *too* large. The fact that (5.32) yields a tridiagonal system which is diagonally dominant can be seen by writing out a few equations and will be proved, in general, in the following theorem. Moreover, when h satisfies inequality (5.32), it is easy to verify by direct substitution in (5.28), that the resulting unique numerical solution is $y_i = x_i - 1$, which agrees exactly with the analytical solution on the R_{n+1} set.

THEOREM 5.1. Let the boundary value problem (5.13)-(5.14) be defined on [a,b] where P(x), Q(x) and R(x) are continuous. Assume that $Q(x) \leq 0$ on [a,b]. Let M be a constant such that $|P(x)| \leq M$ on [a,b]. If

(5.33) $\qquad$ Mh < 2 ,

then the numerical solution obtained with the central difference approximation (5.19) exists and is unique.

PROOF. We need only show that the conditions of Theorem 1.2 are satisfied. To do this observe that (5.33) implies

$$hP(x_i) \leq hM < 2 , \qquad i=1,2,\ldots,n-1$$
$$-hP(x_i) \leq hM < 2 , \qquad i=1,2,\ldots,n-1$$

so that

(5.34) $\qquad$ $2 - hP(x_i) > 0 ,$ $\qquad$ $i=1,2,\ldots,n-1$
(5.35) $\qquad$ $2 + hP(x_i) > 0 ,$ $\qquad$ $i=1,2,\ldots,n-1.$

Thus, from (5.19), the subdiagonal and the superdiagonal elements of the matrix of the resulting tridiagonal system are positive. The main diagonal elements

(5.36) $-4+2h^2Q(x_i),$ $i=1,2,\ldots,n-1$

are all negative because $Q(x)\leq 0$. Finally, let us estabilish diagonal dominance. Consider any one of the equations of the tridiagonal system for which $i=2,3,\ldots,n-2$, that is, any one but the first and the last. The sum of the off diagonal elements is the sum of the left sides of (5.34) and (5.35), which is 4. The absolute value of the main diagonal element (5.36) is $4-h^2Q(x_i)$, which is greater than or equal to 4. Thus, diagonal dominance is not violated by the second through the $(n-2)$nd rows of the matrix of the system. Consider, then, the very first row of the matrix. It is the row vector

$$[\ -4+2h^2Q(x_1) \quad 2+hP(x_1) \quad 0 \quad 0 \quad 0 \quad \ldots \quad 0 \ ].$$

Since, by (5.34), $2-hP(x_1)>0$, then

$$2-hP(x_1) > 2h^2Q(x_1) .$$

Hence,

$$-2+hP(x_1) < -2h^2Q(x_1) .$$

Addition of 4 to both sides of the last inequality yields

$$2+hP(x_1) < 4-2h^2Q(x_1) ,$$

which is equivalent to

$$|2+hP(x_1)| < |-4+2h^2Q(x_1)| .$$

Thus, the first row of the matrix satisfies (1.4) with strict inequality. A similar argument holds for the last row and diagonal dominance is established. Thus, by Theorem 1.2, Theorem 5.1 is proved.

We have now two favorable properties for our numerical method. First, if h satisfies (5.33), the numerical solution exists and is unique. Second, the methods of Chapter I enable us to produce the numerical solution in a constructive fashion. Another important feature of the difference equation (5.19) is that the resulting numerical solution possesses discrete max-min properties, in analogy with continuous solutions. This is proved in the next theorem.

THEOREM 5.2. If Mh<2 and R(x)≡0, then a nonconstant numerical solution y_i of (5.19) cannot assume a positive maximum or a negative minimum at any interior grid point x_i of R_{n+1}. If in addition Q(x)≡0, then a nonconstant numerical solution y_i cannot assume its maximum or minimum at any interior grid point.

PROOF. For Mh<2 and R(x)≡0, assume that there exists an interior grid point, say x_k, 1≤k≤n–1, such that $y_k>0$ is a relative maximum, that is, $y_k \geq \max(y_{k-1}, y_{k+1})$. We will then show that y_i must be constant for i=0,1,2,...,n, which is a contradiction. From (5.19) it follows that

$$[4-2h^2Q(x_k)]y_k = [2-hP(x_k)]y_{k-1}+[2+hP(x_k)]y_{k+1} ,$$

or, equivalently,

(5.37) $$y_k = \frac{2-hP(x_k)}{4-2h^2Q(x_k)} y_{k-1}+ \frac{2+hP(x_k)}{4-2h^2Q(x_k)} y_{k+1} .$$

Now, since Q(x)≤0, and since hP(x)<2, we have

(5.38) $$\frac{2-hP(x_k)}{4-2h^2Q(x_k)} + \frac{2+hP(x_k)}{4-2h^2Q(x_k)} \leq 1 .$$

Note that strict inequality cannot apply in (5.38) since, otherwise, Lemma 1.2, applied to (5.37) would imply that $y_k<\max(y_{k-1},y_{k+1})$ unless $y_k=y_{k-1}=y_{k+1}=0$, which is a contradiction. Thus, it must follow that $Q(x_k)=0$ and Lemma 1.1 applies to (5.37) to yield $y_k \leq \max(y_{k-1},y_{k+1})$, where, if the equal sign applies then $y_k=y_{k-1}=y_{k+1}$. But, $y_k \geq \max(y_{k-1},y_{k+1})$, so that

(5.39) $$y_k = y_{k-1} = y_{k+1} .$$

But (5.39) implies that y_{k-1} and y_{k+1} are also relative maximum. Hence the above analysis can be repeated recursively on y_j, j=k–1,k–2,...,1, and j=k+1,k+2,...,n–1, to yield

$$y_0 = y_1 = y_2 = \cdots = y_n ,$$

Thus, y_i is a constant solution of (5.19) which is a contradiction. The proof that a nonconstant solution of (5.19) cannot assume a negative minimum at any point x_i, $i=1,2,\ldots,n-1$, can be developed in an entirely similar fashion.

The second part of Theorem 5.2 is proved by the fact that when $Q(x_i)=0$, $i=1,2,\ldots,n-1$, equation (5.37) reduces to

$$(5.40) \qquad y_k = \frac{2-hP(x_k)}{4} y_{k-1} + \frac{2+hP(x_k)}{4} y_{k+1} \cdot$$

Thus, since $hP(x_k)<2$, by Lemma 1.1, we have

$$(5.41) \qquad \min(y_{k-1},y_{k+1}) < y_k < \max(y_{k-1},y_{k+1}) \ ,$$

unless $y_k=y_{k-1}=y_{k+1}$. Hence, if y_i is nonconstant, it cannot assume its minimum or maximum for $1 \le i \le n-1$, and the theorem is proved.

As an example, note that in the boundary value problem (5.20)-(5.21) solved in Example 1, one has $R(x)\equiv 0$ and, accordingly, the computed numerical solution has neither a positive maximum, nor a negative minimum for $i=1,2,3,4,5$.

5.3 UPWIND DIFFERENCE METHOD FOR LINEAR BOUNDARY VALUE PROBLEMS

Thus far we have used central difference formula (3.37) to approximate y'. The reason for having chosen this formula instead of either the forward approximation (3.28) or the backward approximation (3.32) is that, in general, (3.37) is the more accurate one. This will be shown in the next example, and, more rigorously, it will be proved later in this chapter. However, there are problems of efficiency which are related to the use of the central difference method. Such problems will be studied in this section. For this purpose, let (5.16) be an approximation for y'', while y' will be approximated either with the forward difference formula

$$(5.42) \qquad y' \approx \frac{y_{i+1}-y_i}{h} \ ,$$

or with the backward difference formula

(5.43) $y' \approx \dfrac{y_i - y_{i-1}}{h}$.

EXAMPLE 1. Consider, again, boundary value problem (5.20)-(5.21):

(5.44) $y'' + \dfrac{x-3}{2} y' - y = 0$

(5.45) $y(0) = 11$, $y(6) = 11$.

Problem (5.44)-(5.45) was solved numerically in Example 1 of the last section by means of a central difference approximation and yielded the exact solution $y_i = x_i^2 - 6x_i + 11$ on the R_7 set. This time, consider again the same R_7 set $x_i = ih$, $i = 0,1,2,3,4,5,6$, so that $y_0 = 11$, $y_6 = 11$ and $h = 1$. Then, let us use the forward difference formula (5.42) to approximate $y'(x_i)$. The resulting difference equation for (5.44) becomes

(5.46) $\dfrac{y_{i+1} - 2y_i + y_{i-1}}{h^2} + \dfrac{x_i - 3}{2} \dfrac{y_{i+1} - y_i}{h} - y_i = 0$.

Rewriting (5.46) so that the coefficients of y_{i-1}, y_i, y_{i+1} are displayed clearly yields

(5.47) $2y_{i-1} - (3 + x_i)y_i + (x_i - 1)y_{i+1} = 0$.

For $i = 1,2,3,4,5$, the difference equation (5.47) is, respectively,

$$2y_0 - 4y_1 \qquad\quad = 0$$
$$2y_1 - 5y_2 + y_3 = 0$$
$$2y_2 - 6y_3 + 2y_4 = 0$$
$$2y_3 - 7y_4 + 3y_5 = 0$$
$$2y_4 - 8y_5 + 4y_6 = 0$$

into which, substitution of the known values y_0 and y_6 yields the system

$$-4y_1 \qquad\qquad\qquad = -22$$
$$2y_1 - 5y_2 + y_3 \qquad\quad = 0$$
$$2y_2 - 6y_3 + 2y_4 \quad\ = 0$$
$$2y_3 - 7y_4 + 3y_5 = 0$$
$$2y_4 - 8y_5 = -44 \ .$$

This system is diagonally dominant with strict inequality satisfied for each equation. Thus, by Theorem 1.3, it has a unique solution which, to two decimal places, is given by

$$y_1 = 5.50 \ , \qquad y_2 = 2.59 \ , \qquad y_3 = 1.95 \ , \qquad y_4 = 3.26 \ , \qquad y_5 = 6.32 \ .$$

Note that, while the central difference method yielded the same results as those of the analytical solution, the above method does not. The central difference formula is to be preferred to the forward or backward difference formulas because the corresponding numerical solution is, usually, *more* accurate. Unfortunately, however, the central difference formula cannot always apply to boundary value problems because the required limitation on h implies

$$Mh = M\frac{b-a}{n} < 2 \ ,$$

that is,

(5.48) $\qquad n > M\dfrac{b-a}{2} \ .$

Thus, for large M, the corresponding linear system to be solved may have too many equations, and hence, it may be *too* large for practical calculation. For example, if M=100 and b–a=500, then n must be greater than 2500. We will show next, then, that if one is willing to sacrifice a degree of accuracy, it will *always* be possible, by employing (5.42) and (5.43) judiciously, to solve a linear boundary value problem numerically for *any* given h with the assurance that the conditions of Theorem 1.2 are valid. Let us start with an illustrative example.

EXAMPLE 2. Consider, again, the boundary value problem (5.24)-(5.25) for which the central difference approximation required h<1/4. We now violate this condition by fixing h=1, so that [0,6] is divided into six equal parts by the R_7 set $x_0=0$, $x_1=1$, $x_2=2$, $x_3=3$, $x_4=4$, $x_5=5$, $x_6=6$. Of course,

$$y_0 = -1, \qquad y_6 = 5.$$

At the points x_i, $i=1,2,3,4,5$, let (5.24) be approximated by

(5.49) $\qquad \dfrac{y_{i+1} - 2y_i + y_{i-1}}{h^2} + 2(2-x_i)v_i = 2(2-x_i),$

where v_i are approximations, as yet unspecified, for y' at x_i, $i=1,2,3,4,5$, respectively. Equations (5.49) are equivalent to

(5.50) $\qquad -2y_1 + y_2 \qquad\qquad\qquad +2v_1 = 3$

(5.51) $\qquad\qquad y_1 - 2y_2 + y_3 \qquad\qquad\qquad\quad = 0$

(5.52) $\qquad\qquad\qquad y_2 - 2y_3 + y_4 \qquad -2v_3 = -2$

(5.53) $\qquad\qquad\qquad\qquad y_3 - 2y_4 + y_5 - 4v_4 = -4$

(5.54) $\qquad\qquad\qquad\qquad\qquad y_4 - 2y_5 - 6v_5 = -11$

In (5.50)-(5.54), we now choose one of the forward or backward formulas (5.42) or (5.43) to approximate y', rather than the more accurate central formula (5.18). We will do this carefully, however, so that system (5.50)-(5.54) satisfies the conditions of Theorem 1.2. Since the coefficient of v_1 is positive, let us choose the forward difference approximation (5.42)

$$v_1 = \frac{y_2 - y_1}{h} = y_2 - y_1,$$

for the approximation v_1 of $y'(x_1)$, so that the main diagonal becomes even more negative and, indeed, (5.50) becomes

$$-4y_1 + 3y_2 = 3.$$

Equation (5.51) requires no approximation for v_2. In (5.52), we have to choose the approximation v_3 for $y'(x_3)$. This time, since the coefficient of v_3 is negative, we choose the backward difference approximation (5.43)

$$v_3 = \frac{y_3 - y_2}{h} = y_3 - y_2,$$

so that (5.52) becomes

$$3y_2 - 4y_3 + y_4 = -2.$$

Similarly, using the backward difference approximation (5.43) at x_4 and x_5, (5.53) and (5.54) become, respectively,

$$5y_3 - 6y_4 + y_5 = -4$$
$$7y_4 - 8y_5 = -11.$$

Thus, system (5.50)-(5.54) becomes

$$
\begin{aligned}
-4y_1 + 3y_2 & & = 3 \\
y_1 - 2y_2 + y_3 & & = 0 \\
3y_2 - 4y_3 + y_4 & & = -2 \\
5y_3 - 6y_4 + y_5 & = -4 \\
7y_4 - 8y_5 & = -11
\end{aligned}
$$

whose solution exists and is unique, by Theorem 1.2. Indeed, the solution is $y_1=0$, $y_2=1$, $y_3=2$, $y_4=3$, $y_5=4$, which agrees exactly with the analytical solution $y=x-1$ of (5.24)-(5.25), and the example is complete.

The above example suggests, then, a second numerical method, called the **upwind difference method**, for general linear boundary value problems. Specifically, it is described as follows. Given the linear boundary value problem

(5.55) $y'' + P(x)y' + Q(x)y = R(x)$ $a < x < b$,

(5.56) $y(a) = \alpha$, $y(b) = \beta$,

divide the interval $[a,b]$ into n equal parts, each of length h, by the R_{n+1} set $x_i = a + ih$, $i = 0,1,2,\ldots,n$, and let $P(x_i)=P_i$, $Q(x_i)=Q_i$, $R(x_i)=R_i$. At each interior grid point x_i, $i=1,2,\ldots,n-1$, the upwind difference approximation for (5.55) is

(5.57) $\quad \dfrac{y_{i+1}-2y_i+y_{i-1}}{h^2}+P_i\dfrac{y_{i+1}-y_i}{h}+Q_iy_i = R_i\,, \qquad \text{if } P_i \geq 0\,,$

(5.58) $\quad \dfrac{y_{i+1}-2y_i+y_{i-1}}{h^2}+P_i\dfrac{y_i-y_{i-1}}{h}+Q_iy_i = R_i\,, \qquad \text{if } P_i < 0\,,$

or, more compactly,

(5.59) $\quad \dfrac{y_{i+1}-2y_i+y_{i-1}}{h^2}+\dfrac{(|P_i|+P_i)y_{i+1}-2|P_i|y_i+(|P_i|-P_i)y_{i-1}}{2h}+Q_iy_i = R_i\,, \quad i=1,2,\dots,n-1.$

If one then inserts the known boundary values y_0 and y_n, there results a tridiagonal linear algebraic system of $n-1$ equations in the $n-1$ unknowns $y_1, y_2, \dots, y_{n-1}$.

The upwind method, though less accurate than the central difference method described in the previous section, has the very same properties, but with *no restriction* on h. In particular, the following theorems are valid.

THEOREM 5.3. **Let the boundary value problem (5.55)-(5.56) be defined on [a,b], where P(x), Q(x) and R(x) are continuous. If Q(x)$\leq$0, then the numerical solution obtained with the upwind difference approximation (5.59) exists and is unique.**

PROOF. The proof follows in a fashion completely analogous to that of Theorem 5.1 by establishing the hypotheses of Theorem 1.2.

THEOREM 5.4. **If Q(x)$\leq$0 and R(x)$\equiv$0, then a nonconstant numerical solution y_i of (5.59) cannot assume a positive maximum or a negative minimum at any interior grid point x_i of R_{n+1}. If, in particular, Q(x)$\equiv$0, then a nonconstant numerical solution y_i cannot assume its minimum or maximum at any interior grid point.**

The proof of this theorem is entirely similar to that of Theorem 5.2.

As an example, note that in the boundary value problem of Example 1, above, one has R(x)$\equiv$0 and, accordingly, the resulting numerical solution has neither a positive maximum, nor a negative minimum at the interior grid points.

The importance of Theorems 5.1-5.4 for the central and upwind difference schemes can be summarized now as follows. Theorems 5.1 and 5.3 prove the *existence* and *uniqueness* of the numerical solution. Theorems 5.2 and 5.4 prove that any numerical solution has the same qualitative behaviour of the analytical solution as relates to *max-min* properties.

5.4 NUMERICAL SOLUTION OF MILDLY NONLINEAR BOUNDARY VALUE PROBLEMS

The methodology developed thus far for the numerical solution of linear boundary value problems extends to mildly nonlinear problems [L.Bers (1953), H.B.Keller (1968)]. A mildly nonlinear boundary value problem is defined as follows. Find a continuous function y(x) on [a,b] which has continuous second derivatives on a<x<b and which satisfies

(5.60) $\qquad y'' + P(x)y' + F(x,y) = 0$, $\qquad a<x<b$,

(5.61) $\qquad y(a) = \alpha$, $\qquad y(b) = \beta$.

Two conditions are sufficient to assure the existence and uniqueness of the analytical solution of (5.60)-(5.61). These are as follows. Assume first that P(x) is continuous on [a,b]. Thus, there exists a constant M such that

(5.62) $\qquad |P(x)| \le M$, $\qquad a \le x \le b$.

Second, we assume that F has continuous partial derivatives with respect to y which satisfy

(5.63) $\qquad \dfrac{\partial F(x,y)}{\partial y} \le 0$, $\qquad a \le x \le b$, $\qquad -\infty < y < \infty$.

Note, in particular, that a linear boundary value problem is a special case of (5.60)-(5.61) in which $F(x,y) = Q(x)y - R(x)$, so that condition (5.63) reduces to $Q(x) \le 0$.

Numerical methods for solving (5.60)-(5.61) can be derived easily from the methods described in the last two sections. Specifically, the interval [a,b] is first divided into n equal parts by the R_{n+1} set $x_i = a + ih$, $i = 0,1,2,\ldots,n$. Then, at each interior grid point x_i, $i = 1,2,\ldots,n-1$, substitution of (5.16) for y'' and any one of (5.17), (5.45) or (5.46) for y' into (5.60) yields a difference equation approximation of the differential equation. For example, use of (5.16) and the central difference formula (5.17) yields

$$\frac{y_{i+1} - 2y_i + y_{i-1}}{h^2} + P(x_i)\frac{y_{i+1} - y_{i-1}}{2h} + F(x_i,y_i) = 0 ,$$

or, equivalently,

(5.64) $\qquad [2-hP(x_i)]y_{i-1}-4y_i+[2+hP(x_i)]y_{i+1}+2h^2F(x_i,y_i) = 0$.

Now, set

$$f_1(y_1) = 2h^2F(x_1,y_1)+(2-hP_1)y_0 ,$$

$$f_i(y_i) = 2h^2F(x_i,y_i) , \qquad i=2,3,...n-2,$$

$$f_{n-1}(y_{n-1}) = 2h^2F(x_{n-1},y_{n-1})+(2+hP_{n-1})y_n .$$

If one then writes (5.64) consecutively for each of $i=1,2,...,n-1$, and inserts the known values (5.61), then there results the following mildly nonlinear system of $n-1$ equations in the $n-1$ unknowns $y_1, y_2, ..., y_{n-1}$:

$$-4y_1+(2+hP_1)y_2 \qquad\qquad\qquad\qquad +f_1(y_1) = 0$$

$$(2-hP_2)y_1-4y_2+(2+hP_2)y_3 \qquad\qquad +f_2(y_2) = 0$$

$$(2-hP_3)y_2-4y_3+(2+hP_3)y_4 \qquad\qquad +f_3(y_3) = 0$$

(5.65)
$$\qquad\qquad \cdot \qquad \cdot \qquad \cdot \qquad\qquad\qquad \cdot \qquad \cdot$$

$$\qquad\qquad\qquad \cdot \qquad \cdot \qquad \cdot \qquad\qquad \cdot \qquad \cdot$$

$$\qquad\qquad\qquad\qquad \cdot \qquad \cdot \qquad \cdot \qquad \cdot \qquad \cdot$$

$$(2-hP_{n-1})y_{n-2}-4y_{n-1}+f_{n-1}(y_{n-1}) = 0 .$$

In general, system (5.65) may not have a solution or may have more than one solution. Note, however, that if h satisfies the inequality

(5.66) $\qquad$ $Mh < 2$,

then system (5.65) satisfies all the assumptions of Theorem 1.4, and hence it has a ***unique*** solution. Such a solution can be found by the generalized Newton's method, which, in this case, converges for all initial guesses and for all ω in a subrange of $0<\omega<2$.

For those boundary value problems in which M is so large that the condition (5.66) is too restrictive, one can use the upwind difference approximation for $y'(x_i)$. In this case the difference approximation for (5.60) at x_i, is

$$\frac{y_{i+1}-2y_i+y_{i-1}}{h^2}+P_i\frac{y_{i+1}-y_i}{h}+F(x_i,y_i)=0\,,\qquad\text{if }\ P_i\geq 0\,,$$

$$\frac{y_{i+1}-2y_i+y_{i-1}}{h^2}+P_i\frac{y_i-y_{i-1}}{h}+F(x_i,y_i)=0\,,\qquad\text{if }\ P_i<0\,,$$

or, more compactly,

(5.67) $$\frac{y_{i+1}-2y_i+y_{i-1}}{h^2}+\frac{(|P_i|+P_i)y_{i+1}-2|P_i|y_i+(|P_i|-P_i)y_{i-1}}{2h}+F(x_i,y_i)=0\,,\qquad i=1,2,\ldots,n-1.$$

Equations (5.67) constitute a mildly nonlinear system which satisfies all the assumptions of Theorem 1.4 for *any* given h. Thus, a unique numerical solution exists and can be determined with the generalized Newton's method.

Of course, if $P(x)\equiv 0$, the central difference method and the upwind difference method are the same.

EXAMPLE. Consider the mildly nonlinear boundary value problem

$$y''-e^y = 0\,,\qquad\qquad y(0) = 0\,,\qquad\qquad y(1) = 0\,.$$

In this case, $F(x,y)=-e^y$, so that $(\partial F/\partial y)=-e^y<0$, and $P(x)\equiv 0$. Thus, the solution exists and is unique. Numerically, then, let $h=0.2$ and divide $[0,1]$ into five equal parts by the R_6 set $x_0=0$, $x_1=0.2$, $x_2=0.4$, $x_3=0.6$, $x_4=0.8$, $x_5=1$. Let $y(x_i)=y_i$, so that $y_0=0$ and $y_5=0$. To approximate y_1, y_2, y_3, y_4, the differential equation is approximated by the following difference equation:

$$\frac{y_{i+1}-2y_i+y_{i-1}}{(0.2)^2}-e^{y_i}=0\,,\qquad i=1,2,3,4.$$

or, equivalently,

(5.68) $$25y_{i-1}-50y_i+25y_{i+1}-e^{y_i}=0\,,\qquad\qquad i=1,2,3,4.$$

For each $i=1,2,3,4$, (5.68) yields the following mildly nonlinear system

$$-50y_1 + 25y_2 \qquad -e^{y_1} = 0$$

$$25y_1 - 50y_2 + 25y_3 \qquad -e^{y_2} = 0$$

$$25y_2 - 50y_3 + 25y_4 - e^{y_3} = 0$$

$$25y_3 - 50y_4 - e^{y_4} = 0$$

For this system the generalized Newton's formulas (1.76) are

$$y_1^{(k+1)} = y_1^{(k)} - \omega \frac{-50y_1^{(k)} + 25y_2^{(k)} - e^{y_1^{(k)}}}{-50 - e^{y_1^{(k)}}}$$

$$y_2^{(k+1)} = y_2^{(k)} - \omega \frac{25y_1^{(k+1)} - 50y_2^{(k)} + 25y_3^{(k)} - e^{y_2^{(k)}}}{-50 - e^{y_2^{(k)}}}$$

$$y_3^{(k+1)} = y_3^{(k)} - \omega \frac{25y_2^{(k+1)} - 50y_3^{(k)} + 25y_4^{(k)} - e^{y_3^{(k)}}}{-50 - e^{y_3^{(k)}}}$$

$$y_4^{(k+1)} = y_4^{(k)} - \omega \frac{25y_3^{(k+1)} - 50y_4^{(k)} - e^{y_4^{(k)}}}{-50 - e^{y_4^{(k)}}} .$$

After making a zero initial guess for the solution and using $\omega=1.3$, these formulas converged in eight iterations to the numerical solution, which, to four decimal places, is

$$y_1 = -0.0731 , \qquad y_2 = -0.1089 , \qquad y_3 = -0.1089 , \qquad y_4 = -0.0731 .$$

*5.5 CONVERGENCE OF DIFFERENCE METHODS FOR BOUNDARY VALUE PROBLEMS

Numerical solutions of *certain* differential equation problems have the following peculiarity. As h converges to zero, the difference equation converges to the differential equation, *but* the numerical solution does *not* converge to the analytical solution [G.E.Forsythe and W.Wasow (1960), p.18; G.I.Marchuk (1975), pp. 30-33, 279-281]. One of our main concerns in this section is to show that, under rather general conditions, this cannot happen when the numerical methods of Sections 5.2-5.4 are applied. This important result will be established next.

Throughout, we will consider mildly nonlinear boundary value problem (5.60)-(5.61) under the assumptions that (5.62) and (5.63) are satisfied, so that the the analytical solution exists and is unique. Of course, the results will apply to linear boundary value problems as well.

In solving (5.60)-(5.61) numerically, consider, first, the central difference approximation

$$(5.69) \qquad \frac{y_{i+1}-2y_i+y_{i-1}}{h^2} + P(x_i)\frac{y_{i+1}-y_{i-1}}{2h} + F(x_i,y_i) = 0 \ .$$

We now prove a lemma about difference operators which will aid in establishing an error bound theorem. This theorem will then be used to prove the convergence theorem.

LEMMA 5.1. Let $P(x_i)=P_i$ and $Q(x_i)=Q_i$ be two discrete functions defined on the R_{n+1} set $x_i=a+ih$, $i=0,1,2,...,n$. Assume that $|P_i|\leq M$. Assume also that there exists a positive number **w** such that

$$(5.70) \qquad Q_i \leq -w < 0 \ , \qquad i=1,2,...,n-1.$$

Set $C=\max(1,1/w)$. Let $E(x_i)=E_i$, $i=0,1,2,...,n$, be an arbitrary discrete function defined on the same R_{n+1} set. At the interior grid points x_1, x_2, ..., x_{n-1}, define L_h, called a difference operator, by

$$(5.71) \qquad L_h E_i = \frac{E_{i+1}-2E_i+E_{i-1}}{h^2} + P_i\frac{E_{i+1}-E_{i-1}}{2h} + Q_i E_i \ .$$

If Mh<2, then

$$(5.72) \qquad |E_i| \leq C\left[\max(|E_0|,|E_n|)+(\max_{0<j<n}|L_h E_j|)\right] \ , \qquad i=0,1,2,...,n.$$

PROOF. Note first that $C\geq 1$. Thus, if $\max|E_i|$ occurs for i=0 or i=n, then the lemma is valid because, for i=0,1,2,...,n,

$$|E_i| \leq \max_{0\leq j\leq n}|E_j| \leq C[\max(|E_0|,|E_n|)] \leq C[\max(|E_0|,|E_n|)+(\max_{0<j<n}|L_h E_j|)] \ .$$

Suppose then that $\max|E_i|$ occurs for one of $i=1,2,\ldots,n-1$. We now rewrite (5.71) as

$$[2-hP_i]E_{i-1}-[4-2h^2Q_i]E_i+[2+hP_i]E_{i+1} = 2h^2L_hE_i$$

so that

(5.73) $\qquad [4-2h^2Q_i]E_i = [2-hP_i]E_{i-1}+[2+hP_i]E_{i+1}-2h^2L_hE_i \; .$

Note that, by (5.70),

$$4-2h^2Q_i \geq 4+2h^2w \; .$$

Thus, by taking the absolute values of both sides of (5.73), one finds

$$[4+2h^2w]|E_i| \leq |2-hP_i||E_{i-1}| + |2 + hP_i||E_{i+1}|+2h^2|L_hE_i|$$

$$\leq |2-hP_i|(\max_{0<j<n}|E_j|)+|2+hP_i|(\max_{0<j<n}|E_j|)+2h^2(\max_{0<j<n}|L_hE_j|) \; ,$$

so that

$$[4+2h^2w]|E_i| \leq 4(\max_{0<j<n}|E_j|)+2h^2(\max_{0<j<n}|L_hE_j|) \; .$$

However, the right-hand side of the last inequality is independent of i, which means that

$$[4+2h^2w](\max_{0<i<n}|E_i|) \leq 4(\max_{0<j<n}|E_j|)+2h^2(\max_{0<j<n}|L_hE_j|) \; ,$$

which, since $\max_{0<i<n}|E_i|=\max_{0<j<n}|E_j|$, implies

$$2h^2w(\max_{0<j<n}|E_j|) \leq 2h^2(\max_{0<j<n}|L_hE_j|) \; .$$

Hence

$$\max_{0<j<n}|E_j| \le \frac{1}{w}(\max_{0<j<n}|L_hE_j|) \le C(\max_{0<j<n}|L_hE_j|) \le C[\max(|E_0|,|E_n|)+\max_{0<j<n}|L_hE_j|)] \ ,$$

and the lemma follows readily.

Now, convergence theorems are usually *very* difficult to prove and, often, one finds it embarassingly convenient to assume more than one actually needs numerically. This will be the case for the theorems which follow, in order to avoid the need for advanced mathematical theory. We will indicate which these additional assumptions are as we proceed.

THEOREM 5.5. **Let $Y(x)$ be the analytical solution of the mildly nonlinear boundary value problem (5.60)-(5.61), where $P(x)$ is continuous on $[a,b]$, so that (5.62) is valid. Assume (this is an additional assumption) that $\partial F/\partial y$ exists and satisfies the following inequality**

$$\frac{\partial F(x,y)}{\partial y} \le -w < 0 \ , \qquad a \le x \le b \ , \quad -\infty < y < \infty \ .$$

Set $C=\max(1,1/w)$. Let y_i, $i=0,1,2,\dots,n$, be the numerical solution of (5.60)-(5.61) obtained by the central difference method, and let $E(x_i)=E_i$ be the error, defined by $E_i=Y(x_i)-y_i$. Assume also (another additional assumption) that $Y(x)$ has continuous derivatives up to and including order four on $[a,b]$. If $Mh<2$, then

(5.74) $$|E_i| \le C\frac{h^2}{12}(M_4+2MM_3) \ , \qquad i=0,1,2,\dots,n,$$

where

$$M_3 = \max\left|\frac{d^3Y}{dx^3}\right| \ , \qquad M_4 = \max\left|\frac{d^4Y}{dx^4}\right| \ .$$

PROOF. For $i=0$ and for $i=n$, $|E_0|=|Y(x_0)-y_0|=|\alpha-\alpha|=0$ and $|E_n|=|Y(x_n)-y_n|=|\beta-\beta|=0$, so that (5.74) is valid. We must, then, prove (5.74) for $i=1,2,\dots,n-1$.

Since $Y(x)$ is the analytical solution of (5.60)-(5.61), we have,

$$Y''(x_i)+P(x_i)Y'(x_i)+F(x_i,Y(x_i)) = 0 \ ,$$

which, by (3.37), (3.38), (3.40) and (3.41), implies

(5.75)
$$\left[\frac{Y(x_{i+1})-2Y(x_i)+Y(x_{i-1})}{h^2}+\frac{h^2}{24}[Y^{iv}(\xi_1)+Y^{iv}(\xi_2)]\right]$$

$$+P(x_i)\left[\frac{Y(x_{i+1})-Y(x_{i-1})}{2h}+\frac{h^2}{12}[Y'''(\xi_3)+Y'''(\xi_4)]\right]+F(x_i,Y(x_i))=0 .$$

Now, by subtracting (5.69) from (5.75), one finds

$$\frac{E_{i+1}-2E_i+E_{i-1}}{h^2}+P(x_i)\frac{E_{i+1}-E_{i-1}}{2h}+F(x_i,Y(x_i))-F(x_i,y_i)$$

$$=-\frac{h^2}{24}[Y^{iv}(\xi_1)+Y^{iv}(\xi_2)]-\frac{h^2}{12}P(x_i)[Y'''(\xi_3)+Y'''(\xi_4)] ,$$

which implies

$$\frac{E_{i+1}-2E_i+E_{i-1}}{h^2}+P(x_i)\frac{E_{i+1}-E_{i-1}}{2h}+\frac{\partial F(x_i,\mu_i)}{\partial y}E_i$$

$$=-\frac{h^2}{24}[Y^{iv}(\xi_1)+Y^{iv}(\xi_2)]-\frac{h^2}{12}P(x_i)[Y'''(\xi_3)+Y'''(\xi_4)] .$$

By setting

$$P_i = P(x_i) , \qquad\qquad Q_i = \frac{\partial F(x_i,\mu_i)}{\partial y} ,$$

the last equation can be written as

$$L_hE_i = -\frac{h^2}{24}[Y^{iv}(\xi_1)+Y^{iv}(\xi_2)]-\frac{h^2}{12}P(x_i)[Y'''(\xi_3)+Y'''(\xi_4)] ,$$

where L_h is defined by (5.71). Thus, we have

$$|L_h E_i| \le \frac{h^2}{12} M_4 + \frac{h^2}{6} M M_3 .$$

Hence, by Lemma 5.1,

$$|E_i| \le C[\max(|E_0|,|E_n|) + (\max_{0<j<n}|L_h E_j|)] = C \max_{0<j<n}(|L_h E_j|) \le C \frac{h^2}{12}(M_4 + 2M M_3) ,$$

and the theorem is proved.

Of course, (5.74) is an error bound for the numerical solution of any mildly nonlinear boundary value problem which satisfies all the prescribed conditions. This error bound is of little direct value because we rarely know M_3 and M_4. However, it is the key to the next theorem, which is the convergence theorem.

THEOREM 5.6. Under the assumptions of Theorem 5.5, the numerical solution obtained with the central difference method converges to the analytical solution as h→0, that is,

$$\lim_{h \to 0} |Y(x_i) - y_i| = 0 , \qquad\qquad i = 0,1,2,...,n.$$

The proof follows directly from (5.74).

Note that the theory developed above tells us almost immediately why Examples 1 and 3 of Section 5.2 yielded numerical results which agreed exactly with the analytical solution. In both examples, $M_3 = M_4 = 0$, so that |E| in (5.74) is zero. Thus, it was the simplicity of the *solutions*, not really the simplicity of the problems, which yielded the *unusually* good numerical results.

Theorems 5.5 and 5.6 provide an error bound and the convergence of the the numerical solution obtained by using the central difference method for mildly nonlinear boundary value problems. An error bound and a convergence theorem for the upwind method can be established in a similar fashion. Specifically, the analogue of Lemma 5.1 is the following.

LEMMA 5.2. Let $P(x_i) = P_i$ and $Q(x_i) = Q_i$ be two discrete functions defined on the R_{n+1} set $x_i = a + ih$, $i = 0,1,2,...,n$. Assume that there exists a positive number w such that $Q_i \le -w < 0$, $i = 1,2,...,n-1$. Set $C = \max(1, 1/w)$. Let $E(x_i) = E_i$, $i = 0,1,2,...,n$ be another discrete function defined on the same R_{n+1} set. At the interior grid points x_1, x_2, ..., x_{n-1}, define the difference operator L_h by

$$L_h E_i = \frac{E_{i+1} - 2E_i + E_{i-1}}{h^2} + \frac{(|P_i| + P_i)E_{i+1} - 2|P_i|E_i + (|P_i| - P_i)E_{i-1}}{2h} + Q_i E_i .$$

Then,

$$|E_i| \le C\left[\max(|E_0|, |E_n|) + (\max_{0 < j < n} |L_h E_j|) \right] , \qquad i = 0, 1, 2, \dots, n.$$

The proof of this lemma is similar to the proof given for Lemma 5.1.

We are now ready for an error bound theorem concerning the upwind difference method.

THEOREM 5.7. Let $Y(x)$ be the analytical solution of the mildly nonlinear boundary value problem (5.60)-(5.61), where $P(x)$ is continuous on $[a,b]$, so that (5.62) is valid. Assume (this is an additional assumption) that $\partial F/\partial y$ exists and satisfies the following inequality:

$$\frac{\partial F(x,y)}{\partial y} \le -w < 0 , \qquad a \le x \le b , \qquad -\infty < y < \infty .$$

Set $C = \max(1, 1/w)$. Let y_i, $i = 0, 1, 2, \dots, n$ be the numerical solution of (5.60)-(5.61) obtained by the upwind difference method, and let $E(x_i) = E_i$ be the error, defined by $E_i = Y(x_i) - y_i$. Assume also (another additional assumption) that $Y(x)$ has continuous derivatives up to and including order four on $[a,b]$. Then

(5.76) $$|E_i| \le C\, \frac{h}{12}\left(h M_4 + 6 M M_2\right) , \qquad i = 0, 1, 2, \dots, n,$$

where

$$M_2 = \max\left|\frac{d^2 Y}{dx^2}\right| , \qquad M_4 = \max\left|\frac{d^4 Y}{dx^4}\right| .$$

The proof of Theorem 5.7 is similar to that of Theorem 5.5.

We state, next, a final theorem, which is the convergence theorem for the upwind difference method.

THEOREM 5.8. Under the assumptions of Theorem 5.7, the numerical solution obtained with the upwind difference method converges to the analytical solution as $h \to 0$, that is,

$$\lim_{h \to 0} |Y(x_i) - y_i| = 0 , \qquad\qquad i = 0, 1, 2, \dots, n.$$

The proof follows directly from (5.76).

Theorem 5.7 suggests why Example 1 of Section 5.3 yielded only an approximation for the analytical solution $y = x^2 - 6x + 11$. In that example, in fact, $M_4 = 0$, but $M \neq 0$ and $M_2 \neq 0$. Example 2 in Section 5.3 yielded numerical results which agreed exactly with the analytical solution $y = x - 1$ since, now, $M_2 = M_4 = 0$.

In general, the reason why the central difference method, when applicable, is to be preferred to the upwind method can be explained by the respective error bounds, for, if h is small, then h^2 in (5.74) is smaller than h in (5.76). Note, however, that no restriction on h is required in Lemma 5.2 and in Theorems 5.7-5.8.

*5.6 THE FINITE ELEMENT METHOD

An important class of boundary value problems arises from the need to minimize a special type of integral called a functional. Minimization of such an integral with the aid of one of the interpolation formulas from Chapter 2 is called a *finite element* method. It is to such methods that this section is directed. For clarity, the discussion begins with a summary of essential background material from the classical mathematical discipline called the *calculus of variations*.

The *fundamental problem* in the calculus of variations is as follows. For a, b, α and β real numbers, with a<b, and for given F(x,y,p) which has continuous first-order partial derivatives, find a function y(x) which is defined and has continuous first-order derivatives on [a,b], which satisfies the boundary conditions

$$(5.77) \qquad y(a) = \alpha , \qquad y(b) = \beta ,$$

and which minimizes the integral

$$(5.78) \qquad J = \int_a^b F(x, y, y')dx .$$

Because the value of J depends on a function, and not just on a real number, the integral (5.78) is in reality a function of a function, and is therefore called a *functional*.

Analytically, the fundamental problem is, in general, exceptionally difficult to solve. As in elementary calculus, where one attempts to find a minimum of a function $y=f(x)$ by solving $f'(x)=0$, so in the calculus of variations one can attempt to find the minimum of J by solving the equation

(5.79) $\dfrac{\partial F}{\partial y} - \dfrac{d}{dx}\left(\dfrac{\partial F}{\partial y'}\right) = 0$.

This can be seen to be correct from the following heuristic argument. Let $y(x)$ minimize (5.78) and satisfy boundary conditions (5.77). Moreover, let $y(x)$ be unique. Let $z(x)$ be a function with continuous first-order derivative which satisfies

$z(a) = z(b) = 0$.

Consider, then, the family of functions

(5.80) $Y(x) = y(x)+\varepsilon z(x)$,

where ε is a real parameter. From (5.78) and (5.80), one has

$$J(Y) = \int_a^b F(x,y+\varepsilon z,y'+\varepsilon z')dx \ .$$

Since y and z are known functions, $J(Y)$ is a function only of ε, that is, $J(Y)=I(\varepsilon)$. Thus,

(5.81) $I(\varepsilon) = \int_a^b F(x,y+\varepsilon z,y'+\varepsilon z')dx \ .$

Now, since $y(x)$ minimizes (5.78), it follows that $(dI/d\varepsilon)=0$ when $\varepsilon=0$, so that, from (5.81),

(5.82) $\dfrac{dI}{d\varepsilon}(0) = \displaystyle\int_a^b \left(z\dfrac{\partial F}{\partial y} + z'\dfrac{\partial F}{\partial y'} \right) dx = 0$.

However, using integration by parts, one finds

$$\int_a^b z'\frac{\partial F}{\partial y'}\,dx = z\frac{\partial F}{\partial y'}\Big|_a^b - \int_a^b z\frac{d}{dx}\left(\frac{\partial F}{\partial y'}\right)dx = -\int_a^b z\frac{d}{dx}\left(\frac{\partial F}{\partial y'}\right)dx\ ,$$

so that, from (5.82),

(5.83) $\displaystyle\int_a^b \left[\dfrac{\partial F}{\partial y} - \dfrac{d}{dx}\left(\dfrac{\partial F}{\partial y'}\right)\right] z\, dx = 0$.

Finally, since (5.83) is valid for any function z(x), it follows readily that

$$\frac{\partial F}{\partial y} - \frac{d}{dx}\left(\frac{\partial F}{\partial y'}\right) = 0\ ,$$

which is (5.79).

Equation (5.79) is called the Euler differential equation of functional (5.78).

EXAMPLE. The Euler equation of the functional

$$\int_0^1 [xy^3 - (y')^2 + 3xyy']\,dx$$

is

$$3xy^2 + 3xy' - \frac{d}{dx}(-2y' + 3xy) = 0\ ,$$

or, equivalently,

$$2y'' - 3y + 3xy^2 = 0 \ .$$

As in the above example, the Euler equation (5.79) is, in general, a nonlinear, second-order, ordinary differential equation, and, although such equations are very difficult to solve, still they seem to be more accessible analytically than the functionals from which they are derived. For this reason, in precomputer days the fundamental problem (5.77)-(5.78) was usually studied by replacing it with the two-point boundary value problem defined by (5.77) and the second-order differential equation (5.79). Nevertheless, *numerically,* it is now often easier to approximate a solution of the original variational problem. Hence, the approach in this section will be to consider boundary value problems which are usually stated in terms of (5.77) and (5.79) by considering their primitive variational formulation. For this purpose, we develop, next, a numerical method which is called a finite element method.

Consider the fundamental problem of the calculus of variations. Divide the interval $[a,b]$ into n parts by the R_{n+1} set $a=x_0<x_1<x_2<\ldots<x_n=b$. Each such subdivision is called an element and, in general, the length of the elements need *not* be equal, though, for simplicity, we will take them all to be equal. Set $h=(b-a)/n$. Let $y_i=y(x_i)$, $i=0,1,2,\ldots,n$, so that $y_0=\alpha$, $y_n=\beta$, while $y_1, y_2, \ldots, y_{n-1}$ are unknowns. Next, rewrite the functional as

$$(5.84) \qquad J = \int_{x_0}^{x_1} F(x,y,y')dx + \int_{x_1}^{x_2} F(x,y,y')dx + \ldots + \int_{x_{n-1}}^{x_n} F(x,y,y')dx$$

and choose any piecewise interpolating function from Chapter 2. For illustrative purposes, let us choose the piecewise linear interpolating function $L(x)$ determined by the points $(x_0,y_0),(x_1,y_1), (x_2,y_2), \ldots, (x_n,y_n)$, so that

$$(5.85) \qquad L(x) = y_i + \frac{y_{i+1}-y_i}{h}(x-x_i) \ , \qquad x_i \leq x \leq x_{i+1} \ , \qquad i=0,1,2,\ldots,n-1.$$

Substitution of L for y and L' for y' in (5.84), and assuming that the integrals can be carried out exactly, yields

$$J_{n-1} = J_{n-1}(y_1,y_2,\ldots,y_{n-1}) \ .$$

One next minimizes J_{n-1} by solving the system

$$\frac{\partial J_{n-1}}{\partial y_i} = 0 , \qquad i=1,2,\ldots,n-1,$$

the solution of which, when substituted into (5.85), yields a continuous, piecewise linear approximation for the exact solution y(x).

EXAMPLE. Consider the problem of minimizing the functional

$$(5.86) \qquad J = \int_0^6 [(y')^2 + y^2 - 2y - 2xy] dx ,$$

subject to the boundary conditions

$$(5.87) \qquad y(0) = 1 , \qquad y(6) = 7 .$$

Let [0,6] be divided into three equal parts of length h=2 by the R_4 set $x_0=0$, $x_1=2$, $x_2=4$, $x_3=6$. On this R_4 set let $y_i=y(x_i)$ so that $y_0=1$, $y_3=7$, while y_1 and y_2 are unknowns. Setting

$$(5.88) \qquad F(x,y,y') = (y')^2 + y^2 - 2y - 2xy$$

implies

$$(5.89) \qquad J = \int_0^2 F(x,y,y') dx + \int_2^4 F(x,y,y') dx + \int_4^6 F(x,y,y') dx .$$

Taking

$$(5.90) \qquad L(x) = \begin{cases} y_0 + (y_1 - y_0)x/2, & 0 \le x \le 2 \\ y_1 + (y_2 - y_1)(x-2)/2, & 2 \le x \le 4 \\ y_2 + (y_3 - y_2)(x-4)/2, & 4 \le x \le 6 \end{cases}$$

yields

$$
(5.91) \qquad L'(x)=
\begin{cases}
(y_1-y_0)/2, & 0\le x\le 2 \\[2mm]
(y_2-y_1)/2, & 2\le x\le 4 \\[2mm]
(y_3-y_2)/2, & 4\le x\le 6 .
\end{cases}
$$

Substitution of (5.90) for y and (5.91) for y' into (5.89) yields by means of (5.88)

$$
J = \int_0^2 \left[\left(\tfrac{y_1-y_0}{2}\right)^2+\left(y_0+\tfrac{y_1-y_0}{2}x\right)^2-2\left(y_0+\tfrac{y_1-y_0}{2}x\right)-2x\left(y_0+\tfrac{y_1-y_0}{2}x\right)\right]dx
$$

$$
+\int_2^4 \left[\left(\tfrac{y_2-y_1}{2}\right)^2+\left(y_1+\tfrac{y_2-y_1}{2}(x-2)\right)^2-2\left(y_1+\tfrac{y_2-y_1}{2}(x-2)\right)-2x\left(y_1+\tfrac{y_2-y_1}{2}(x-2)\right)\right]dx
$$

$$
+\int_4^6 \left[\left(\tfrac{y_3-y_2}{2}\right)^2+\left(y_2+\tfrac{y_3-y_2}{2}(x-4)\right)^2-2\left(y_2+\tfrac{y_3-y_2}{2}(x-4)\right)-2x\left(y_2+\tfrac{y_3-y_2}{2}(x-4)\right)\right]dx ,
$$

which, after substitution of the known values for y_0 and y_3, yields by termwise integration

$$
J_2 = \tfrac{7}{3}(y_1^2+y_2^2) - \tfrac{1}{3}y_1y_2 - \tfrac{37}{3}y_1 - \tfrac{67}{3}y_2 - \tfrac{101}{3} .
$$

Thus, the equations for y_1 and y_2 are

$$
\frac{\partial J_2}{\partial y_1} = \tfrac{14}{3}y_1 - \tfrac{1}{3}y_2 - \tfrac{37}{3} = 0
$$

$$
\frac{\partial J_2}{\partial y_2} = -\tfrac{1}{3}y_1 + \tfrac{14}{3}y_2 - \tfrac{67}{3} = 0 ,
$$

the solution of which is

$$
(5.92) \qquad y_1 = 3 , \qquad\qquad y_2 = 5 .
$$

Substitution of (5.92) into (5.90) yields for the approximate solution

(5.93) $L(x) = 1+x$, $0 \le x \le 6$,

Note that the Euler equation of (5.86) is

(5.94) $y'' - y = -(1+x)$

and the solution of (5.94) subject to boundary conditions (5.87) is, indeed,

(5.95) $y = 1+x$.

The fact that (5.93) and (5.95) are identical is a consequence only of the simplicity of the problem, for, in general, these functions will differ.

5.7 DIFFERENTIAL EIGENVALUE PROBLEMS

Eigenvalue and eigenvector problems for matrices were discussed in Section 1.7. Problems with the same names arise also in the study of differential equations. For example, in the study of atoms and molecules, the following type problem arises and is called an eigenvalue problem. For what positive value of λ does the boundary value problem

(5.96) $y'' + \lambda^2 y = 0$, $y(0) = y(\pi) = 0$ $(\lambda > 0)$

have nonzero solutions? We will call such a problem a differential eigenvalue problem. The nonzero, positive values λ are called the eigenvalues and the associated solutions of the differential equation are called the eigenvectors. Note, in particular, that uniqueness condition (5.15) is not valid in (5.96).

Because of its simplicity, (5.96) can be solved exactly as follows. The general solution of the differential equation is

(5.97) $y(x) = c_1 \sin(\lambda x) + c_2 \cos(\lambda x)$.

However, the boundary conditions imply only that $c_2 = 0$, so that

(5.98) $y(x) = c_1 \sin(\lambda x)$.

The boundary conditions imply also, with respect to (5.98), that λ can take on any of the values $\lambda=1,2,3,\ldots$, and only these values, since $\sin(\lambda x)$ must be zero when $x=\pi$. Thus the eigenvalues are $\lambda=1,2,3,\ldots$, and the eigenvectors for each λ are given by (5.98), where c_1 is any nonzero constant.

If the differential equation in (5.96) were, however, more complex, then one might not be able to carry through an exact analysis, as was done above. In such cases, one could approximate differential eigenvalue problem solutions by adapting one of the numerical methods already described, the choice being dependent on the physical results desired. To illustrate, suppose one were interested primarily in the minimum eigenvalue of (5.96), which is often of interest in atomic and molecular studies. One could then apply a difference method as follows.

Divide the interval $[0,\pi]$ into four equal parts, each of length $h=\pi/4$, by the R_5 set $x_0=0$, $x_1=\pi/4$, $x_2=\pi/2$, $x_3=3\pi/4$, $x_4=\pi$. At each interior grid point approximate (5.96) by

$$y_{i+1}-2y_i+y_{i-1}+\lambda^2(\pi/4)^2 y_i = 0 .$$

Setting $\mu=(\lambda\pi/4)^2$ and $i=1,2,3$, yields the three equations

$$
\begin{aligned}
(\mu-2)y_1 \quad + \; y_2 \qquad\qquad &= 0 \\
y_1+(\mu-2)y_2 \quad + \; y_3 &= 0 \\
y_2+(\mu-2)y_3 &= 0 .
\end{aligned}
$$

For this system to have a nonzero solution, the determinant of the system must be zero, or, equivalently,

$$(\mu-2)^3-2(\mu-2) = 0 ,$$

which yields

$$\mu = 2-\sqrt{2}, \; 2, \; 2+\sqrt{2} .$$

The minimum root μ is then $2-\sqrt{2}$, which yields, approximately, a minimum λ given by

(5.99) $\lambda = 0.974$.

The exact solution, as described above, is $\lambda=1$.

In order to improve on (5.99), one need only decrease the grid size. If, for example, one halves the grid size, so that $h=\pi/8$, and then repeats the procedure with $\mu=(\lambda\pi/8)^2$, then the equation one has to solve for μ is

(5.100) $\mu^7-14\mu^6+78\mu^5-220\mu^4+330\mu^3-252\mu^2+84\mu-8 = 0$.

Computer evaluation of the left side of (5.100) for $\mu=0$, 0.1, 0.2, ..., 9.9, 10, reveals quickly by the sign changes that all seven roots of (5.100) are real, positive and lie in the interval [0.1,3.9]. In particular, the smallest root lies in [0.1,0.2]. Application of Newton's method on this interval yields $\mu=0.152$, from which it follows that the minimum λ is

$\lambda = 0.993$,

which is a significant improvement over (5.99). Note that use of the grid size $\pi/4$ yields only three approximate eigenvalues, while use of the grid size $\pi/8$ yields seven approximations. In general, division of $[0,\pi]$ into n parts results in n−1 approximate eigenvalues, even though the number of exact eigenvalues is infinite.

EXERCISES

Basic Exercises

1. Show that each of the following boundary value problems has a unique solution by showing that the sufficient conditions described in Section 5.2 are satisfied. Then, solve each exactly. Finally, using h=0.2 and the central difference method, solve each numerically to three decimal places.

(a) $y'' = x$, $y(0) = 0$, $y(1) = 1$
(b) $y''-4y'-3y = 0$, $y(0) = 0$, $y(1) = 1$
(c) $y''-y = x$, $y(0) = 0$, $y(1) = -1$.

2.   Give an example to show that the condition $Q(x) \le 0$ is not necessary for boundary value problem (5.13)-(5.14) to have a unique solution.

3.   Determine what choice of h will assure, a priori, that the numerical solution of

$$y'' + 4\sin(x)y' - 4\cos(x)y = -\sin(x), \qquad y(0) = 0, \qquad y(\pi/2) = 1,$$

with the central difference method exists and is unique.

4.   Using $h = \pi/100$ and the central difference method, find to four decimal places the numerical solution of

$$y'' + 4\sin(x)y' - 4\cos(x)y = -\sin(x), \qquad y(0) = 0, \qquad y(\pi/2) = 1.$$

Then, compare your result with the exact solution $y = \sin(x)$.

5.   Using $h = 1/3$, find a numerical solution of the boundary value problem

$$y'' + 8xy' - y = 7x, \qquad y(0) = 0, \qquad y(1) = 1.$$

Be sure that the tridiagonal system which you set up has a unique solution. Compare your result with the exact solution $y = x$.

6.   Give a complete proof of Theorem 5.3.

7.   Give a complete proof of Theorem 5.4.

8.   Using $h = 0.2$ and calculating to four decimal places, find an approximate solution for each of the following boundary value problems.

(a) $y'' = e^y$, $y(0) = 0$, $y(1) = 1$
(b) $y'' + y' = e^y$, $y(0) = 0$, $y(1) = 1$
(c) $y'' - 100y' = e^y$, $y(0) = 0$, $y(1) = 1$
(d) $y'' = y^3$, $y(0) = 0$, $y(1) = 1$.

9.   For the following choices of E(x), P(x) and Q(x), determine $L_h E(x_i)$ from (5.71).

(a) $E(x) = 1$, $P(x) = 1$, $Q(x) = -1$
(b) $E(x) = x$, $P(x) = 1$, $Q(x) = -1$
(c) $E(x) = x^2$, $P(x) = x$, $Q(x) = -1$.

10. For each of the following boundary value problems, find an estimate for the constant
     C of Lemma 5.1.

(a) $y'' + y' - y = 0$, $y(0) = y(1) = 2$
(b) $y'' - 2y' - 4y = 0$, $y(0) = y(1) = 3$
(c) $y'' + 3y' - (1 + x^2)y = 0$, $y(0) = y(1) = 4$
(d) $y'' - 2y' - 2e^x y = 0$, $y(0) = y(1) = 5$.

11. Prove Lemma 5.2.

12. Prove Theorem 5.7.

13. Using the finite element method with piecewise parabolic interpolating function P(x),
     approximate the minimum of the functional

$$J = \int_0^1 \left[\frac{1}{2}(y')^2 + \frac{1}{2}y^2 + 12x^2 y - x^4 y \right] dx$$

subject to the boundary conditions

$$y(0) = 0, \qquad y(1) = 1.$$

Find the exact solution of the problem and compare your numerical result with it.

14. Repeat Exercise 13 but use cubic splines.

15. Approximate the smallest eigenvalue of (5.96) using $h=\pi/6$.

Supplementary Exercises

16. With $h=0.1$, find an approximate solution of each of the following boundary value problems:

(a)	$y''+y'-y = 2+2x-x^2,$	$y(0) = 0,$	$y(1) = 1$
(b)	$y'' = y^3-x^6+2,$	$y(0) = 0,$	$y(1) = 1$
(c)	$y'' = y^5-x^{10}+2,$	$y(0) = 0,$	$y(1) = 1$
(d)	$y'' = y^2,$	$y(0) = y(1) = 1$	
(e)	$y'' = 2y-2y'-e^{-2x},$	$y(0) = 1,$	$y(\infty) = 0$
(f)	$y'' = (1/2)(1+x+y)^3,$	$y(0) = 0,$	$y(1) = 0$
(g)	$(0.01)y''+y' = 0,$	$y(0) = 0,$	$y(\infty) = 1$
(h)	$(0.0001)y''+y' = 0,$	$y(0) = 0,$	$y'(\infty) = 1.$

17. Show that all solutions of

$$y_{n+1}-2\alpha y_n+y_{n-1} = 0$$

are bounded as $n \to \infty$ if $-1<\alpha<1$.

18. The Fibonacci numbers are defined by

$$y_n = y_{n-1}+y_{n-2}, \qquad y_0 = 0, \qquad y_1 = 1.$$

Show that

$$\lim_{n \to \infty} \frac{y_{n+1}}{y_n} = \frac{1}{2}(1+\sqrt{5}).$$

19. With h=0.4, approximate the minimum eigenvalue of

$$\frac{d}{dx}\left[(1+x^2)\frac{dy}{dx}\right]+\lambda y = 0, \qquad y(-1) = y(1) = 0.$$

20. Approximate the minimum eigenvalue of

$$y''+\lambda xy = 0, \qquad y(0) = y(1) = 0.$$

21. Approximate the maximum eigenvalue of

$$y''+\lambda xy = 0, \qquad y(0) = y(1) = 0.$$

22. Find an approximate solution of the differential equation

$$y''-y = -4xe^x$$

subject to the conditions

$$y(0) = y'(0)-1, \qquad y(1) = -[y'(1)+e].$$

Compare your results with the exact solution $y=x(1-x)e^x$.

23. Find an approximate solution of the boundary value problem

$$\frac{d^4y}{dx^4}+\frac{d^2y}{dx^2} = 2, \qquad y(0) = y'(0) = 0, \qquad y(1) = 1, \qquad y'(1) = 2.$$

Compare your results with the exact solution $y=x^2$.

Elliptic Equations

6.1 INTRODUCTION

A very large portion of mathematical physics is devoted to the study of the class of partial differential equations

$$(6.1) \qquad a(x,y,u, \frac{\partial u}{\partial x}, \frac{\partial u}{\partial y}) \frac{\partial^2 u}{\partial x^2} + 2b(x,y,u, \frac{\partial u}{\partial x}, \frac{\partial u}{\partial y}) \frac{\partial^2 u}{\partial x \partial y} + c(x,y,u, \frac{\partial u}{\partial x}, \frac{\partial u}{\partial y}) \frac{\partial^2 u}{\partial y^2}$$

$$+ f(x,y,u, \frac{\partial u}{\partial x}, \frac{\partial u}{\partial y}) = 0 ,$$

where a, b, c and f can, as indicated, be functions of x, y, u, $\partial u/\partial x$ and $\partial u/\partial y$. There is no term $\partial^2 u/\partial y \partial x$ in (6.1) because it is assumed, generally, that $\partial^2 u/\partial x \partial y = \partial^2 u/\partial y \partial x$. In the notation $u_x = \partial u/\partial x$, $u_y = \partial u/\partial y$, $u_{xx} = \partial^2 u/\partial x^2$, $u_{yy} = \partial^2 u/\partial y^2$, $u_{xy} = \partial^2 u/\partial x \partial y$, ..., (6.1) can, and will, be written more simply as

$$(6.2) \qquad a(x,y,u,u_x,u_y)u_{xx} + 2b(x,y,u,u_x,u_y)u_{xy} + c(x,y,u,u_x,u_y)u_{yy}$$

$$+ f(x,y,u,u_x,u_y) = 0 .$$

Finally, we assume that at any point of definition

$$(6.3) \qquad a^2 + b^2 + c^2 \neq 0 ,$$

so that at least one of the second-order derivatives in (6.2) is present. Equation (6.2) is called the general, second-order, *quasilinear* partial differential equation in two independent variables x and y.

At each point of definition, (6.2) is called *linear* if it has the particular form

(6.4) $a(x,y)u_{xx}+2b(x,y)u_{xy}+c(x,y)u_{yy}+d(x,y)u_x+e(x,y)u_y+f(x,y)u+g(x,y) = 0$,

while it is called *mildly nonlinear* if it has the particular form

(6.5) $a(x,y)u_{xx}+2b(x,y)u_{xy}+c(x,y)u_{yy}+d(x,y)u_x+e(x,y)u_y+f(x,y,u) = 0$.

EXAMPLE 1. At each point of the XY plane, denoted by E^2, the equation

$$u_{xx}+u_{yy}-3xu_x+x^2u_y+e^{xy}u = 0$$

is linear.

EXAMPLE 2. At each point of E^2, the equation

$$u_{xx}+u_{yy}-e^u = 0$$

is mildly nonlinear.

EXAMPLE 3. At each point of E^2, the equation

$$(1+u_y^2)u_{xx}-2u_xu_yu_{xy}+(1+u_x^2)u_{yy} = 0$$

is quasilinear.

      Though attention will be restricted largely to linear and mildly nonlinear equations, we will continue, at present, with definitions that are relatively general.

      For both practical and theoretical reasons, it is convenient to classify the possible forms of (6.2) as follows. At any point (x,y) of definition, equation (6.2) is said to be *elliptic, parabolic* or *hyperbolic*, according as b^2-ac is *less* than, *equal* to, or *greater* than zero. This is in analogy with the classification of the general, second-order algebraic equation

$$ax^2+2bxy+cy^2+dx+ey+f = 0$$

as being elliptic, parabolic, or hyperbolic according as the discriminant b^2-ac is less than, equal to, or greater than zero. The analogy, however, ends here, since no actual relationship exists between a conic section and a partial differential equation with the same name type.

EXAMPLE 1. At each point of E^2, the equation

$$\frac{\partial^2 u}{\partial x^2} + \frac{\partial^2 u}{\partial y^2} = 0 \,,$$

or, equivalently,

(6.6) $u_{xx} + u_{yy} = 0$

is elliptic, since $a=c=1$, $b=0$, and $b^2-ac=-1<0$. This equation is called the *potential*, or *Laplace's*, equation and it is the prototype elliptic partial differential equation.

EXAMPLE 2. At each point of E^2, the equation

$$\frac{\partial^2 u}{\partial x^2} - \frac{\partial u}{\partial y} = 0 \,,$$

or, equivalently,

(6.7) $u_{xx} - u_y = 0$

is parabolic, since $a=1$, $b=c=0$, and $b^2-ac=0$. This equation is called the *heat*, or *diffusion*, equation and it is the prototype parabolic partial differential equation.

EXAMPLE 3. At each point of E^2, the equation

$$\frac{\partial^2 u}{\partial x^2} - \frac{\partial^2 u}{\partial y^2} = 0 \,,$$

or, equivalently,

(6.8) $u_{xx} - u_{yy} = 0$

is hyperbolic, since $a=1$, $b=0$, $c=-1$ and $b^2-ac=1>0$. This equation is called the *wave* equation and it is the prototype hyperbolic partial differential equation.

EXAMPLE 4. At each point of E^2, the equation

$$u_{xx} + u_{yy} - e^u = 0$$

is elliptic.

Elliptic equations will be studied in this chapter, parabolic equations will be studied in Chapter VII and hyperbolic equations will be studied in Chapter VIII.

6.2 BOUNDARY VALUE PROBLEMS FOR THE LAPLACE EQUATION

Let us begin the study of elliptic equations by considering the simplest such equation, that is, Laplace's equation (6.6). Other notations for (6.6) are

$$\Delta u = 0$$

and

$$\nabla^2 u = 0 .$$

Any particular solution of (6.6) is called a **harmonic** function. Examples of functions which are harmonic at each point of E^2 are $u=1$, $u=x$, $u=y$, $u=x-y$, $u=x^2-y^2$, $u=3xy^2-x^3$, $u=xy^3-x^3y$. Special properties of harmonic functions, because of their importance in the study of gravitation and potential theory, have been studied in great detail. One such property is the **max-min** property, which can be stated as follows: if Ω is a bounded, simply connected region whose boundary is Γ, and if u is harmonic on Ω and continuous on $\Omega \cup \Gamma$, then u takes on its maximum and its minimum values on Γ.

The simplest and most basic problem of physical interest associated with Laplace's equation is the **Dirichlet problem**, which is formulated as follows. Let Ω be an open, bounded point set which is simply connected and whose boundary Γ is piecewise regular, that is, is piecewise continuously differentiable. If f(x,y) is given and continuous on Γ, then the Dirichlet problem for the Laplace's equation is that of finding a function u(x,y) on $\Omega \cup \Gamma$ which satisfies the following three conditions:

(a) it is continuous on $\Omega \cup \Gamma$,

(b) it is identical with f(x,y) on Γ, and

(c) it is harmonic on Ω.

EXAMPLE. The problem of finding a function u(x,y) such that:

(a) u is continuous at each point (x,y) whose coordinates satisfy $x^2+y^2 \leq 25$,

(b) u is identical with $f(x,y)=x-y^2$ at each point (x,y) which satisfies $x^2+y^2 = 25$,

(c) u is harmonic at each point (x,y) whose coordinates satisfy $x^2+y^2<25$,

is a Dirichlet problem.

Geometrically, the Dirichlet problem can be described as follows. Since $f(x,y)$ is defined only on Γ, the graph of $f(x,y)$ is a closed, three dimensional space curve (see Figure 6.1). Then one is being asked to find a function $u(x,y)$ which is a solution of Laplace's equation at each point of Ω and whose graph is a continuous surface on $\Omega \cup \Gamma$ which has the given space curve for its boundary. For this reason, the Dirichlet problem is called a boundary value problem, and it is an extension, dimensionwise, of the types of boundary value problems studied in Chapter V.

We come now to an interesting, and often initially perplexing, mathematical situation. We can prove in a variety of ways that the Dirichlet problem has one and only one solution. Yet, in only very few cases can we actually give the solution constructively. Thus, the methods of finite differences, Green's functions, integral equations, Dirichlet's principle, conformal mapping, and subharmonic-superharmonic functions all have been applied to show the same result, namely, that the Dirichlet problem has exactly one solution. However, only in such simple cases as when Γ is a circle, an ellipse, a rectangle, or an elementary transformation of any of these curves, called a conformal transformation, can the solution be found explicitly. These explicit solutions are usually given as series or integrals. Unfortunately, if one wishes to know the solution at a specific point, these integrals may require numerical integration and the series may require truncation and computer approximation. To make mathematical endeavor even more frustrating, there is no general way to construct the solution of any Dirichlet problem in which the Laplace equation is replaced by even the simplest mildly nonlinear elliptic equation, and such equations are often basic to refined modeling.

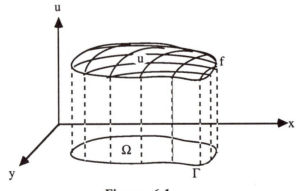

Figure 6.1

For all the above reasons, we turn now to a numerical method for approximating the solution of the Dirichlet problem. As will be indicated later, the method will be so general that it will extend in a natural way to Dirichlet problems for mildly nonlinear elliptic equations.

6.3 APPROXIMATE SOLUTION OF THE DIRICHLET PROBLEM ON A RECTANGLE

In developing a numerical method for the Dirichlet problem, it will be of value to have a definition of discrete functions in two independent variables, and, for clarity only, we will consider, first, Dirichlet problems defined on a rectangle. Later, we will see how all of the methodology extends readily to apply to Dirichlet problems on more complex domains. Throughout, we assume that a_1, b_1, a_2, b_2 are four real constants which satisfy $a_1 < b_1$ and $a_2 < b_2$.

Consider now four points $D=(a_1,a_2)$, $E=(b_1,a_2)$, $F=(b_1,b_2)$ $G=(a_1,b_2)$, as shown in Figure 6.2. Let $[a_1,b_1]$ be the interval $a_1 \le x \le b_1$, and let $[a_2,b_2]$ be the interval $a_2 \le y \le b_2$. Subdivide $[a_1,b_1]$ into n equal parts, each of length $h=(b_1-a_1)/n$, by an R_{n+1} set $x_i=a_1+ih$, $i=0,1,2,...,n$, and subdivide $[a_2,b_2]$ into m equal parts, each of length $k=(b_2-a_2)/m$, by an R_{m+1} set $y_j=a_2+jk$, $j=0,1,2,...,m$. Then the set of number couples (x_i,y_j), $x_i \in R_{n+1}$, $y_j \in R_{m+1}$ is called an $R_{n+1,m+1}$ set. An $R_{n+1,m+1}$ set is also called a set of planar grid points, and h is called the x-grid size while k is called the y-grid size. The points (x_0,y_j), (x_n,y_j), $j=1,2,...m-1$, and (x_i,y_0), (x_i,y_m), $i=1,2,...n-1$, are called *boundary* grid points, while (x_i,y_j), $i=1,2,...,n-1$, $j=1,2,...,m-1$ are called *interior* grid points. The remaining points are called the *corner* points. Any subrectangular domain with four grid points (x_i,y_j), (x_{i+1},y_j), (x_i,y_{j+1}), (x_{i+1},y_{j+1}) for vertices is called a *cell*.

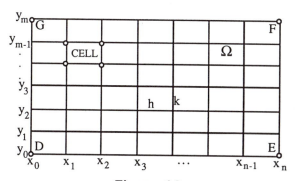

Figure 6.2.

EXAMPLE 1. Consider a rectangle with vertices (0,0), (1,0), (1,4), (0,4). Divide the interval [0,1] into two equal parts, each of length h=0.5, by the R_3 set $x_0=0$, $x_1=0.5$, $x_2=1$. Next, divide the interval [1,4] into three equal parts, each of length k=1, by the R_4 set $y_0=1$, $y_1=2$, $y_2=3$, $y_3=4$. The set $R_{3,4}$ is the set of points

(0.0,4)	(0.5,4)	(1.0,4)
(0.0,3)	(0.5,3)	(1.0,3)
(0.0,2)	(0.5,2)	(1.0,2)
(0.0,1)	(0.5,1)	(1.0,1) .

The interior points are (0.5,2), (0.5,3). The boundary points are (0.0,2), (0.0,3), (0.5,1), (0.5,4), (1.0,2), (1.0,3). The corner points are (0.0,1), (0.0,4), (1.0,1), (1.0,4).

DEFINITION 6.1. A function z=f(x,y) that is defined only on an $R_{n+1,m+1}$ set is called a discrete function of *two* variables.

The notation $z_{i,j}=f(x_i,y_j)$, $(x_i,y_j) \in R_{n+1,m+1}$ will be commonly used when dealing with discrete functions of two variables. However, when no confusion can exist, we will often delete the comma and use $z_{ij}=z_{i,j}$. This will hold also for other items with double subscripts.

EXAMPLE 2. On the $R_{3,4}$ set given in Example 1, the equation $z_{ij}=2x_i+y_j^2$, $(x_i,y_j) \in R_{3,4}$, defines a discrete function of two variables.

We formulate now the basic algorithm for approximating the solution of the Dirichlet problem on a rectangle. Let Γ be the rectangle with vertices $D=(a_1,a_2)$, $E=(b_1,a_2)$, $F=(b_1,b_2)$, $G=(a_1,b_2)$ and let Ω be its interior, as shown in Figure 6.2. Divide $[a_1,b_1]$ into n equal parts, each of length h, by the R_{n+1} set $a_1=x_0<x_1<x_2<...<x_n=b_1$, and divide $[a_2,b_2]$ into m equal parts, each of length k, by the R_{m+1} set $a_2=y_0<y_1<y_2<...<y_m=b_2$. One says, then, that Ω is covered by the lattice of points (x_i,y_j) which belong to the $R_{n+1,m+1}$ set. One also says that Ω is covered by the cells which are determined by these grid points. On this $R_{n+1,m+1}$ set, let $u_{i,j}=u(x_i,y_j)$ be a discrete function which satisfies

(6.9)

$$u_{0,j} = f(x_0,y_j) , \qquad u_{n,j} = f(x_n,y_j) , \qquad j=0,1,2,...,m,$$

$$u_{i,0} = f(x_i,y_0) , \qquad u_{i,m} = f(x_i,y_m) , \qquad i=0,1,2,...,n,$$

where f(x,y) is given by the Dirichlet problem. The values $u_{i,j}$, i=1,2,...,n−1, j=1,2,...,m−1 are unknown. It is these that one has to determine. This will be accomplished by approximating Laplace's equation at *each* interior grid point (x_i,y_j), i=1,2,...,n−1,

$j=1,2,\ldots,m-1$, as follows.

Consider, from Section 3.6, the approximations

(6.10) $$u_{xx}(x_i,y_j) \approx \frac{u_{i+1,j}-2u_{i,j}+u_{i-1,j}}{h^2}$$

(6.11) $$u_{yy}(x_i,y_j) \approx \frac{u_{i,j+1}-2u_{i,j}+u_{i,j-1}}{k^2} ,$$

so that the Laplace differential equation is approximated at each interior grid point (x_i,y_j) by the difference equation

(6.12) $$\frac{u_{i+1,j}-2u_{i,j}+u_{i-1,j}}{h^2} + \frac{u_{i,j+1}-2u_{i,j}+u_{i,j-1}}{k^2} = 0 .$$

One then writes (6.12) *consecutively* as follows. First, fix $j=1$ and, in order, set $i=1,2,\ldots,n-1$. Then fix $j=2$ and, in order, set $i=1,2,\ldots,n-1$. Continue in this fashion until one has, finally, $j=m-1$ and, in order, $i=1,2,\ldots,n-1$. Since u is known at all boundary grid points, there results a linear algebraic system of $(n-1)(m-1)$ equations in the $(n-1)(m-1)$ unknowns $u_{i,j}$, $i=1,2,\ldots,n-1$, $j=1,2,\ldots,m-1$. A solution of this system with the known values (6.9) is called a numerical solution on the given $R_{n+1,m+1}$ set.

EXAMPLE. Consider the Dirichlet problem defined on the rectangle Γ with vertices $(0,0)$, $(3,0)$, $(3,4)$, $(0,4)$ and let Ω be the interior of Γ. On Γ let $f(x,y)=(x-1)^2-(y-2)^2$. Set $n=3$, $m=4$, so that $h=k=1$ and the points of $R_{4,5}$ are simply given by (i,j), $i=0,1,2,3$, $j=0,1,2,3,4$. At the boundary and corner grid points one has $u(x_i,y_j)=f(x_i,y_j)$, that is,

(6.13)

$$
\begin{array}{llll}
u_{04} = -3, & u_{14} = -4, & u_{24} = -3, & u_{34} = 0, \\
u_{03} = 0, & & & u_{33} = 3, \\
u_{02} = 1, & & & u_{32} = 4, \\
u_{01} = 0, & & & u_{31} = 3, \\
u_{00} = -3, & u_{10} = -4, & u_{20} = -3, & u_{30} = 0.
\end{array}
$$

Application of (6.12) consecutively on the interior grid points yields

$$u_{21} - 2u_{11} + u_{01} + u_{12} - 2u_{11} + u_{10} = 0$$

$$u_{31} - 2u_{21} + u_{11} + u_{22} - 2u_{21} + u_{20} = 0$$

(6.14) $\qquad u_{22} - 2u_{12} + u_{02} + u_{13} - 2u_{12} + u_{11} = 0$

$$u_{32} - 2u_{22} + u_{12} + u_{23} - 2u_{22} + u_{21} = 0$$

$$u_{23} - 2u_{13} + u_{03} + u_{14} - 2u_{13} + u_{12} = 0$$

$$u_{33} - 2u_{23} + u_{13} + u_{24} - 2u_{23} + u_{22} = 0 .$$

Substituting known values from (6.13) into (6.14) then implies

$$
\begin{aligned}
-4u_{11} + u_{21} + u_{12} &= 4 \\
u_{11} - 4u_{21} \qquad + u_{22} &= 0 \\
u_{11} \qquad -4u_{12} + u_{22} + u_{13} &= -1 \\
u_{21} + u_{12} - 4u_{22} \qquad + u_{23} &= -4 \\
u_{12} \qquad -4u_{13} + u_{23} &= 4 \\
u_{22} + u_{13} - 4u_{23} &= 0 ,
\end{aligned}
$$

the solution of which is

$$
\begin{aligned}
u_{13} &= -1 , & u_{23} &= 0 , \\
u_{12} &= 0 , & u_{22} &= 1 , \\
u_{11} &= -1 , & u_{21} &= 0 ,
\end{aligned}
$$

which, agrees exactly with the analytical solution of the given Dirichlet problem, that is, $u = (x-1)^2 - (y-2)^2$. Note finally that the values of $u_{i,j}$ at the corner points in (6.13) played *no* role in the simplification of system (6.14), which is why we have separated boundary and corner points in the considerations.

It is worth reemphasizing that the only set on which one actually approximates values of u is the set of *interior* grid points, since $u \equiv f$ at each grid point which is in Γ. Moreover, the only known function values which are utilized in the approximation come from the boundary grid points. In this chapter, known boundary values at corner points have no practical value, but will be of theoretical convenience, as seen in Section 6.7.

6.4 APPROXIMATE SOLUTION OF THE DIRICHLET PROBLEM ON A GENERAL DOMAIN

The ideas developed in Section 6.3 to solve the Dirichlet problem on a rectangle will be extended next to Dirichlet problems on more complex domains.

Let Ω be an open, simply connected region of E^2 whose boundary Γ is piecewise regular. Let $f(x,y)$ be given and continuous on Γ. Then choose any rectangle with vertices $D=(a_1,a_2)$, $E=(b_1,a_2)$, $F=(b_1,b_2)$, $G=(a_1,b_2)$, which contains Ω entirely (see Figure 6.3). Divide $[a_1,b_1]$ into n equal parts, each of length h, by an R_{n+1} set $x_i=a_1+ih$, $i=0,1,2,\ldots,n$, and divide $[a_2,b_2]$ into m equal parts, each of length k, by an R_{m+1} set $y_j=a_2+jk$, $j=0,1,2,\ldots,m$. The region is said to be covered by the lattice of points (x_i,y_j) which belong to the $R_{n+1,m+1}$ set. The subset of points of $R_{n+1,m+1}$ which are also points of Ω is denoted by Ω_{hk} and is called the set of *interior* grid points. Each interior grid point (x_i,y_j) has four grid point *neighbors*, namely, $(x_{i\pm1},y_j)$ and $(x_i,y_{j\pm1})$, which are the grid points nearest to the given interior point in the east, west, north and south directions. A neighbor of an interior grid point which does not belong to Ω_{hk} is called a *boundary* grid point. The set of boundary grid points is denoted with Γ_{hk}. A *corner* grid point is a point which is not an interior grid point and not a boundary grid point, but which is the vertex of a cell which has at least one other vertex which is an interior grid point. As the concept of a corner point plays no role in approximating u at an interior grid point, as indicated in the last illustrative example, we will concentrate in this section only on the interior and boundary grid points.

Note that we are assuming tacitly that if (x_i,y_j) is an interior grid point and if a neighbor, for example, (x_{i+1},y_j), is also an interior grid point, then the entire segment joining these two points is contained in Ω. This is usually valid for domains which occur in applied problems, like convex and L-shaped domains. In other domains, it can often be achieved by choosing the grid sizes to be sufficiently small or by rotation of axes.

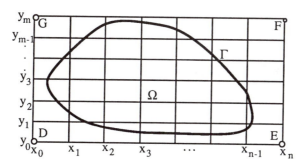

Figure 6.3

Our problem is, again, that of determining u_{ij} at each interior grid point. For this purpose, a value to u_{ij} will be assigned at each boundary grid point by

(6.15) $u_{i,j} = f(x_i, y_j)$, $(x_i, y_j) \in \Gamma_{hk}$.

Observe also that at each interior grid point $(x_i, y_j) \in \Omega_{hk}$, where u_{ij} is unknown, the Laplace difference equation (6.12) is:

(6.16) $$\frac{u_{i+1,j} - 2u_{i,j} + u_{i-1,j}}{h^2} + \frac{u_{i,j+1} - 2u_{i,j} + u_{i,j-1}}{k^2} = 0 .$$

The basic algorithm for the Dirichlet problem in a complex domain is then formulated as follows.

Fix an $R_{n+1,m+1}$ set which covers Ω. Determine Ω_{hk} and Γ_{hk}. Suppose that Ω_{hk} consists of K points, $K \leq (n-1)(m-1)$. At any *boundary* grid point (x_i, y_j) a value to u_{ij} is determined by (6.15). If one then writes (6.16) consecutively for all $(x_i, y_j) \in \Omega_{hk}$, *and* inserts the approximated values of u at the boundary grid points, there results a linear algebraic system of K equations with K unknowns. As indicated in the last section, *consecutively* implies proceeding from left to right on each row of grid points, starting with the lowest row and proceeding in order to the highest. A solution of this system is the numerical solution on Ω_{hk}.

EXAMPLE 1. Let Γ be the quadrilateral with vertices (0,0), (7,0), (2,5), and (0,4), which is shown in Figure 6.4. Let Ω be the interior of Γ. On $\Omega \cup \Gamma$ consider the Dirichlet problem with $f(x,y)=x^2-y^2$. A rectangle which contains Ω has vertices (0,0), (8,0), (8,6) and (0,6). By setting h=k=2, the points of $R_{5,3}$ are $(x_i, y_j)=(2i,2j)$, i=0,1,2,3,4, j=0,1,2,3. The points of Ω_{hk} are (2,2), (4,2) and (2,4) and are shown as crossed points in Figure 6.4. The points of Γ_{hk} are (2,0), (4,0), (6,2), (4,4), (2,6), (0,4) and (0,2), and are shown as circled points in Figure 6.4. At the boundary grid points, (6.15) yields

(6.17)
$$u_{10} = f(2,0) = 4 , \qquad u_{20} = f(4,0) = 16 ,$$
$$u_{31} = f(6,2) = 32 , \qquad u_{22} = f(4,4) = 0 , \qquad u_{13} = f(2,6) = -32 ,$$
$$u_{01} = f(0,2) = -4 , \qquad u_{02} = f(0,4) = -16 .$$

Application of (6.16) at the interior grid points and substitution of (6.17) yields

$$\frac{u_{21}-2u_{11}-4}{4} + \frac{u_{12}-2u_{11}+4}{4} = 0$$

$$\frac{32-2u_{21}+u_{11}}{4} + \frac{0-2u_{21}+16}{4} = 0$$

$$\frac{0-2u_{12}-16}{4} + \frac{-32-2u_{12}+u_{11}}{4} = 0 ,$$

or, equivalently,

$$-4u_{11} + u_{21} + u_{12} = 0$$
$$u_{11} - 4u_{21} = -48$$
$$u_{11} \qquad -4u_{12} = 48 ,$$

whose unique solution is

(6.18) $\qquad u_{11} = 0 , \qquad u_{21} = 12 , \qquad u_{12} = -12 .$

Thus, the numerical solution agrees exactly with the analytical solution $u = x^2 - y^2$ in Ω_{hk}.

It should be noted, finally, that methods other than the one developed in this section are available for nonrectangular domains [Greenspan (1974)]. Moreover, different methods often possess different advantages. The advantages of the method we have developed are programming simplicity and broad applicability.

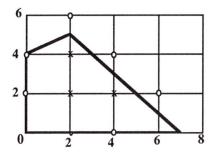

Figure 6.4

6.5 The General Linear Elliptic Equation

The methods developed in Sections 6.3 and 6.4 for solving the Dirichlet problem numerically extend to boundary value problems for a wide class of linear elliptic problems. For clarity only, in this and in the following sections, we will restrict attention to boundary value problems defined on *rectangles*. The ideas and methods extend in a natural way to more complex regions.

Consider a rectangle Γ with vertices (a_1,a_2), (b_1,a_2), (b_1,b_2) and (a_1,b_2) and let Ω be the interior of Γ. If A, B and C are constants and P(x,y), Q(x,y), R(x,y) and S(x,y) are continuous on $\Omega \cup \Gamma$, it is known [Courant and Hilbert (1962), Greenspan (1961)] that the partial differential equation

$$(6.19) \qquad Au_{xx} + 2Bu_{xy} + Cu_{yy} + P(x,y)u_x + Q(x,y)u_y + R(x,y)u + S(x,y) = 0 ,$$

$$A^2 + B^2 + C^2 \neq 0 ,$$

can be simplified by a rotation of axes. When (6.19) is elliptic, one can, for example, eliminate the u_{xy} term. Thus, without loss of generality, we can assume that B=0:

$$Au_{xx} + Cu_{yy} + P(x,y)u_x + Q(x,y)u_y + R(x,y)u + S(x,y) = 0 , \qquad AC > 0 .$$

Moreover, by an appropriate scale change, this equation can be further simplified so that A=C=1. Consider then, the following Dirichlet problem

$$(6.20) \qquad u_{xx} + u_{yy} + P(x,y)u_x + Q(x,y)u_y + R(x,y)u + S(x,y) = 0 , \qquad (x,y) \in \Omega$$

$$(6.21) \qquad u(x,y) = f(x,y) , \qquad (x,y) \in \Gamma .$$

Further, for both practical and theoretical reasons, it will be convenient to assume that

$$(6.22) \qquad R(x,y) \leq 0 , \qquad (x,y) \in (\Omega \cup \Gamma)$$

which will assure [Courant and Hilbert (1962)] that any solution of (6.20) has certain properties, like a general *max-min* property, in common with harmonic functions. Specifically, for $S(x,y) \equiv 0$ and $R(x,y) \leq 0$, any nonconstant solution u(x,y) of (6.20) has the following (weak) max-min property: u(x,y) cannot assume a positive maximum or a negative minimum on Ω. Also, if $S(x,y)\equiv0$ and $R(x,y)\equiv0$ then any nonconstant solution $u(x,y)$ of equation (6.20) has the following (strong) max-min property: $u(x,y)$ cannot assume its maximum or minimum on Ω [Protter and Weinberger (1984)]. Moreover, inequality (6.22) is *sufficient* to insure that the solution of the Dirichlet problem (6.20)-(6.21) *exists* and is *unique*. Of course, these results are direct extensions of those given for boundary value problems for ordinary differential equations.

Numerically, one proceeds as follows. Subdivide $[a_1,b_1]$ into n equal parts, each of length $h=(b_1-a_1)/n$, by the R_{n+1} set $x_i=a_1+ih$, i=0,1,2,...,n. Next, subdivide $[a_2,b_2]$ into m equal parts, each of length $k=(b_2-a_2)/m$, by the R_{m+1} set $y_j=a_2+jk$, j=0,1,2,...,m, so that Ω is covered by the set of planar grid points $R_{n+1,m+1}$. On this $R_{n+1,m+1}$ set define the discrete functions P_{ij}, Q_{ij}, R_{ij} and S_{ij} as $P_{ij}=P(x_i,y_j)$, $Q_{ij}=Q(x_i,y_j)$, $R_{ij}=R(x_i,y_j)$ and $S_{ij}=S(x_i,y_j)$. As the region is rectangular, interior, boundary and corner grid points are defined exactly as in Section 7.3. Then, at the boundary and corner grid points set

$$(6.23) \qquad u_{ij} = f(x_i,y_j) ,$$

while at each interior grid point (x_i,y_j) substitution of

$$u_{xx} \approx \frac{u_{i+1,j}-2u_{i,j}+u_{i-1,j}}{h^2} , \qquad u_x \approx \frac{u_{i+1,j}-u_{i-1,j}}{2h} ,$$

$$u_{yy} \approx \frac{u_{i,j+1}-2u_{i,j}+u_{i,j-1}}{k^2} , \qquad u_y \approx \frac{u_{i,j+1}-u_{i,j-1}}{2k} ,$$

into (6.20) yields the following difference equation approximation of the differential equation:

$$\frac{u_{i+1,j}-2u_{i,j}+u_{i-1,j}}{h^2} + \frac{u_{i,j+1}-2u_{i,j}+u_{i,j-1}}{k^2} +P_{ij}\frac{u_{i+1,j}-u_{i-1,j}}{2h} + Q_{ij}\frac{u_{i,j+1}-u_{i,j-1}}{2k} +R_{ij}u_{i,j}+S_{ij}=0,$$

or, equivalently,

$$(6.24) \qquad h^2(2-kQ_{ij})u_{i,j-1}+k^2(2-hP_{ij})u_{i-1,j}-2(2h^2+2k^2-h^2k^2R_{ij})u_{i,j}$$

$$+k^2(2+hP_{ij})u_{i+1,j}+h^2(2+kQ_{ij})u_{i,j+1}+2h^2k^2S_{ij} = 0 .$$

If one then writes (6.24) consecutively at the interior grid points, *and* inserts the known values (6.23), there results a linear algebraic system of $(n-1)(m-1)$ equations in the $(n-1)(m-1)$ unknowns u_{ij}, $i=1,2,...,n-1$, $j=1,2,...,m-1$. A solution of this system, together, with the boundary and corner values (6.23), is called a numerical solution on the given $R_{n+1,m+1}$ set. This method is called the **central difference method** because central difference formulas have been used to approximate u_x and u_y.

EXAMPLE. Consider the rectangle Γ with vertices $(0,0)$, $(2,0)$, $(2,1.5)$ and $(0,1.5)$. Let Ω be the interior of Γ. On $\Omega \cup \Gamma$ consider the boundary value problem

(6.25)
$$u_{xx}+u_{yy}-yu_x-xu_y-u = 0, \quad (x,y) \in \Omega$$

$$u(x,y) = x-y, \quad (x,y) \in \Gamma.$$

Fix $h=k=0.5$ so that Ω is covered by the $R_{5,4}$ set $(0.5i,0.5j)$, $i=0,1,2,3,4$, $j=0,1,2,3$. On the boundary and corner grid points one has

$$u_{03} = -1.5, \quad u_{13} = -1, \quad u_{23} = -0.5, \quad u_{33} = 0, \quad u_{43} = 0.5,$$
$$u_{02} = -1, \quad\quad\quad\quad\quad\quad\quad\quad\quad\quad\quad\quad\quad\quad\quad u_{42} = 1,$$
$$u_{01} = -0.5, \quad\quad\quad\quad\quad\quad\quad\quad\quad\quad\quad\quad\quad u_{41} = 1.5,$$
$$u_{00} = 0, \quad u_{10} = 0.5, \quad u_{20} = 1, \quad u_{30} = 1.5, \quad u_{40} = 2,$$

while $u_{11}, u_{21}, u_{31}, u_{12}, u_{22}$ and u_{32} are unknown. Application of (6.24) at each interior grid point yields

$$0.25[2+0.5(0.5)]u_{10}+0.25[2+0.5(0.5)]u_{01}-2(0.25)(2+2+0.25)u_{11}$$
$$+ 0.25[2-0.5(0.5)]u_{21}+0.25[2-0.5(0.5)]u_{12} = 0$$

$$0.25[2+0.5(1)]u_{20}+0.25[2+0.5(0.5)]u_{11}-2(0.25)(2+2+0.25)u_{21}$$
$$+ 0.25[2-0.5(0.5)]u_{31}+0.25[2-0.5(1)]u_{22} = 0$$

$$0.25[2+0.5(1.5)]u_{30}+0.25[2+0.5(0.5)]u_{21}-2(0.25)(2+2+0.25)u_{31}$$
$$+ 0.25[2-0.5(0.5)]u_{41}+0.25[2-0.5(1.5)]u_{32} = 0$$

$$0.25[2+0.5(0.5)]u_{11}+0.25[2+0.5(1)]u_{02}-2(0.25)(2+2+0.25)u_{12}$$
$$+ 0.25[2-0.5(1)]u_{22}+0.25[2-0.5(0.5)]u_{13} = 0$$

$$0.25[2+0.5(1)]u_{21}+0.25[2+0.5(1)]u_{12}-2(0.25)(2+2+0.25)u_{22}$$
$$+ 0.25[2-0.5(1)]u_{32}+0.25[2-0.5(1)]u_{23} = 0$$

$$0.25[2+0.5(1.5)]u_{31}+0.25[2+0.5(1)]u_{22}-2(0.25)(2+2+0.25)u_{32}$$
$$+ 0.25[2-0.5(1)]u_{42}+0.25[2-0.5(1.5)]u_{33} = 0,$$

or, equivalently,

$$2.25u_{10} + 2.25u_{01} - 8.5u_{11} + 1.75u_{21} + 1.75u_{12} = 0$$
$$2.5u_{20} + 2.25u_{11} - 8.5u_{21} + 1.75u_{31} + 1.5u_{22} = 0$$
$$2.75u_{30} + 2.25u_{21} - 8.5u_{31} + 1.75u_{41} + 1.25u_{32} = 0$$
$$2.25u_{11} + 2.5u_{02} - 8.5u_{12} + 1.5u_{22} + 1.75u_{13} = 0$$
$$2.5u_{21} + 2.5u_{12} - 8.5u_{22} + 1.5u_{32} + 1.5u_{23} = 0$$
$$2.75u_{31} + 2.5u_{22} - 8.5u_{32} + 1.5u_{42} + 1.25u_{33} = 0 \ .$$

From the known boundary values, this system yields

$$u_{11} = 0 \ , \qquad u_{21} = 0.5 \ , \qquad u_{31} = 1 \ ,$$
$$u_{12} = -0.5 \ , \qquad u_{22} = 0 \ , \qquad u_{32} = 0.5 \ ,$$

which is in perfect agreement with the analytical solution $u = x - y$.

An important feature of the central difference approximation (6.24) is that the resulting numerical solution possesses discrete max-min properties, in analogy with continuous solutions.

THEOREM 6.1. Let $P(x,y)$, $Q(x,y)$ and $R(x,y)$ be continuous on $\Omega \cup \Gamma$. Assume that $R(x,y) \leq 0$. Let N and M be two constants such that $|P(x,y)| \leq N$ and $|Q(x,y)| \leq M$. If $S(x,y) \equiv 0$, and if h and k satisfy the following inequalities

(6.26)   **Nh < 2 ,   Mk < 2 ,**

then a nonconstant numerical solution u_{ij} of (6.24) cannot assume a positive maximum or a negative minimum at any interior grid point (x_i, y_j) of $R_{n+1,m+1}$. If, in addition, $R(x,y) \equiv 0$, then a nonconstant numerical solution u_{ij} cannot assume its maximum or minimum at any interior grid point.

The proof is entirely analogous to that of Theorem 5.2.

As an example, note that in the boundary value problem (6.25) solved in the Example above, one has $R(x,y) = -1 < 0$, $S(x,y) \equiv 0$ and, accordingly, the computed numerical solution has neither a positive maximum, nor a negative minimum at the interior grid points.

We prove, next, existence and uniqueness of a numerical solution of the Dirichlet problem. Note, however, that one cannot invoke the proof of Theorem 5.1 because the linear algebraic system is no longer tridiagonal.

THEOREM 6.2. Consider the Dirichlet problem (6.20)-(6.21) on a rectangular domain. Let P(x,y), Q(x,y), R(x,y) and S(x,y) be continuous on $\Omega \cup \Gamma$, where R(x,y)≤0. If Nh<2 and Mk<2, then the numerical solution obtained with the central difference approximation (6.24) exists and is unique.

PROOF. For the boundary value problem (6.20)-(6.21) consider the corresponding linear algebraic system of finite difference equations (6.24), that is,

$$(6.27) \qquad h^2(2-kQ_{ij})u_{i,j-1}+k^2(2-hP_{ij})u_{i-1,j}-2(2h^2+2k^2-h^2k^2R_{ij})u_{i,j}+k^2(2+hP_{ij})u_{i+1,j}$$

$$+h^2(2+kQ_{ij})u_{i,j+1}+2h^2k^2S_{ij} = 0\ , \qquad i=1,2,...,n-1, \qquad j=1,2,...,m-1,$$

where $u_{i,0}$, $u_{i,m}$, i=0,1,2,...,n, and $u_{0,j}$, $u_{n,j}$, j=1,2,...,m-1, are known by (6.23). To show that the solution of (6.23), (6.27) exists and is unique, one need only show that the unique solution of the related homogeneous system is $u_{ij}=0$, i=1,2,...,n-1, j=1,2,...,m-1. Note, in fact, that the homogeneous algebraic system related to (6.27) is given by

$$(6.28) \qquad h^2(2-kQ_{ij})u_{i,j-1}+k^2(2-hP_{ij})u_{i-1,j}-2(2h^2+2k^2-h^2k^2R_{ij})u_{i,j}+k^2(2+hP_{ij})u_{i+1,j}$$

$$+h^2(2+kQ_{ij})u_{i,j+1} = 0\ , \qquad i=1,2,...,n-1, \qquad j=1,2,...,m-1,$$

where

$$u_{i,0} = u_{i,m} = 0\ , \qquad i=0,1,2,...,n,$$
$$(6.29)$$
$$u_{0,j} = u_{n,j} = 0\ , \qquad j=1,2,...,m-1.$$

Assume that (6.28) has a solution for which at least one of u_{ij}, i=1,2,...,n-1, j=1,2,...,m-1, is not zero. Without loss of generality, assume that at least one of these values is positive (a completely analogous proof follows under the assumption that at least one is negative). Now, if one of u_{ij}, i=1,2,...,n-1, j=1,2,...,m-1 is positive, then there is a maximum positive number, say u_{pq}, such that $u_{pq}=\max(u_{ij})>0$. But, since $u_{ij}=0$ at the boundary grid points, system (6.28) has a nonconstant numerical solution with a positive maximum at (x_p,y_q), which contradicts Theorem 6.1. Thus, by contradiction, system

(6.28) has only the zero solution, and the theorem is proved.

Note that Theorems 6.1 and 6.2 also apply, as a particular case, to the numerical method for the Dirichlet problem for Laplace's equation, in which case, inequalities (6.26) are always valid, since N=M=0. Note also that, when inequalities (6.26) are satisfied, equations (6.24) yield a linear system which is *diagonally dominant* with negative elements on the main diagonal and nonnegative ones elsewhere. This is the case since one starts at (x_1,y_1) and then proceeds consecutively, which is from left to right, and from bottom to top. A system of this form can be solved by the generalized Newton's method with convergence *assured* for any initial guess and for any ω in the range $0<\omega<2$. In this case the Generalized Newton's formulas (1.76) reduce to

$$(6.30) \qquad u_{i,j}^{(r+1)} = u_{i,j}^{(r)} - \omega \left[\frac{h^2(2-kQ_{ij})u_{i,j-1}^{(r+1)}+k^2(2-hP_{ij})u_{i-1,j}^{(r+1)}-2(2h^2+2k^2-h^2k^2R_{ij})u_{i,j}^{(r)}}{-2(2h^2+2k^2-h^2k^2R_{ij})} \right.$$

$$\left. + \frac{k^2(2+hP_{ij})u_{i+1,j}^{(r)}+h^2(2+kQ_{ij})u_{i,j+1}^{(r)}+2h^2k^2S_{ij}}{-2(2h^2+2k^2-h^2k^2R_{ij})} \right], \qquad \begin{array}{l} i=1,2,\ldots,n-1 \\ j=1,2,\ldots,m-1, \end{array}$$

where r denotes the iteration index.

6.6 UPWIND DIFFERENCE METHOD FOR GENERAL LINEAR ELLIPTIC EQUATIONS

In the previous section central difference formulas have been used to approximate u_x and u_y. These formulas are, of course, more accurate than either the forward or the backward finite difference approximations. However, when inequalities (6.26) are not satsfied, a numerical solution may not be unique or may not have the max-min property.

EXAMPLE 1. Consider the square Γ with vertices $(0,0)$, $(3,0)$, $(3,3)$ and $(0,3)$, and let Ω be its interior. On $\Omega \cup \Gamma$ consider the boundary value problem

$$(6.31) \qquad u_{xx}+u_{yy}+6u_x+6u_y = 0 , \qquad\qquad (x,y)\in\Omega$$

$$(6.32) \qquad u(x,y) = (10/9)(x-3)^2(y-3)^2, \qquad\qquad (x,y)\in\Gamma$$

Fix h=k=1 so that, as shown in Figure 6.5, $\Omega \cup \Gamma$ is covered by the $R_{4,4}$ set (i,j), i=0,1,2,3, j=0,1,2,3. At the boundary grid points (6.32) yields

(6.33) $u_{10} = u_{01} = 40$, $u_{20} = u_{02} = 10$, $u_{31} = u_{32} = u_{13} = u_{23} = 0$.

At the interior grid points the difference equation (6.24) becomes

(6.34) $-4u_{i,j-1} - 4u_{i-1,j} - 8u_{i,j} + 8u_{i+1,j} + 8u_{i,j+1} = 0$.

For i=1,2 and j=1,2, equation (6.34) yields in a consecutive fashion

$$-4u_{10} - 4u_{01} - 8u_{11} + 8u_{21} + 8u_{12} = 0$$
$$-4u_{20} - 4u_{11} - 8u_{21} + 8u_{31} + 8u_{22} = 0$$
$$-4u_{11} - 4u_{02} - 8u_{12} + 8u_{22} + 8u_{13} = 0$$
$$-4u_{21} - 4u_{12} - 8u_{22} + 8u_{32} + 8u_{23} = 0 ,$$

into which, substitution of the known boundary values (6.33), yields

$$-8u_{11} + 8u_{21} + 8u_{12} = 320$$
$$-4u_{11} - 8u_{21} + 8u_{22} = 40$$
$$-4u_{11} - 8u_{12} + 8u_{22} = 40$$
$$-4u_{21} - 4u_{12} - 8u_{22} = 0 .$$

The solution of this system is given by

$$u_{11} = -30 , \qquad u_{12} = u_{21} = 5 , \qquad u_{22} = -5 .$$

Thus, since $u_{11} < 0$ and $u_{22} < 0$, the numerical solution on $R_{4,4}$ does not possess the max-min property. In order for a numerical solution of (6.31)-(6.32) to possess the max-min property one must choose h and k small enough so as to satisfy inequalities (6.24), which, in this example, imply

$$h < \frac{2}{6} = \frac{1}{3} \quad \text{and} \quad k < \frac{2}{6} = \frac{1}{3} .$$

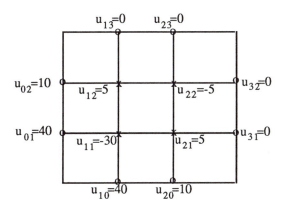

Figure 6.5

But, this requires a covering for $\Omega \cup \Gamma$ by an $R_{n+1,m+1}$ set with n>9 and m>9. Thus, one has to solve a linear system with at least 81 equations and 81 unknowns.

In general, if N or M, or both, are large, the corresponding linear algebraic system to be solved may have too many equations, and hence, it may be *too* large for practical calculation. Let us show next that if one is willing to sacrifice a degree of accuracy, then it will *always* be possible, by using backward or forward finite differences for approximating u_x and u_y, to solve a linear elliptic boundary value problem numerically for *any* given h and k with the assurance that the resulting numerical solution exists, is unique, and possesses the discrete max-min property.

In this section, then, let us introduce the so-called *upwind difference method* for general linear elliptic boundary value problems. The development is a direct extension of that given in Section 5.3. Specifically, it is described as follows. Given the linear boundary value problem

$$(6.35) \qquad u_{xx}+u_{yy}+P(x,y)u_x+Q(x,y)u_y+R(x,y)u+S(x,y) = 0 , \qquad (x,y)\in \Omega$$

$$(6.36) \qquad u(x,y) = f(x,y) , \qquad (x,y)\in \Gamma ,$$

let u_{xx} and u_{yy} be approximated at an interior grid point (x_i,y_j) by

$$(6.37) \qquad u_{xx} \approx \frac{u_{i+1,j}-2u_{i,j}+u_{i-1,j}}{h^2} , \qquad u_{yy} \approx \frac{u_{i,j+1}-2u_{i,j}+u_{i,j-1}}{k^2} ,$$

while u_x and u_y are approximated by

$$(6.38) \qquad u_x \approx \frac{u_{i+1,j}-u_{i,j}}{h}, \quad \text{if } P_{ij} \geq 0; \qquad u_x \approx \frac{u_{i,j}-u_{i-1,j}}{h}, \quad \text{if } P_{ij} < 0;$$

$$(6.39) \qquad u_y \approx \frac{u_{i,j+1}-u_{i,j}}{k}, \quad \text{if } Q_{ij} \geq 0; \qquad u_y \approx \frac{u_{i,j}-u_{i,j-1}}{k}, \quad \text{if } Q_{ij} < 0.$$

Substitution of (6.37)-(6.39) into (6.35) yields the following upwind difference approximation:

$$(6.40) \qquad \frac{u_{i+1,j}-2u_{i,j}+u_{i-1,j}}{h^2} + \frac{(|P_{ij}|+P_{ij})u_{i+1,j}-2|P_{ij}|u_{i,j}+(|P_{ij}|-P_{ij})u_{i-1,j}}{2h}$$

$$+ \frac{u_{i,j+1}-2u_{i,j}+u_{i,j-1}}{k^2} + \frac{(|Q_{ij}|+Q_{ij})u_{i,j+1}-2|Q_{ij}|u_{i,j}+(|Q_{ij}|-Q_{ij})u_{i,j-1}}{2k}$$

$$+ R_{ij}u_{i,j}+S_{ij} = 0, \qquad i=1,2,\ldots,n-1, \qquad j=1,2,\ldots,m-1.$$

If one then inserts the known boundary values, there results a linear algebraic system of $(n-1)(m-1)$ equations in the $(n-1)(m-1)$ unknowns u_{ij}, $i=1,2,\ldots,n-1$, $j=1,2,\ldots,m-1$.

The upwind method, though less accurate than the central difference method described in the previous section, has the same properties as the other, but with *no restrictions* on h and k. In particular, the following theorems are valid.

THEOREM 6.3. If $S(x,y) \equiv 0$, then a nonconstant solution u_{ij} of (6.40) cannot assume a positive maximum or a negative minimum at any interior grid point (x_i,y_j) of $R_{n+1,m+1}$. If, in addition, $R(x,y) \equiv 0$, then a nonconstant numerical solution u_{ij} cannot assume its maximum or minimum at any interior grid point.

The proof follows in a fashion completely analogous to that of Theorem 5.2.

THEOREM 6.4. Let the Dirichlet problem (6.35)-(6.36) be defined on a rectangular domain. Let $P(x,y)$, $Q(x,y)$, $R(x,y)$ and $S(x,y)$ be continuous on $\Omega \cup \Gamma$. If $R \leq 0$ then the numerical solution obtained with the upwind difference method exists and is unique.

The proof is a direct consequence of Theorem 6.3.

The importance of Theorems 6.1-6.4 for the central and upwind difference methods can be summarized now as follows. Theorems 6.2 and 6.4 provide the *existence* and *uniqueness* of the numerical solution. Theorems 6.1 and 6.3 establish that any numerical solution has the same qualitative behaviour of the analytical solution relative to *max-min* properties.

EXAMPLE 2. Let us solve numerically the boundary value problem (6.31)-(6.32) with the upwind method. For h=k=1, since P=Q=6>0, equation (6.40) reduces to

(6.41) $\qquad u_{i+1,j} - 2u_{i,j} + u_{i-1,j} + 6(u_{i+1,j} - u_{i,j}) + u_{i,j+1} - 2u_{i,j} + u_{i,j-1} + 6(u_{i,j+1} - u_{i,j}) = 0$.

For i=1,2 and j=1,2, (6.41) implies

$$u_{10} + u_{01} - 16u_{11} + 7u_{21} + 7u_{12} = 0$$
$$u_{20} + u_{11} - 16u_{21} + 7u_{31} + 7u_{22} = 0$$
$$u_{11} + u_{02} - 16u_{12} + 7u_{22} + 7u_{13} = 0$$
$$u_{21} + u_{12} - 16u_{22} + 7u_{32} + 7u_{23} = 0 ,$$

into which, substitution of the boundary values (6.33) yields

$$-16u_{11} + 7u_{21} + 7u_{12} \qquad\qquad = -80$$
$$u_{11} - 16u_{21} \qquad + 7u_{22} = -10$$
$$u_{11} \qquad -16u_{12} + 7u_{22} = -10$$
$$u_{21} + u_{12} - 16u_{22} = 0 .$$

This system has a unique solution given by

$$u_{11} = 5.92105 , \qquad u_{12} = u_{21} = 1.05263 , \qquad u_{22} = 0.131579 .$$

Thus, when the upwind method is used for (6.31)-(6.32), the numerical solution obtained does possess the max-min property.

*6.7 THEORY FOR THE NUMERICAL SOLUTION OF LINEAR BOUNDARY VALUE PROBLEMS

In this section let us show that, under rather general conditions, as h and k converge to zero, the numerical solution of an elliptic boundary value problem obtained with either the central or the upwind method converges to the analytical solution.

Consider the Dirichlet problem (6.20)-(6.21) and assume that (6.22) is satisfied, so that the the analytical solution exists and is unique. In solving (6.20)-(6.21) numerically, consider, first, the central difference approximation

$$(6.42) \qquad \frac{u_{i+1,j}-2u_{i,j}+u_{i-1,j}}{h^2} + \frac{u_{i,j+1}-2u_{i,j}+u_{i,j-1}}{k^2} + P_{ij}\frac{u_{i+1,j}-u_{i-1,j}}{2h}$$

$$+ Q_{ij}\frac{u_{i,j+1}-u_{i,j-1}}{2k} + R_{ij}u_{i,j}+S_{ij} = 0 .$$

Consider first a lemma about elliptic difference operators which will aid in establishing an error bound theorem.

LEMMA 6.1. Let P_{ij}, Q_{ij} and R_{ij} be three discrete functions defined on an $R_{n+1,m+1}$ set. Assume that $|P_{ij}|\leq N$ and $|Q_{ij}|\leq M$. Assume also that there exists a positive number w such that $R_{ij}\leq-w<0$. Set $C=1/w$. Let E_{ij} be a discrete function, defined on the same $R_{n+1,m+1}$ set, such that $E_{i,0}=E_{i,m}=0$, $i=0,1,2,...,n$, and $E_{0,j}=E_{n,j}=0$, $j=1,2,...,m-1$. At the interior grid points, define L_{hk}, called an elliptic difference operator, by

$$(6.43) \qquad L_{hk}E_{i,j} = \frac{E_{i+1,j}-2E_{i,j}+E_{i-1,j}}{h^2} + \frac{E_{i,j+1}-2E_{i,j}+E_{i,j-1}}{k^2}$$

$$+ P_{ij}\frac{E_{i+1,j}-E_{i-1,j}}{2h} + Q_{ij}\frac{E_{i,j+1}-E_{i,j-1}}{2k} + R_{ij}E_{i,j} .$$

If Nh<2 and Mk<2, then

$$(6.44) \qquad |E_{ij}| \leq C[\max_{p,q}(|L_{hk}E_{pq}|)] , \qquad i=0,1,2,...,n, \qquad j=0,1,2,...,m.$$

The proof is a direct extension of that of Lemma 5.1. Note, however, that the condition $C=\max(1,1/w)$ has been replaced by $C=1/w$ because the E_{ij} are now assumed to be zero on the boundary grid points.

Again, in order to avoid the need for advanced mathematical theory, let us formulate convergence theorems under some additional assumptions which are not always required numerically.

THEOREM 6.5. Let $U(x,y)$ be the analytical solution of the Dirichlet problem (6.20)-(6.21). Assume that $P(x,y)$, $Q(x,y)$, $R(x,y)$ and $S(x,y)$ are continuous on $\Omega \cup \Gamma$, with $R(x,y) \leq -w < 0$. Set $C=1/w$. Let u_{ij}, $i=0,1,2,...,n$, $j=0,1,2,...,m$ be the numerical solution of (6.20)-(6.21) obtained from the central difference approximation (6.24), and let $E_{ij}=E(x_i,y_j)$, be the error, defined by $E_{ij}=U(x_i,y_j)-u_{ij}$. Assume also that $U(x,y)$ has continuous partial derivatives up to and including order four on $\Omega \cup \Gamma$. If $Nh<2$ and $Mk<2$, then

$$(6.45) \qquad |E_{ij}| \leq C\left[\frac{h^2}{12}(N_4+2NN_3)+\frac{k^2}{12}(M_4+2MM_3)\right] ,$$

where

$$N_3= \max_{\Omega \cup \Gamma}|U_{xxx}| , \quad N_4= \max_{\Omega \cup \Gamma}|U_{xxxx}| , \quad M_3= \max_{\Omega \cup \Gamma}|U_{yyy}| , \quad M_4= \max_{\Omega \cup \Gamma}|U_{yyyy}| .$$

The proof is a direct extension of Theorem 5.5.

THEOREM 6.6. Under the assumptions of Theorem 6.5, the numerical solution obtained with the central difference method converges to the analytical solution as $h \to 0$ and $k \to 0$, that is,

$$\lim_{h,k \to 0} |U(x_i,y_j)-u_{ij}| = 0 , \qquad i=0,1,2,...,n, \qquad j=0,1,2,...,m.$$

The proof follows directly from (6.45).

Theorems 6.5 and 6.6 provide an error bound and the convergence of the the numerical solution obtained by using the central difference method for a linear boundary value problem. An error bound and a convergence theorem for the upwind method can also be established readily. Specifically, the analogue of Lemma 6.1 is the following.

LEMMA 6.2. Let P_{ij}, Q_{ij} and R_{ij} be three discrete functions defined on an $R_{n+1,m+1}$ set, and assume that there exists a positive number w such that $R_{ij} \leq -w < 0$. Set $C=1/w$. Let E_{ij} be a discrete function, defined on the same $R_{n+1,m+1}$ set, such that $E_{i,0}=E_{i,m}=0$, $i=0,1,2,...,n$, and $E_{0,j}=E_{n,j}=0$, $j=1,2,...,m-1$. At the interior grid points, define the difference operator L_{hk} by

$$L_{hk}E_{i,j} = \frac{E_{i+1,j}-2E_{i,j}+E_{i-1,j}}{h^2} + \frac{(|P_{ij}|+P_{ij})E_{i+1,j}-2|P_{ij}|E_{i,j}+(|P_{ij}|-P_{ij})E_{i-1,j}}{2h}$$

$$+\frac{E_{i,j+1}-2E_{i,j}+E_{i,j-1}}{k^2} + \frac{(|Q_{ij}|+Q_{ij})E_{i,j+1}-2|Q_{ij}|E_{i,j}+(|Q_{ij}|-Q_{ij})E_{i,j-1}}{2k} + R_{ij}E_{i,j}.$$

Then,

$$|E_{ij}| \le C[\max_{p,q}(|L_{hk}E_{pq}|)] , \qquad i=0,1,2,...,n, \qquad j=0,1,2,...,m.$$

The proof of this Lemma is similar to the proof of Lemma 5.1.

The error bound theorem concerning the upwind difference method is as follows.

THEOREM 6.7. Let $U(x,y)$ be the analytical solution of the Dirichlet problem (6.20)-(6.21). Assume that $P(x,y)$, $Q(x,y)$, $R(x,y)$ and $S(x,y)$ are continuous on $\Omega \cup \Gamma$, with $R(x,y) \le -w < 0$. Set $C=1/w$. Let u_{ij}, $i=0,1,2,...,n$, $j=0,1,2,...,m$ be the numerical solution of (6.20)-(6.21) obtained from the upwind difference approximation (6.40), and let $E_{ij}=E(x_i,y_j)$ be the error, defined by $E_{ij}=U(x_i,y_j)-u_{ij}$. Assume also that $U(x,y)$ has continuous partial derivatives up to and including order four on $\Omega \cup \Gamma$. Then

$$(6.46) \qquad |E_{ij}| \le C\left[\frac{h}{12}\left(hN_4+6NN_2\right)+\frac{k}{12}\left(kM_4+6MM_2\right)\right] ,$$

where

$$N_2= \max_{\Omega\cup\Gamma}|U_{xx}| , \qquad N_4= \max_{\Omega\cup\Gamma}|U_{xxxx}| , \qquad M_2= \max_{\Omega\cup\Gamma}|U_{yy}| , \qquad M_4= \max_{\Omega\cup\Gamma}|U_{yyyy}| .$$

The proof of Theorem 6.7 is similar to that of Theorem 5.5.

Consider finally a convergence theorem for the upwind difference method.

THEOREM 6.8. Under the assumptions of Theorem 6.7, the numerical solution obtained with the upwind difference method converges to the analytical solution as $h \to 0$ and $k \to 0$, that is,

$$\lim_{h,k\to 0} |U(x_i,y_j)-u_{ij}| = 0 , \qquad i=0,1,2,...,n, \qquad j=0,1,2,...,m.$$

The proof follows directly from (6.46).

In general, the reason why the central difference method, when applicable, is to be preferred to the upwind method can be explained by the respective error bounds, for, if h and k are small, then h^2 and k^2 in (6.45) are smaller than h and k, respectively, in (6.46). Note, however, that no restrictions on h and k are required in Lemma 6.2 and Theorems 6.7-6.8.

6.8 NUMERICAL SOLUTION OF MILDLY NONLINEAR PROBLEMS

The methodology developed thus far for the numerical solution of linear boundary value problems extends to mildly nonlinear problems. A mildly nonlinear boundary value problem is defined as follows. Find a continuous function u(x,y) such that

$$(6.47) \qquad u_{xx}+u_{yy}+P(x,y)u_x+Q(x,y)u_y+F(x,y,u) = 0 , \qquad (x,y)\in \Omega$$

$$(6.48) \qquad u(x,y) = f(x,y) , \qquad (x,y)\in \Gamma ,$$

under the assumption

$$(6.49) \qquad \frac{\partial F(x,y,u)}{\partial y} \leq 0 , \qquad (x,y)\in (\Omega\cup\Gamma) , \qquad -\infty < u < \infty .$$

Condition (6.49) assures that solution of the Dirichlet problem (6.47)-(6.48) exists, is unique, and has certain general properties in common with harmonic functions.

Numerical methods for solving (6.47)-(6.48) can be derived easily from the methods described previously. Specifically, central difference approximations for (6.47) yield

$$\frac{u_{i+1,j}-2u_{i,j}+u_{i-1,j}}{h^2} + \frac{u_{i,j+1}-2u_{i,j}+u_{i,j-1}}{k^2} + P_{ij}\frac{u_{i+1,j}-u_{i-1,j}}{2h} + Q_{ij}\frac{u_{i,j+1}-u_{i,j-1}}{2k} + F(x_i,y_j,u_{i,j}) = 0 .$$

or, equivalently,

$$(6.50) \qquad h^2(2-kQ_{ij})u_{i,j-1}+k^2(2-hP_{ij})u_{i-1,j}-4(h^2+k^2)u_{i,j}+k^2(2+hP_{ij})u_{i+1,j}$$

$$+h^2(2+kQ_{ij})u_{i,j+1}+2h^2k^2F(x_i,y_j,u_{i,j}) = 0 .$$

If one then writes (6.50) consecutively and inserts the known boundary values, there results a mildly nonlinear system of $(n-1)(m-1)$ equations in the $(n-1)(m-1)$ unknowns u_{ij}, $i=1,2,\ldots,n-1$, $j=1,2,\ldots,m-1$.

In general, the system of equations (6.50) may not have a solution or may have more than one solution. However, note that if h and k satisfy the inequalities

$$(6.51) \qquad Nh < 2, \qquad Mk < 2,$$

then this system has a unique solution [Ortega and Rheinboldt (1970)]. Such a solution, can be found by the generalized Newton's method which, in this case, converges for all initial guesses and for all ω in a subrange of $0<\omega<2$. The specific generalized Newton's formulas for (6.50) are given by

$$(6.52) \qquad u_{i,j}^{(r+1)} = u_{i,j}^{(r)} - \omega\left[\frac{h^2(2-kQ_{ij})u_{i,j-1}^{(r+1)}+k^2(2-hP_{ij})u_{i-1,j}^{(r+1)}-4(h^2+k^2)u_{i,j}^{(r)}}{-4(h^2+k^2)+2h^2k^2[\partial F(x_i,y_j,u_{i,j}^{(r)})/\partial u]}\right.$$

$$\left.+\frac{k^2(2+hP_{ij})u_{i+1,j}^{(r)}+h^2(2+kQ_{ij})u_{i,j+1}^{(r)}+2h^2k^2F(x_i,y_j,u_{i,j}^{(r)})}{-4(h^2+k^2)+2h^2k^2[\partial F(x_i,y_j,u_{i,j}^{(r)})/\partial u]}\right],$$

$$i=1,2,3,\ldots,n-1, \qquad j=1,2,3\ldots,m-1.$$

For those boundary value problems for which conditions (6.51) are *too* restrictive, one can use the upwind difference approximation for u_x and u_y. In this case the difference approximation for (6.47) at (x_i,y_j) is

$$(6.53) \qquad \frac{u_{i+1,j}-2u_{i,j}+u_{i-1,j}}{h^2} + \frac{(|P_{ij}|+P_{ij})u_{i+1,j}-2|P_{ij}|u_{i,j}+(|P_{ij}|-P_{ij})u_{i-1,j}}{2h}$$

$$+\frac{u_{i,j+1}-2u_{i,j}+u_{i,j-1}}{k^2} + \frac{(|Q_{ij}|+Q_{ij})u_{i,j+1}-2|Q_{ij}|u_{i,j}+(|Q_{ij}|-Q_{ij})u_{i,j-1}}{2k}$$

$$+ F(x_i,y_j,u_{i,j}) = 0 .$$

Equations (6.53) constitute a mildly nonlinear system which has a unique solution for *any* given h and k. This solution can be determined with the generalized Newton's method.

Of course, if $P\equiv Q\equiv 0$, the central difference method and the upwind difference method are equivalent.

EXAMPLE. Let Γ be the square with vertices (0,0), (0,1), (1,1), (1,0) and let Ω be its interior. On Γ, set $f(x,y)=0$. For $h=k=1/3$, the points of $R_{4,4}$, are shown in Figure 6.6. If the differential equation defined on Ω is

$$u_{xx}+u_{yy} = e^u,$$

then application of (6.50) at each interior point of $R_{4,4}$ yields the following mildly nonlinear system

$$\frac{u_{i+1,j}-2u_{i,j}+u_{i-1,j}}{(1/3)^2} + \frac{u_{i,j+1}-2u_{i,j}+u_{i,j-1}}{(1/3)^2} - e^{u_{i,j}} = 0, \qquad i=1,2, \qquad j=1,2.$$

or, equivalently,

(6.54)

$$-36u_{11}+9u_{21}+9u_{12} \qquad -e^{u_{11}} = 0$$

$$9u_{11}-36u_{21} \qquad +9u_{22}-e^{u_{21}} = 0$$

$$9u_{11} \qquad -36u_{12}+9u_{22}-e^{u_{12}} = 0$$

$$9u_{21}+9u_{12}-36u_{22}-e^{u_{22}} = 0$$

This system can be solved readily by the generalized Newton's method (see Exercise 18).

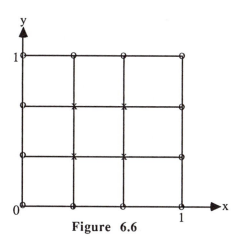

Figure 6.6

EXERCISES

Basic Exercises

1. Classify each of the following partial differential equations as elliptic, parabolic, or hyperbolic at the point (0,0):

   (a) $u_{xx} + 2u_{yy} = 0$

   (b) $u_{xx} - 2u_{yy} = 0$

   (c) $u_{xx} - 2u_y = 0$

   (d) $u_{xx} - 4u_{xy} + u_{yy} = 0$

   (e) $3u_{xx} - 4u_{xy} - 5u_{yy} = 0$

   (f) $3u_{xx} - 4u_{xy} - 5u_{yy} + 8u_x - 9yu_y + 6u = 27xy$.

2. Determine, if possible, in which portions of the plane each of the following is elliptic, parabolic or hyperbolic:

   (a) $yu_{xx} - u_{yy} = 0$

   (b) $u_{xx} + 2xu_{xy} + (1-y^2)u_{yy} = 0$

   (c) $(1+u_y^2)u_{xx} - 2u_xu_yu_{xy} + (1+u_x^2)u_{yy} = 0$.

3. Let Γ be the square whose vertices are (1/2,1/2), (1/2,–1/2), (–1/2,1/2), (–1/2,–1/2) and let Ω be its interior. Show that each of the following continuous functions is harmonic on Ω:

   (a) $u = 5$ (b) $u = 4y-7$ (c) $u = 7x-4y-2$

   (d) $u = x^2-y^2$ (e) $u = (1/2)xy^2-x^3/6$ (f) $u = (xy^3-x^3y)/6$.

4. Repeat Exercise 3 but let Γ be the circle whose equation is $x^2+y^2=1$.

5. With h=k=2 find the numerical solution of the Dirichlet problem for Laplace's

equation for which $f(x,y)=x^2-y^2$ and Γ is the rectangle whose vertices are (0,0), (5,0), (5,4), (0,4).

6. With $h=k=2$ find the numerical solution of the Dirichlet problem for Laplace's equation for which $f(x,y)=x-2y$ and Γ is the triangle whose vertices are (0,0), (7,0), (0,7).

7. With $h=k=1/2$ find the numerical solution of the Dirichlet problem for Laplace's equation for which $f(x,y)=x^2-y$ and Γ is the circle of unit radius whose center is (1,1).

8. Let Γ be the square whose vertices are (1,1), (2,1), (2,2), (1,2) and let Ω be the interior of Γ. Let $f(x,y)=x-y^3$ on Γ and consider the resulting Dirichlet problem for Laplace's equation. By a change of variables, transform the problem into one in which the origin lies interior to the region of interest.

9. Let Γ be the rectangle whose vertices are (1,1), (9,1), (9,4), (1,4) and let Ω be the interior of Γ. On Γ let $f(x,y)=x^2-y^2$. By using the central difference method, find a numerical solution of the Dirichlet problem for each of the following elliptic equations:

(a) $u_{xx}+u_{yy}+10(yu_x+xu_y) = 0$

(b) $u_{xx}+u_{yy}+0.1(yu_x+xu_y) = 0$

(c) $u_{xx}+u_{yy}+xu_x+yu_y-2u = 0$.

Be sure that the numerical solution exists, is unique, and possesses the discrete max-min property. Then, compare your result with the exact solution $u=x^2-y^2$.

10. Prove Theorem 6.1.

11. Repeat Exercise 9 by using the upwind difference method with $h=2$ and $k=1$.

12. Prove Theorem 6.4.

13. Prove Lemma 6.1.

14. Prove Theorem 6.5.

15. For each of the boundary value problems in Exercise 9, find an estimate for the constant C of Lemma 6.1.

16. Knowing that the exact solution for each of the boundary value problems in Exercise 9 is $u=x^2-y^2$, determine an error bound for the numerical solutions when the central difference method is used for each of these problems.

17. Repeat Exercise 16 for the upwind difference method.

18. Find the solution of system (6.54).

Supplementary Exercises

19. Let Γ be the rectangle whose vertices are A(0,0), B(5,0), C(5,4), D(0,4). On AB, BC, and AD, let $f(x,y)=x^2-y^2$. If, in the statement of the Dirichlet problem for Laplace's equation one does not give f(x,y) on CD, but instead gives $\partial u/\partial n$, that is the derivative of u in the direction normal to CD, then the resulting problem is called a *Mixed Type* problem. Suppose, then, one is given $(\partial u/\partial y)=-8$ on CD. Devise a numerical method for this mixed problem and apply it with h=k=1. Compare your answer with the exact solution $u=x^2-y^2$.

20. Let Γ be the square whose vertices are (0,0), (1,0), (1,1), (0,1). Let Ω be the interior of Γ and let $f(x,y)=x^2-y^2$ on Γ. Using h=k=0.1, approximate the solution of the Dirichlet problem for Laplace's equation and compare your answer with the exact solution $u=x^2-y^2$.

21. Repeat Exercise 20, but replace Γ with the ellipse whose equation is $x^2+2y^2=1$.

22. Repeat Exercise 20, but replace Γ with the convex pentagon whose consecutive vertices are (0,0), (1,0), (2,3), (1,4), (–1,1).

23. Let Γ be the square whose vertices are (0,0), (1,0), (1,1), (0,1). Let Ω be the interior of Γ and let $f(x,y)=x+3$ on Γ. Using $h=k=0.2$, approximate a solution of the Dirichlet problem for each of the following elliptic equations:

(a) $u_{xx}+u_{yy}-4u_y = 0$

(b) $u_{xx}+u_{yy}+u_x+u_y+u-(x+4) = 0$

(c) $u_{xx}+u_{yy}-e^u = -e^{x+3}$

(d) $u_{xx}+u_{yy}-(1/y)u_y = 0$.

In each case, compare the numerical results with the exact solution $u=x+3$. Does the fact that the coefficient of u_y in (d) is unbounded near $y=0$ cause practical problems?

7

Parabolic Equations

7.1 INTRODUCTION

The prototype parabolic differential equation is the heat equation

$$u_{xx} - u_y = 0 \; ,$$

and we will examine it first. Because, physically, the variable y represents *time* in the problem to be studied, we will set y=t and examine the heat equation in its more customary form

(7.1) $u_t = u_{xx} \; .$

Two kinds of problems are of fundamental interest both mathematically and physically with regard to (7.1). These are the *initial value* problem and the *initial-boundary* problem, which are defined as follows. In an initial value problem for (7.1), one is given a function f(x) which is continuous for all values of x and one is asked to find a function u(x,t) which is

(a) defined and continuous for $-\infty < x < \infty$, $0 \leq t$;

(b) satisfies (7.1) for $-\infty < x < \infty$, $0 < t$; and

(c) satisfies the initial conditions u(x,0)=f(x) at time t=0 for $-\infty < x < \infty$.

As shown geometrically in Figure 7.1, initial value problems are defined on an upper-half plane.

In an initial-boundary problem, one is given a constant a>0 and three continuous functions

214

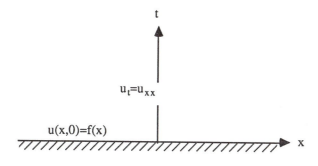

Figure 7.1

$$g_1(t), \quad t \geq 0; \qquad g_2(t), \quad t \geq 0; \qquad f(x), \quad 0 \leq x \leq a,$$

which satisfy $g_1(0) = f(0)$, $g_2(0) = f(a)$, and one is asked to find a function $u(x,t)$ which is

(a) defined and continuous for $t \geq 0$, $0 \leq x \leq a$;

(b) satisfies (7.1) on $0 < x < a$, $t > 0$; and

(c) satisfies the initial and boundary conditions

(7.2) $u(x,0) = f(x)$, $0 \leq x \leq a$,   (initial condition)

(7.3)
$$u(0,t) = g_1(t),$$
$$\qquad\qquad\qquad t \geq 0, \qquad\qquad \text{(boundary conditions).}$$
$$u(a,t) = g_2(t),$$

     As shown geometrically in Figure 7.2, initial-boundary problems are defined on a semi-infinite strip.

     The unique solution of an initial value problem can be given in terms of the Fourier integral, while that for an initial-boundary problem can be given in terms of series. However, again, because analytical solutions so given are not evaluated easily at particular points of interest, and, because methods for generating such solutions do not extend to nonlinear problems, we turn to numerical methods. For clarity, we will concentrate on initial-boundary problems, though most of the ideas extend to the initial value problem as well.

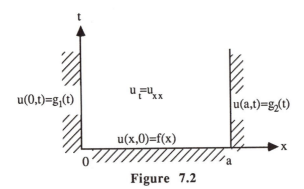

Figure 7.2

7.2 AN EXPLICIT NUMERICAL METHOD FOR THE HEAT EQUATION

Let us first develop a simple numerical method to solve the initial-boundary problem for the heat equation. Because in the initial-boundary problem the time variable t can vary in the unbounded range $0 \leq t < \infty$, and since the computer cannot calculate forever, we will first replace $t \geq 0$ by $0 \leq t \leq T$. The constant value T is usually determined by the physics of the phenomenon under study. Thus, in fast reaction type problems, like those related, for example, to the release of nuclear energy, it is often sufficient to choose T relatively small. On the other hand, slow reaction type problems, like those related, for example, to radioactive decay, may require the choice of a relatively large T. In any case, T is some constant, positive value.

To begin with, it is important to know that, like harmonic functions, solutions of the heat equation possess the *max-min* property. Specifically [Forsythe and Wasow (1960)], in the rectangle $0 \leq x \leq a$, $0 \leq t \leq T$, *any solution of (7.1) is bounded from above by the maximum and from below by the minimum of its initial and boundary values.* (The line t=T is not considered part of the boundary.)

With the above considerations in mind, let us begin by subdividing the interval [0,a] into n equal parts, each of length h=a/n, by the R_{n+1} set $x_i = ih$, $i=0,1,2,\ldots,n$. Next, the interval [0,T] is divided into m equal parts, each of length k=T/m, by the R_{m+1} set $t_j = jk$, $j=0,1,2,\ldots,m$. Thus, as indicated in Figure 7.3, on the rectangle with vertices (0,0), (a,0), (a,T) and (0,T), we have defined a set of planar grid points $(x_i, t_j) \in R_{n+1,m+1}$. The points with coordinates $(x_i, 0)$, $i=0,1,2,\ldots,n$, are called the *initial* grid points, the points with coordinates (x_0, t_j) and (x_n, t_j), $j=1,2,\ldots,m$, are called the *boundary* grid points, and the points with coordinates (x_i, t_j), $i=1,2,\ldots,n-1$, $j=1,2,\ldots,m$, are called the *interior* grid points. Those grid points whose coordinates are (x_i, t_j), $i=0,1,2,\ldots,n$, are called the jth row of grid points, or, the grid points at *time level* t_j.

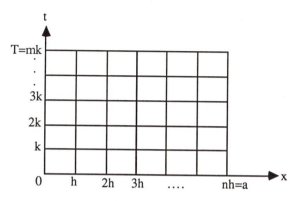

Figure 7.3

A simple numerical method for approximating the solution of initial-boundary problem (7.1)-(7.3) can be formulated now as follows. At the 0th row of grid points the initial condition (7.2) implies

(7.4) $u_{i,0} = f(x_i)$, $i=0,1,2,\ldots,n.$

At x=0 and at x=a, the boundary conditions (7.3) yield

(7.5) $u_{0,j} = g_1(t_j)$, $u_{n,j} = g_2(t_j)$, $j=1,2,\ldots,m.$

At each interior grid point, for the point arrangement shown in Figure 7.4, consider from Section 3.6 the approximations

(7.6) $u_t(x_i,t_j) \approx \dfrac{u_{i,j+1} - u_{i,j}}{k}$

(7.7) $u_{xx}(x_i,t_j) \approx \dfrac{u_{i+1,j} - 2u_{i,j} + u_{i-1,j}}{h^2}$,

substitution of which into (7.1) yields the approximation

(7.8)
$$\frac{u_{i,j+1}-u_{i,j}}{k} = \frac{u_{i+1,j}-2u_{i,j}+u_{i-1,j}}{h^2},$$

or, equivalently,

(7.9)
$$u_{i,j+1} = u_{i,j} + \frac{k}{h^2}(u_{i+1,j}-2u_{i,j}+u_{i-1,j}).$$

In (7.9), setting

$$\alpha = \frac{k}{h^2}$$

yields finally

(7.10)
$$u_{i,j+1} = \alpha u_{i+1,j} + (1-2\alpha)u_{i,j} + \alpha u_{i-1,j}.$$

Now, on the first row of grid points, apply (7.10) with j=0 to approximate $u_{i,1}$ explicitly, for each of i=1,2,...,n–1. Then, using the numerical results generated for row 1, set j=1 and approximate u explicitly on the second row by means of (7.5) and (7.10). Continue in the indicated fashion to approximate u explicitly at each grid point of row j+1, j=2,3,...,m-1, by applying (7.10), and by making use of (7.5) and the numerical approximation generated on row j. The values u_{ij}, i=0,1,2,...,n, j=0,1,2,...,m, are called the numerical solution on the $R_{n+1,m+1}$ set.

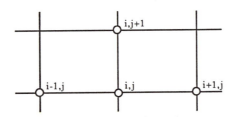

Figure 7.4

EXAMPLE 1. Consider the initial-boundary problem defined by (7.1), a=3 and

(7.11) $u(x,0) = f(x) = x^2,$ $0 \leq x \leq 3$

(7.12)
$$u(0,t) = g_1(t) = 0 ,$$
$$u(3,t) = g_2(t) = 9 ,$$
              $t > 0 .$

Set T=3, n=3, m=6 so that h=1, k=0.5, and the $R_{4,7}$ set is (i,0.5j), i=0,1,2,3, j=0,1,2,3,4,5,6. Then, at the initial time $t_0=0$ the initial condition (7.11) implies

(7.13) $u_{00} = 0 ,$ $u_{10} = 1 ,$ $u_{20} = 4 ,$ $u_{30} = 9 .$

At the boundary grid points the boundary conditions (7.12) yield

(7.14)
$$u_{01} = u_{02} = u_{03} = u_{04} = u_{05} = u_{06} = 0 ,$$
$$u_{31} = u_{32} = u_{33} = u_{34} = u_{35} = u_{36} = 9 .$$

At the interior grid points, since $\alpha=0.5$, equation (7.10) reduces to

(7.15) $u_{i,j+1} = 0.5(u_{i-1,j} + u_{i+1,j}) ,$ i=1,2.

Thus, for j=0 and by use of (7.13), equation (7.15) yields $u_{11}=2$ and $u_{21}=5$. Hence, the numerical solution at time level $t_1=0.5$ is

(7.16) $u_{01} = 0 ,$ $u_{11} = 2 ,$ $u_{21} = 5 ,$ $u_{31} = 9 .$

Next, for j=1 and by use of (7.16), equation (7.15) yields $u_{12}=2.5$ and $u_{22}=5.5$. Hence the numerical solution at time level $t_2=1$ is

(7.17) $u_{02} = 0 ,$ $u_{12} = 2.5 ,$ $u_{22} = 5.5 ,$ $u_{32} = 9 .$

Continuing in the indicated fashion for j=2,3,4,5 one finds

(7.18)

$$
\begin{array}{llll}
u_{03} = 0 , & u_{13} = 2.75 , & u_{23} = 5.75 , & u_{33} = 9 , \\
u_{04} = 0 , & u_{14} = 2.875 , & u_{24} = 5.875 , & u_{34} = 9 , \\
u_{05} = 0 , & u_{15} = 2.9375 , & u_{25} = 5.9375 , & u_{35} = 9 , \\
u_{06} = 0 , & u_{16} = 2.96875 , & u_{26} = 5.96875 , & u_{36} = 9 .
\end{array}
$$

The numerical solution on the $R_{4,7}$ set is given finally by (7.13), (7.16), (7.17) and (7.18).

We say that the numerical solution possesses the max-min property when the discrete function u_{ij}, i=0,1,2,…,n, j=0,1,2,…,m, takes on its maximum and its minimum on the union of of the initial and the boundary grid points. Note that the discrete function u_{ij} in the above example possesses the max-min property.

EXAMPLE 2. Let us reconsider the initial-boundary problem defined by (7.1), (7.11) and (7.12), and fix T=30. Set h=1. Since k=0.5 will then result in 60 rows of grid points, let us set k=5 so that only 6 rows will result. Then, $\alpha=5$, and equation (7.10) reduces to

(7.19) $\qquad u_{i,j+1} = 5u_{i-1,j} - 9u_{i,j} + 5u_{i+1,j} ,$ $\qquad\qquad\qquad$ i=1,2.

Thus, for j=0 and by use of (7.13), equation (7.19) yields $u_{11}=11$ and $u_{21}=14$. Hence, the numerical solution at time level $t_1=5$ is

(7.20) $\qquad u_{01} = 0 , \qquad u_{11} = 11 , \qquad u_{21} = 14 , \qquad u_{31} = 9 .$

Next, for j=1,2,3,4,5 and by use of the numerical solution generated for row j, equation (7.19) yields

(7.21)

$$
\begin{array}{llll}
u_{02} = 0 , & u_{12} = -29 , & u_{22} = -26 , & u_{32} = 9 , \\
u_{03} = 0 , & u_{13} = 131 , & u_{23} = 134 , & u_{33} = 9 , \\
u_{04} = 0 , & u_{14} = -509 , & u_{24} = -506 , & u_{34} = 9 , \\
u_{05} = 0 , & u_{15} = 2051 , & u_{25} = 1964 , & u_{35} = 9 , \\
u_{06} = 0 , & u_{16} = -8639 , & u_{26} = -7376 , & u_{36} = 9 .
\end{array}
$$

Thus, the numerical solution on the $R_{4,7}$ set is now given by (7.13), (7.20) and (7.21). Note, however, that this solution has a nonphysical behavior, since it oscillates with positive and negative values. Indeed it does not possess the max-min property, and, if one proceeds further with j=6,7,8,..., overflow will occur readily.

To develop a condition for (7.10) to yield a numerical solution which possesses the max-min property, consider the following simple argument. For a given initial-boundary problem, let h=a/2, k=T/m, so that the points of $R_{3,m+1}$ are shown in Figure 7.5. At the point (a/2,0), set $u_{10}=\varepsilon>0$. At the remaining initial and boundary grid points, set $u_{0j}=u_{2j}=0$, j=0,1,2,...,m. Then, application of (7.10) at the interior grid points yields

$$u_{11} = (1-2\alpha)\varepsilon, \quad u_{12} = (1-2\alpha)u_{11} = (1-2\alpha)^2\varepsilon, \quad u_{13} = (1-2\alpha)^3\varepsilon, \quad \ldots,$$

and, in general,

(7.22) $\qquad u_{1j} = (1-2\alpha)^j\varepsilon, \qquad\qquad$ j=0,1,2,...,m.

Now, to have the max-min property, we require from (7.22) that $0 \le u_{1j} \le \varepsilon$, j=1,2,...,m, that is,

$$0 \le (1-2\alpha)^j\varepsilon \le \varepsilon.$$

or

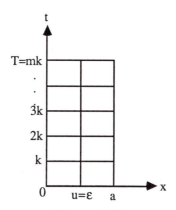

Figure 7.5

$$0 \leq 1-2\alpha \leq 1 \ .$$

Thus, since $\alpha > 0$, this implies

(7.23) $\alpha \leq \dfrac{1}{2} \ ,$

which is equivalent to

(7.24) $k \leq \dfrac{h^2}{2} \ .$

Inequality (7.24) is, indeed, the correct mathematical condition for the finite difference approximation (7.8) to possess the max-min property. This condition was violated in Example 2 where we selected $\alpha = 5 > 1/2$. A rigorous development will be given later.

7.3 THE GENERAL LINEAR PARABOLIC EQUATION

The ideas developed thus far will be extended next to general linear parabolic equations as follows.

Let $a > 0$ be a fixed constant and let $P(x,t)$, $Q(x,t)$, $R(x,t)$ and $S(x,t)$ be bounded and continuous for $0 \leq x \leq a$, $t \geq 0$. Then consider the initial-boundary problem

(7.25) $u_t = P(x,t)u_{xx} + Q(x,t)u_x + R(x,t)u + S(x,t) \ ,$ $0 < x < a \ ,$ $t > 0$

(7.26) $u(x,0) = f(x) \ ,$ $0 \leq x \leq a$

(7.27) $u(0,t) = g_1(t) \ ,$ $u(a,t) = g_2(t) \ ,$ $t \geq 0,$

in which $g_1(0)=f(0)$ and $g_2(0)=f(a)$. We will assume that $P(x,t) \geq v > 0$ and $R(x,t) \leq 0$, in order to be assured that the solution of the initial-boundary problem (7.25)-(7.27) exists, is unique, and has a general *max-min* property, which can be stated as follows. If $S(x,t) \equiv 0$, then in the rectangle $0 \leq x \leq a$, $0 \leq t \leq T$, any positive maximum of a solution of (7.25) is bounded above by the maximum of its initial and boundary values. Similarly, any negative minimum is bounded below by the minimum of its initial and boundary values. If, in addition, $R(x,t) \equiv 0$, then any solution of (7.25) is bounded above by the maximum and below by the minimum of its initial and boundary values. (The line $t=T$ is not considered part of the boundary.)

A numerical method for (7.25)-(7.27) can be formulated as follows. Select a final time $T>0$. Subdivide the interval $[0,a]$ into n equal parts, each of length $h=a/n$, by an R_{n+1} set $x_i=ih$, $i=0,1,2,\ldots,n$. Next, subdivide the interval $[0,T]$ into m equal parts, each of length $k=T/m$, by an R_{m+1} set $t_j=jk$, $j=0,1,2,\ldots,m$. At the initial grid points, from (7.26), set

$$(7.28) \qquad u_{i,0} = f(x_i) , \qquad\qquad i=0,1,2,\ldots,n.$$

At the boundary grid points, from (7.27), set

$$(7.29) \qquad u_{0,j} = g_1(t_j) , \qquad u_{n,j} = g_2(t_j) , \qquad\qquad j=1,2,\ldots,m.$$

In order to compute u_{ij} at the interior grid points, consider the approximations

$$(7.30) \qquad u_t(x_i,t_j) \approx \frac{u_{i,j+1}-u_{i,j}}{k} ,$$

$$(7.31) \qquad u_x(x_i,t_j) \approx \frac{u_{i+1,j}-u_{i-1,j}}{2h} ,$$

$$(7.32) \qquad u_{xx}(x_i,t_j) \approx \frac{u_{i+1,j}-2u_{i,j}+u_{i-1,j}}{h^2} .$$

Substitution of (7.30)-(7.32) into (7.25) then yields

$$(7.33) \qquad \frac{u_{i,j+1}-u_{i,j}}{k} = P_{ij}\frac{u_{i+1,j}-2u_{i,j}+u_{i-1,j}}{h^2} + Q_{ij}\frac{u_{i+1,j}-u_{i-1,j}}{2h} + R_{ij}u_{i,j} + S_{ij} .$$

It follows that (7.33) can be rewritten, equivalently, as

$$(7.34) \qquad u_{i,j+1} = (\alpha P_{ij}-\beta Q_{ij})u_{i-1,j}+(1+kR_{ij}-2\alpha P_{ij})u_{i,j}+(\alpha P_{ij}+\beta Q_{ij})u_{i+1,j}+kS_{ij} ,$$

where

$$\alpha = \frac{k}{h^2}, \qquad \beta = \frac{k}{2h}.$$

Equation (7.34) can now be used to generate explicitly the numerical solution at each interior grid point at time t_{j+1}. The numerical method so obtained is called the *explicit central difference* method.

EXAMPLE. Consider the initial-boundary problem defined by

(7.35) $u_t = u_{xx} + (x-2)u_x - 3u$

(7.36) $u(x,0) = x^2 - 4x + 5$, $0 \le x \le 4$

(7.37) $u(0,t) = u(4,t) = 5e^{-t}$, $t > 0$

the exact solution of which is $(x^2 - 4x + 5)e^{-t}$. To generate the numerical solution, let $T=1$, $n=4$, $m=10$ so that $h=1$, $k=0.1$, and the $R_{5,11}$ set is $(i, 0.1j)$, $i=0,1,2,3,4$, $j=0,1,2,\ldots,10$. Then, at the initial time $t_0=0$ the initial condition (7.36) implies

(7.38) $u_{0,0} = 5$, $u_{1,0} = 2$, $u_{2,0} = 1$, $u_{3,0} = 2$, $u_{4,0} = 5$.

At the boundary grid points the boundary conditions (7.37), to three decimal places, yield

$u_{0,0} = 5.000$	$u_{4,0} = 5.000$
$u_{0,1} = 4.524$	$u_{4,1} = 4.524$
$u_{0,2} = 4.094$	$u_{4,2} = 4.094$
$u_{0,3} = 3.704$	$u_{4,3} = 3.704$
$u_{0,4} = 3.352$	$u_{4,4} = 3.352$
$u_{0,5} = 3.033$	$u_{4,5} = 3.033$
$u_{0,6} = 2.744$	$u_{4,6} = 2.744$
$u_{0,7} = 2.483$	$u_{4,7} = 2.483$
$u_{0,8} = 2.247$	$u_{4,8} = 2.247$
$u_{0,9} = 2.033$	$u_{4,9} = 2.033$
$u_{0,10} = 1.839$	$u_{4,10} = 1.839$.

(7.39)

At the interior grid points, since $\alpha=0.1$ and $\beta=0.05$, equation (7.34) becomes

(7.40) $u_{i,j+1} = 0.1[1-0.5(x_i-2)]u_{i-1,j}+0.5u_{i,j}+0.1[1+0.5(x_i-2)]u_{i+1,j}$.

Thus, by using (7.38), (7.39), and calculating to three decimal places, equation (7.40), for $j=0,1,2,...,9$, yields

$u_{1,1} = 1.800$	$u_{2,1} = 0.900$	$u_{3,1} = 1.800$
$u_{1,2} = 1.624$	$u_{2,2} = 0.810$	$u_{3,2} = 1.624$
$u_{1,3} = 1.466$	$u_{2,3} = 0.730$	$u_{3,3} = 1.466$
$u_{1,4} = 1.325$	$u_{2,4} = 0.658$	$u_{3,4} = 1.325$
$u_{1,5} = 1.198$	$u_{2,5} = 0.594$	$u_{3,5} = 1.198$
$u_{1,6} = 1.084$	$u_{2,6} = 0.537$	$u_{3,6} = 1.084$
$u_{1,7} = 0.980$	$u_{2,7} = 0.485$	$u_{3,7} = 0.980$
$u_{1,8} = 0.887$	$u_{2,8} = 0.439$	$u_{3,8} = 0.887$
$u_{1,9} = 0.802$	$u_{2,9} = 0.397$	$u_{3,9} = 0.802$
$u_{1,10} = 0.726$	$u_{2,10} = 0.359$	$u_{3,10} = 0.726$,

which is the numerical solution at the interior grid points on the given $R_{5,11}$ set.

With regard to the max-min property of the numerical solution of parabolic equation (7.25), we proceed as follows.

DEFINITION 7.1. For $S(x,t)\equiv0$, a numerical solution u_{ij} of (7.25) is said to possess the weak max-min property if u_{ij} satisfies the following inequalities for $j=1,2,...,m$:

(7.41) $u_{ij} \geq \min[0, \min_{1\leq q\leq j}(u_{0q}), \min_{0\leq p\leq n}(u_{p0}), \min_{1\leq q\leq j}(u_{nq})]$

(7.42) $u_{ij} \leq \max[0, \max_{1\leq q\leq j}(u_{0q}), \max_{0\leq p\leq n}(u_{p0}), \max_{1\leq q\leq j}(u_{nq})]$.

Note that the numerical solution obtained from the example above possesses the weak max-min property since inequalities (7.41) and (7.42) are satisfied.

DEFINITION 7.2. For $S(x,t) \equiv R(x,t) \equiv 0$, a numerical solution u_{ij} of (7.25) is said to possess the **strong max-min property** if u_{ij} satisfies the following inequalities:

(7.43) $u_{ij} \geq \min[\min_{1 \leq q \leq j}(u_{0q}), \min_{0 \leq p \leq n}(u_{p0}), \min_{1 \leq q \leq j}(u_{nq})]$

(7.44) $u_{ij} \leq \max[\max_{1 \leq q \leq j}(u_{0q}), \max_{0 \leq p \leq n}(u_{p0}), \max_{1 \leq q \leq j}(u_{nq})]$.

    If either of Definition 7.1 or 7.2 is applicable, one says that the numerical solution has the max-min property. Note also that only when $S(x,t) \equiv 0$ is the max-min property defined.

    Before continuing, it is important to make the following observation. The explicit finite difference scheme (7.34) may yield a numerical solution which does not possess the max-min property. In fact, for $S(x,t) \equiv 0$, equation (7.34) reduces to

(7.45) $u_{i,j+1} = (\alpha P_{ij} - \beta Q_{ij})u_{i-1,j} + (1 + kR_{ij} - 2\alpha P_{ij})u_{i,j} + (\alpha P_{ij} + \beta Q_{ij})u_{i+1,j}$.

    Now, if any one of the coefficients of $u_{i-1,j}$, $u_{i,j}$ $u_{i+1,j}$ in the right-hand side of (7.45) is negative, then it is easy to construct an example of an initial-boundary problem whose numerical solution does not possess the max-min property. Assume, for instance, that for $j=0$ and $0 < i < n$ one has $(\alpha P_{i0} - \beta Q_{i0}) < 0$. At the point (x_{i-1}, t_0), set $u_{i-1,0} = \varepsilon > 0$. At the remaining initial and boundary grid points of $R_{n+1,m+1}$, set $u = 0$ so that the max-min property implies $0 \leq u \leq \varepsilon$ at each interior grid point. However, application of (7.45) at (x_i, t_0) yields

$u_{i,1} = (\alpha P_{i0} - \beta Q_{i0})\varepsilon < 0$.

    In order to be assured, *a priori*, that the finite difference scheme (7.45) does imply the max-min property, it is necessary and sufficient that the space step h and the time step k satisfy appropriate conditions. The sufficiency is proved in the following theorem. The necessity appears later as an exercise.

THEOREM 7.1. For $S(x,t) \equiv 0$, let the initial-boundary problem (7.25)-(7.27) be defined on $[0,a]$, $0 \leq t \leq T$, where $P(x,t)$, $Q(x,t)$ and $R(x,t)$ are continuous, and hence bounded. Assume that $V \geq P(x,t) \geq v > 0$, $|Q(x,t)| \leq M$ and $-N \leq R(x,t) \leq 0$ on $0 \leq x \leq a$, $0 \leq t \leq T$. If

(7.46) $Mh \leq 2v$

and

(7.47) $k \leq \dfrac{h^2}{Nh^2 + 2V}$,

then the numerical solution of (7.25)-(7.27), obtained with the central difference formula (7.34), possesses the **max-min property**.

PROOF. Note first that inequality (7.46) implies

$$(\alpha P_{ij} - \beta Q_{ij}) \geq (\alpha P_{ij} - \beta |Q_{ij}|) \geq (\alpha v - \beta M) \geq 0 ,$$

and

$$(\alpha P_{ij} + \beta Q_{ij}) \geq (\alpha P_{ij} - \beta |Q_{ij}|) \geq (\alpha v - \beta M) \geq 0 ,$$

while inequality (7.47) implies

$$(1 + kR_{ij} - 2\alpha P_{ij}) \geq (1 - kN - 2\alpha V) \geq 0 .$$

Moreover,

(7.48) $(\alpha P_{ij} - \beta Q_{ij}) + (1 + kR_{ij} - 2\alpha P_{ij}) + (\alpha P_{ij} + \beta Q_{ij}) = 1 + kR_{ij} \leq 1 .$

Thus, (7.45) implies

(7.49) $u_{i,j+1} \leq (\alpha P_{ij} - \beta Q_{ij}) \max_{0 \leq p \leq n} (u_{pj}) + (1 + kR_{ij} - 2\alpha P_{ij}) \max_{0 \leq p \leq n} (u_{pj})$

$$+ (\alpha P_{ij} + \beta Q_{ij}) \max_{0 \leq p \leq n} (u_{pj}) = (1 + kR_{ij}) \max_{0 \leq p \leq n} (u_{pj}) ,$$

which, since $0 \leq (1 + kR_{ij}) \leq 1$, implies

(7.50) $u_{i,j+1} \leq \max[0, \max_{0 \leq p \leq n} (u_{pj})] = \max[0, u_{0,j}, \max_{0 < p < n} (u_{p,j}), u_{n,j}]$.

Now, because (7.50) is valid for i=1,2,...,n−1, it follows that

(7.51) $u_{i,j} \leq \max[0, u_{0,j-1}, \max_{0 < p < n} (u_{p,j-1}), u_{n,j-1}]$

$\leq \max[0, u_{0,j-1}, u_{0,j-2}, \max_{0 < p < n} (u_{p,j-2}), u_{n,j-2}, u_{n,j-1}]$

$\leq \max[0, u_{0,j-1}, u_{0,j-2}, u_{0,j-3}, \max_{0 < p < n} (u_{p,j-3}), u_{n,j-3}, u_{n,j-2}, u_{n,j-1}]$

.
.
.

$\leq \max[0, \max_{0 \leq q < j} (u_{0,q}), \max_{0 < p < n} (u_{p,0}), \max_{0 \leq q < j} (u_{n,q})]$,

which implies (7.42). Inequality (7.41) can be proved in an entirely similar way.
When $R_{ij}=0$, i=0,1,2,...,n, j=0,1,2,...,m, inequality (7.49) reduces to

$u_{i,j+1} \leq \max_{0 \leq p \leq n} (u_{pj}) = \max[u_{0,j}, \max_{0 < p < n} (u_{p,j}), u_{n,j}]$,

which can be used to yield

(7.52) $u_{i,j} \leq \max[u_{0,j-1}, \max_{0 < p < n} (u_{p,j-1}), u_{n,j-1}]$

$\leq \max[u_{0,j-1}, u_{0,j-2}, \max_{0 < p < n} (u_{p,j-2}), u_{n,j-2}, u_{n,j-1}]$

$\leq \max[u_{0,j-1}, u_{0,j-2}, u_{0,j-3}, \max_{0 < p < n} (u_{p,j-3}), u_{n,j-3}, u_{n,j-2}, u_{n,j-1}]$

.
.
.

$\leq \max[\max_{0 \leq q < j} (u_{0,q}), \max_{0 < p < n} (u_{p,0}), \max_{0 \leq q < j} (u_{n,q})]$,

which implies (7.44). Inequality (7.43) can be proved in a similar fashion.
Note that under the hypotheses of Theorem 7.1, if f(x), $g_1(t)$ and $g_2(t)$ are bounded,

the numerical solution of initial-boundary problem (7.25)-(7.27), obtained with the explicit central difference scheme (7.41), remains bounded for any T. Hence, inequalities (7.46), (7.47) can also be regarded as the *sufficient* (not necessary) *stability conditions* for the method. Note also that (7.46) is a restriction on the space step h. Since h=a/n, it implies that the number of parts n in which the interval [0,a] is to be subdivided must satisfy the following inequality:

$$(7.53) \qquad n \geq \frac{aM}{2v} .$$

Once h has been fixed, one then should choose a time step k in such a way that inequality (7.47) is also satisfied. Since h=a/n and k=T/m, inequality (7.47) is equivalent to

$$(7.54) \qquad m \geq T\left[N + \frac{2V}{h^2}\right] = T\left[N + \frac{2Vn^2}{a^2}\right] .$$

From (7.53), (7.54) implies that if the space subdivision n is to be taken large, the corresponding time subdivision m will be required to be much larger, in which case the numerical method may no longer be convenient.

7.4 AN EXPLICIT UPWIND METHOD

If one wishes to eliminate the restriction (7.46) on the space step h, one can use an upwind, rather than a central, difference approximation to discretize the term u_x in (7.25). Specifically, one proceeds as follows. The terms u_t and u_{xx} are approximated by (7.30) and (7.32), respectively, while the term u_x in (7.25) is approximated by

$$(7.55) \qquad u_x \approx \frac{u_{i+1,j} - u_{i,j}}{h}, \quad \text{if} \quad Q_{ij} \geq 0; \qquad u_x \approx \frac{u_{i,j} - u_{i-1,j}}{h}, \quad \text{if} \quad Q_{ij} < 0;$$

Substitution of (7.30), (7.32) and (7.55) into (7.25) then yields

$$\frac{u_{i,j+1} - u_{i,j}}{k} = P_{ij}\frac{u_{i+1,j} - 2u_{i,j} + u_{i-1,j}}{h^2} + Q_{ij}\frac{u_{i+1,j} - u_{i,j}}{h} + R_{ij}u_{i,j} + S_{ij}, \quad \text{if} \quad Q_{ij} \geq 0;$$

$$\frac{u_{i,j+1} - u_{i,j}}{k} = P_{ij}\frac{u_{i+1,j} - 2u_{i,j} + u_{i-1,j}}{h^2} + Q_{ij}\frac{u_{i,j} - u_{i-1,j}}{h} + R_{ij}u_{i,j} + S_{ij}, \quad \text{if} \quad Q_{ij} < 0,$$

or, equivalently,

$$(7.56) \qquad u_{i,j+1} = [\alpha P_{ij} + \beta(|Q_{ij}| - Q_{ij})]u_{i-1,j} + [1 + kR_{ij} - 2(\alpha P_{ij} + \beta|Q_{ij}|)]u_{i,j}$$

$$+ [\alpha P_{ij} + \beta(|Q_{ij}| + Q_{ij})]u_{i+1,j} + kS_{ij},$$

where, of course, $\alpha = k/(h^2)$ and $\beta = k/(2h)$. The algorithm to be used to generate the numerical solution at each interior grid point is the same as described in the previous section but with (7.56) replacing (7.34). The method so obtained is called the *explicit upwind* method.

EXAMPLE. Let us reconsider the initial-boundary problem (7.35)-(7.37). Let T=1, n=4, m=10 so that h=1, k=0.1, and the $R_{5,11}$ set is (i,0.1j), i=0,1,2,3,4, j=0,1,2,...,10. Then, by using (7.38), equation (7.56) yields for j=0

$$u_{1,1} = 1.900 \qquad u_{2,1} = 0.900 \qquad u_{3,1} = 1.900 .$$

Next, by use of (7.38) and the numerical solution just generated, for j=1, equation (7.56) yields

$$u_{1,2} = 1.758 \qquad u_{2,2} = 0.830 \qquad u_{3,2} = 1.758 .$$

Continuing in the indicated fashion, for j=2,3,...,9, one finds, to three decimal places,

$$u_{1,3} = 1.604 \qquad u_{2,3} = 0.766 \qquad u_{3,3} = 1.604$$

$$u_{1,4} = 1.459 \qquad u_{2,4} = 0.704 \qquad u_{3,4} = 1.459$$

$$u_{1,5} = 1.324 \qquad u_{2,5} = 0.647 \qquad u_{3,5} = 1.324$$

$$u_{1,6} = 1.201 \qquad u_{2,6} = 0.587 \qquad u_{3,6} = 1.201$$

$$u_{1,7} = 1.088 \qquad u_{2,7} = 0.533 \qquad u_{3,7} = 1.088$$

$$u_{1,8} = 0.985 \qquad u_{2,8} = 0.484 \qquad u_{3,8} = 0.985$$

$$u_{1,9} = 0.892 \qquad u_{2,9} = 0.439 \qquad u_{3,9} = 0.892$$

$$u_{1,10} = 0.807 \qquad u_{2,10} = 0.398 \qquad u_{3,10} = 0.807 ,$$

Table 7.1 - Explicit methods.

exact	$u_{1,10}=0.736$	$u_{2,10}=0.368$	$u_{3,10}=0.736$
central	$u_{1,10}=0.726$	$u_{2,10}=0.359$	$u_{3,10}=0.726$
upwind	$u_{1,10}=0.807$	$u_{2,10}=0.398$	$u_{3,10}=0.807$

which is another approximation to the exact solution $u=(x^2-4x+5)e^{-t}$. Note that the present numerical solution possesses the discrete max-min property. It, however, is not as accurate as the one which was obtained in the previous section by the central difference method. For clarity in this comparison, we have listed in Table 7.1 the values of the exact solution at time T=1, the results by the central difference method from the example of the previous section, and the results just generated.

Although the explicit upwind method is not very accurate, it yields a numerical solution which possesses the max-min property for any choice of the space step h. This is proved in the next theorem.

THEOREM 7.2. For $S(x,t)\equiv0$, let the initial-boundary problem (7.25)-(7.27) be defined on $0\leq x\leq a$, $0\leq t\leq T$, where $P(x,t)$, $Q(x,t)$ and $R(x,t)$ are continuous, and hence bounded. Assume that $0<v\leq P(x,t)\leq V$, $|Q(x,t)|\leq M$ and $-N\leq R(x,t)\leq0$ on $0\leq x\leq a$, $0\leq t\leq T$. If

$$(7.57)\qquad k \leq \frac{h^2}{Nh^2+Mh+2V},$$

then the numerical solution of (7.25)-(7.27), obtained with the upwind difference formula (7.56), possesses the max-min property.

PROOF. Note first that, for $S(x,t)\equiv0$, equation (7.56) reduces to

$$(7.58)\qquad u_{i,j+1} = [\alpha P_{ij}+\beta(|Q_{ij}|-Q_{ij})]u_{i-1,j} + [1+kR_{ij}-2(\alpha P_{ij}+\beta|Q_{ij}|)]u_{i,j}$$

$$+ [\alpha P_{ij}+\beta(|Q_{ij}|+Q_{ij})]u_{i+1,j},$$

where

$$\alpha P_{ij}+\beta(|Q_{ij}|-Q_{ij}) \geq 0, \qquad \alpha P_{ij}+\beta(|Q_{ij}|+Q_{ij}) \geq 0,$$

and, by inequality (7.57),

$$1+kR_{ij}-2(\alpha P_{ij}+\beta|Q_{ij}|) \geq 1-kN-2(\alpha V+\beta M) \geq 0 \ .$$

Moreover,

(7.59) $[\alpha P_{ij}+\beta(|Q_{ij}|-Q_{ij})]+[1+kR_{ij}-2(\alpha P_{ij}+\beta|Q_{ij}|)]+[\alpha P_{ij}+\beta(|Q_{ij}|+Q_{ij})]=1+kR_{ij} \leq 1 \ .$

The proof now follows directly as in Theorem 7.1.

Note that under the hypotheses of Theorem 7.2, if f(x), $g_1(t)$ and $g_2(t)$ are bounded, the numerical solution of initial-boundary problem (7.25)-(7.27) remains bounded for any fixed T, and hence, inequality (7.57) also constitutes a *stability condition* for the method.

7.5 NUMERICAL SOLUTION OF MILDLY NONLINEAR PROBLEMS

The explicit central difference and upwind method extend in a natural way to mildly nonlinear initial-boundary problems defined by (7.26), (7.27) and

(7.60) $u_t = P(x,t)u_{xx}+Q(x,t)u_x+F(x,t,u) \ ,$ $P(x,t) > 0 \ .$

To assure that the solution of (7.26), (7.27) and (7.60) exists, is unique, and has properties in common with the solution of the corresponding linear equation (7.25), it will be assumed that F is bounded and F_u exists and satisfies

(7.61) $-N \leq F_u \leq 0 \ ,$ $0 \leq x \leq a,$ $0 \leq t \leq T,$ $-\infty < u < \infty.$

The only modifications necessary in the algorithms for the central difference and upwind method when (7.60) replaces (7.25) are that the difference equations must be modified appropriately. For the *central difference* method, one need only replace (7.34) with

(7.62) $u_{i,j+1} = (\alpha P_{ij}-\beta Q_{ij})u_{i-1,j}+(1-2\alpha P_{ij})u_{i,j}+(\alpha P_{ij}+\beta Q_{ij})u_{i+1,j}+kF(x_i,t_j,u_{i,j}) \ .$

For the **upwind** method, one need only replace (7.56) with

$$(7.63) \qquad u_{i,j+1} = [\alpha P_{ij} + \beta(|Q_{ij}| - Q_{ij})]u_{i-1,j} + [1 - 2(\alpha P_{ij} + \beta|Q_{ij}|)]u_{i,j}$$

$$+ [\alpha P_{ij} + \beta(|Q_{ij}| + Q_{ij})]u_{i+1,j} + kF(x_i, t_j, u_{i,j}) \ .$$

With regard to stability, the central difference and the upwind methods continue to remain stable under the constraints of Theorems 7.1 and 7.2, respectively, with F_u replacing R.

EXAMPLE. Consider the initial-boundary problem defined by

$$(7.64) \qquad u_t = u_{xx} - [u + \cos(u)] \ , \qquad 0 < x < 5, \qquad 0 < t \le 1$$

$$(7.65) \qquad u(x,0) = 1 \ , \qquad 0 \le x \le 5$$

$$(7.66) \qquad u(0,t) = u(5,t) = 1 \ , \qquad 0 < t \le 1.$$

Note first that $F(x,t,u) = -[u + \cos(u)]$. Thus, $F_u = -1 + \sin(u)$, which implies $-2 \le F_u \le 0$. Moreover, since in this example $Q(x,t) \equiv 0$, one has M=0 and the central difference method and the upwind method are equivalent. Now, since M=0, h, and hence n, can be chosen arbitrarily. Let us then fix n=5, so that h=1. The stability restriction (7.47) implies $k \le 1/4$. Hence, fix m=5, so that k=1/5=0.2, and the finite difference formula (7.62), in this example, reduces to

$$u_{i,j+1} = \alpha u_{i-1,j} + (1 - 2\alpha)u_{i,j} + \alpha u_{i+1,j} - k[u_{i,j} + \cos(u_{i,j})] \ ,$$

that is,

$$(7.67) \qquad u_{i,j+1} = (0.2)u_{i-1,j} + (0.4)u_{i,j} + (0.2)u_{i+1,j} - (0.2)\cos(u_{i,j}) \ .$$

Now, at $t_0 = 0$, (7.65) yields

$$u_{00} = u_{10} = u_{20} = u_{30} = u_{40} = u_{50} = 1 \ .$$

At x=0 and at x=5, (7.66) yields

$$u_{01} = u_{02} = u_{03} = u_{04} = u_{05} = 1$$
$$u_{51} = u_{52} = u_{53} = u_{54} = u_{55} = 1 \; .$$

By setting $j=0$, $i=1,2,3,4$, and calculating to four decimal places, (7.67) yields, explicitly, the following numerical solution at time level $t_1=k$:

$$u_{11} = 0.6919 \qquad u_{21} = 0.6919 \qquad u_{31} = 0.6919 \qquad u_{41} = 0.6919 \; .$$

Next, by setting $j=1$ and $i=1,2,3,4$, the numerical solution that one obtains from (7.67) at $t_2=2k$ is

$$u_{12} = 0.4611 \qquad u_{22} = 0.3995 \qquad u_{32} = 0.3995 \qquad u_{42} = 0.4611 \; .$$

Continuing in the indicated fashion for $j=2,3,4$, (7.67) yields

$$u_{13} = 0.2852 \qquad u_{23} = 0.1477 \qquad u_{33} = 0.1477 \qquad u_{43} = 0.2852$$
$$u_{14} = 0.1517 \qquad u_{24} = -0.0521 \qquad u_{34} = -0.0521 \qquad u_{44} = 0.1517$$
$$u_{15} = 0.0525 \qquad u_{25} = -0.2006 \qquad u_{35} = -0.2006 \qquad u_{45} = 0.0525 \; ,$$

which gives the numerical solution.

*7.6 CONVERGENCE OF EXPLICIT FINITE DIFFERENCE METHODS

Throughout this section, we will consider mildly nonlinear boundary value problems of the form (7.26), (7.27), (7.60) under the assumptions that F is bounded, $-N \leq F_u \leq 0$, $0 < v \leq P(x,t) \leq V$, $|Q(x,t)| \leq M$, so that the analytical solution exists and is unique.

In developing convergence analyses, let us assume that the numerical solution u_{ij} is free from roundoff errors, so that the only difference between u_{ij} and the exact solution $U(x_i,t_j)$ is the error made by replacing (7.60) by the difference equation, that is, the local truncation error.

In solving (7.26), (7.27), (7.60) numerically, consider, first, the central difference approximation (7.62), for which we will prove the following theorem.

THEOREM 7.3. Let $U(x,t)$ be the analytical solution of the mildly nonlinear initial-boundary problem (7.26), (7.27), (7.60). Let u_{ij}, $i=0,1,2,...,n$, $j=0,1,2,...,m$ be the numerical solution obtained on $0 \leq t \leq T$ with the central difference approximation (7.62), and let $E(x_i,t_j)=E_{ij}$, be the error, defined by $E_{ij}=U(x_i,t_j)-u_{ij}$. Assume that $U(x,t)$ has continuous partial derivatives up to and including order two in t and order four in x. If $Mh \leq 2v$ and $k \leq h^2/(Nh^2+2V)$, then

$$(7.68) \qquad |E_{ij}| \leq T\left[\frac{h^2}{12}\left(VM_4+2MM_3\right)+\frac{k}{2} N_2\right] \ ,$$

where $N_2=\max(|U_{tt}|)$, $M_3=\max(|U_{xxx}|)$ and $M_4=\max(|U_{xxxx}|)$.

PROOF. For $j=0$, for $i=0$ and for $i=n$, one has, respectively,

$$E_{i0} = |U(x_i,0)-u_{i0}| = |f(x_i)-f(x_i)| = 0 \ , \qquad i=0,1,2,...,n,$$

$$E_{0j} = |U(0,t_j)-u_{0j}| = |g_1(t_j)-g_1(t_j)| = 0 \ , \qquad j=0,1,2,...,m,$$

$$E_{nj} = |U(a,t_j)-u_{nj}| = |g_2(t_j)-g_2(t_j)| = 0 \ , \qquad j=0,1,2,...,m,$$

so that (7.68) is valid. One must then prove (7.68) for $j=1,2,...,m$, and $i=1,2,...,n-1$.

Since $U(x,t)$ is the analytical solution of (7.60), it follows that

$$U_t(x_i,t_j) = P_{ij}U_{xx}(x_i,t_j)+Q_{ij}U_x(x_i,t_j)+F(x_i,t_j,U(x_i,t_j)) \ ,$$

which, by (3.30), (3.31), (3.37), (3.38) and (3.40), (3.41), is equivalent to

$$\left[\frac{U(x_i,t_{j+1})-U(x_i,t_j)}{k} + \frac{k}{2} U_{tt}(x_i,\tau)\right] = P_{ij}\left[\frac{U(x_{i+1},t_j)-2U(x_i,t_j)+U(x_{i-1},t_j)}{h^2}\right.$$

$$\left. + \frac{h^2}{24} [U_{xxxx}(\xi_1,t_j)+U_{xxxx}(\xi_2,t_j)]\right]+Q_{ij}\left[\frac{U(x_{i+1},t_j)-U(x_{i-1},t_j)}{2h}\right.$$

$$\left. + \frac{h^2}{12} [U_{xxx}(\xi_3,t_j)+U_{xxx}(\xi_4,t_j)]\right]+F(x_i,t_j,U(x_i,t_j)) \ ,$$

that is,

(7.69)
$$U(x_i,t_{j+1}) = (\alpha P_{ij} - \beta Q_{ij})U(x_{i-1},t_j) + (1-2\alpha P_{ij})U(x_i,t_j) + (\alpha P_{ij}+\beta Q_{ij})U(x_{i+1},t_j)$$

$$+ kF(x_i,t_j,U(x_i,t_j)) + k\frac{h^2}{24}\left\{ P_{ij}[U_{xxxx}(\xi_1,t_j) + U_{xxxx}(\xi_2,t_j)]\right.$$

$$+ 2Q_{ij}[U_{xxx}(\xi_3,t_j) + U_{xxx}(\xi_4,t_j)]\left.\right\} - \frac{k^2}{2}U_{tt}(x_i,\tau) .$$

Now, by subtracting (7.62) from (7.69), one finds

$$E_{i,j+1} = (\alpha P_{ij} - \beta Q_{ij})E_{i-1,j} + (1-2\alpha P_{ij})E_{i,j} + (\alpha P_{ij}+\beta Q_{ij})E_{i+1,j} + k\frac{\partial F(x_i,t_j,\mu)}{\partial u}E_{i,j}$$

$$+ k\frac{h^2}{24}\left\{ P_{ij}[U_{xxxx}(\xi_1,t_j) + U_{xxxx}(\xi_2,t_j)] + 2Q_{ij}[U_{xxx}(\xi_3,t_j) + U_{xxx}(\xi_4,t_j)]\right\}$$

$$- \frac{k^2}{2}U_{tt}(x_i,\tau) .$$

By replacing $\partial F(x_i,t_j,\mu)/\partial u$ by R_{ij}, which is an extension of the notation R_{ij}, one has $-N \leq R_{ij} \leq 0$ and the last equation can be written as

$$E_{i,j+1} = (\alpha P_{ij} - \beta Q_{ij})E_{i-1,j} + (1+kR_{ij}-2\alpha P_{ij})E_{i,j} + (\alpha P_{ij}+\beta Q_{ij})E_{i+1,j}$$

$$+ k\frac{h^2}{24}\left\{ P_{ij}[U_{xxxx}(\xi_1,t_j) + U_{xxxx}(\xi_2,t_j)] + 2Q_{ij}[U_{xxx}(\xi_3,t_j) + U_{xxx}(\xi_4,t_j)]\right\}$$

$$- \frac{k^2}{2}U_{tt}(x_i,\tau) .$$

From the hypotheses, $(\alpha P_{ij} - \beta Q_{ij}) \geq 0$, $(1+kR_{ij}-2\alpha P_{ij}) \geq (1-kN-2\alpha P_{ij}) \geq 0$ and $(\alpha P_{ij}+\beta Q_{ij}) \geq 0$. Thus,

$$|E_{i,j+1}| \leq (\alpha P_{ij} - \beta Q_{ij})|E_{i-1,j}| + (1+kR_{ij}-2\alpha P_{ij})|E_{i,j}| + (\alpha P_{ij}+\beta Q_{ij})|E_{i+1,j}|$$

$$+ k[\frac{h^2}{12}(VM_4+2MM_3) + \frac{k}{2}N_2] ,$$

which implies

$$\max_i(|E_{i,j+1}|) \leq [(\alpha P_{ij}-\beta Q_{ij})+(1+kR_{ij}-2\alpha P_{ij})+(\alpha P_{ij}+\beta Q_{ij})]\max_i(|E_{i,j}|)$$

$$+k[\frac{h^2}{12}(VM_4+2MM_3)+\frac{k}{2}N_2]$$

$$= (1+kR_{ij})\max_i(|E_{i,j}|)+k[\frac{h^2}{12}(VM_4+2MM_3)+\frac{k}{2}N_2]$$

$$\leq \max_i(|E_{i,j}|)+k[\frac{h^2}{12}(VM_4+2MM_3)+\frac{k}{2}N_2] \; .$$

By using this recurrence relation one obtains

$$\max_i(|E_{i,j+1}|) \leq \max_i(|E_{i,j}|)+k[\frac{h^2}{12}(VM_4+2MM_3)+\frac{k}{2}N_2]$$

$$\leq \max_i(|E_{i,j-1}|)+2k[\frac{h^2}{12}(VM_4+2MM_3)+\frac{k}{2}N_2]$$

$$\leq \max_i(|E_{i,j-2}|)+3k[\frac{h^2}{12}(VM_4+2MM_3)+\frac{k}{2}N_2]$$

$$\vdots$$

$$\leq \max_i(|E_{i,0}|)+(j+1)k[\frac{h^2}{12}(VM_4+2MM_3)+\frac{k}{2}N_2] \; .$$

But, $(j+1)k=t_{k+1}\leq T$ and $\max_i(|E_{i,0}|)=0$. Thus,

$$\max_i(|E_{i,j+1}|) \leq T[\frac{h^2}{12}(VM_4+2MM_3)+\frac{k}{2}N_2] \; ,$$

which implies (7.68) and the Theorem is proved.

Theorem 7.3 allows us to give a convergence theorem which is as follows.

THEOREM 7.4. Under the assumptions of Theorem 7.3, the numerical solution of a mildly nonlinear initial-boundary problem obtained with the central difference method converges to the analytical solution as $h \to 0$, that is,

$$(7.70) \qquad \lim_{h \to 0} |U(x_i,t_j) - u_{ij}| = 0 , \qquad i=0,1,2,...,n, \qquad j=0,1,2,...,m.$$

PROOF. Since, by hypotheses, k is required to satisfy the inequality

$$k \le \frac{h^2}{Nh^2+2V} ,$$

error bound (7.68) implies (7.70).

Analogous results to those of Theorems 7.3 and 7.4 are given, next, for the explicit upwind method (7.63).

THEOREM 7.5. Let $U(x,t)$ be the analytical solution of the mildly nonlinear initial-boundary problem (7.26), (7.27), (7.60). Let u_{ij} be the numerical solution obtained on $0 \le t \le T$ with the upwind difference formula (7.63), and let $E_{ij}=E(x_i,t_j)$, $i=0,1,2,...,n$, $j=0,1,2,...,m$ be the error. Assume that $U(x,t)$ has continuous partial derivatives up to and including order two in t and order four in x. If $k \le h^2/(Nh^2+Mh+2V)$, then

$$(7.71) \qquad |E_{ij}| \le T\left[\frac{h}{12}\left(hVM_4+6MM_2\right)+\frac{k}{2}N_2\right] ,$$

where $N_2=\max(|U_{tt}|)$, $M_2=\max(|U_{xx}|)$ and $M_4=\max(|U_{xxxx}|)$.

The proof of Theorem 7.5 is similar to that of Theorem 7.3.

From error bound (7.71) a convergence theorem follows readily.

THEOREM 7.6. Under the assumptions of Theorem 7.5, the numerical solution of a mildly nonlinear initial-boundary problem obtained with the upwind difference method converges to the analytical solution as $h \to 0$, that is,

$$\lim_{h \to 0} |U(x_i,t_j) - u_{ij}| = 0 , \qquad i=0,1,2,...,n, \qquad j=0,1,2,...,m.$$

In general, the reason why the central difference method, when applicable, is to be preferred to the upwind method is explained by the respective error bounds, for, if h is small, the error bound of the central difference method is, in general, smaller than the error bound of the upwind method. In the latter case, for small h, the error bound is dominated by the term $(h/2)TMM_2$, which is called the *artificial* diffusion term.

Note, finally, that the results given in this section also apply to linear initial-boundary problems as a particular case when $F(x,t,u)=R(x,t)u+S(x,t)$.

7.7 IMPLICIT CENTRAL DIFFERENCE METHOD

Suppose now that one wishes to construct a method which possesses the max-min property for *all* choices of time step k. Such a method is desirable, for example, if one has to calculate for long periods of time. If, say, one wishes to have a numerical approximation of the simple heat equation up to T=100 and one has to choose h=0.01, then (7.24) implies that one must choose k to satisfy

$$k \le \frac{1}{2}\left(\frac{1}{100}\right)^2 = 0.00005 .$$

To generate a numerical solution at t=100 would therefore require, using the explicit method, computation on a minimum of two million rows of points of $R_{n+1,m+1}$.

Interestingly enough, conditions (7.43) and (7.57) for the central difference method and for the upwind method, respectively, can be eliminated simply by replacing the point pattern shown in Figure 7.4 by the one shown in Figure 7.6, or, more precisely, by approximating u_t with a backward rather than a forward finite difference, that is,

(7.72) $u_t \approx \dfrac{u_{i,j} - u_{i,j-1}}{k}$.

To begin with, let us consider the central difference method for the mildly nonlinear initial-boundary problem. Substitution of (7.31), (7.32), (7.72) into (7.60) yields

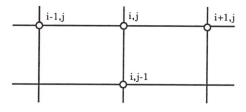

Figure 7.6

$$\frac{u_{i,j}-u_{i,j-1}}{k} = P_{ij}\frac{u_{i+1,j}-2u_{i,j}+u_{i-1,j}}{h^2} + Q_{ij}\frac{u_{i+1,j}-u_{i-1,j}}{2h} + F(x_i,t_j,u_{i,j}) ,$$

or, equivalently,

(7.73) $(\alpha P_{ij}-\beta Q_{ij})u_{i-1,j}-(1+2\alpha P_{ij})u_{i,j}+(\alpha P_{ij}+\beta Q_{ij})u_{i+1,j}+kF(x_i,t_j,u_{i,j})+u_{i,j-1}=0 .$

Equation (7.73) can now be used to generate the numerical solution at each interior grid point at time t_j. In fact, by setting

$$f_1(u_{1,j}) = u_{1,j-1}+kF(x_1,t_j,u_{1,j})+(\alpha P_{1j}-\beta Q_{1j})u_{0,j}$$

(7.74) $f_i(u_{i,j}) = u_{i,j-1}+kF(x_i,t_j,u_{i,j}) ,$ $i=2,3,...,n-2$

$$f_{n-1}(u_{n-1,j}) = u_{n-1,j-1}+kF(x_{n-1},t_j,u_{n-1,j})+(\alpha P_{n-1,j}+\beta Q_{n-1,j})u_{n,j} ,$$

equation (7.73), written consecutively for $i=1,2,...,n-1$, yields

$$-(1+2\alpha P_{1j})u_{1,j}+(\alpha P_{1j}+\beta Q_{1j})u_{2,j} \qquad\qquad +f_1(u_{1,j}) = 0$$

$$(\alpha P_{2j}-\beta Q_{2j})u_{1,j}-(1+2\alpha P_{2j})u_{2,j}+(\alpha P_{2j}+\beta Q_{2j})u_{3,j} \qquad +f_2(u_{2,j}) = 0$$

(7.75)

$$(\alpha P_{n-1,j}-\beta Q_{n-1,j})u_{n-2,j}-(1+2\alpha P_{n-1,j})u_{n-1,j}+f_{n-1}(u_{n-1,j}) = 0 .$$

Starting with $j=1$, on the first row of grid points, if one inserts the known initial and boundary values in (7.74), equations (7.75) result in a mildly nonlinear linear system of $n-1$ equations in the $n-1$ unknowns $u_{1,1}, u_{2,1}, u_{3,1}, ..., u_{n-1,1}$. A solution of this system, together with the boundary values, is called the numerical solution at time t_1. Once a numerical result for row 1 has been generated, set $j=2$, and solve a new mildly nonlinear system (7.75) of $n-1$ equations in the $n-1$ unknowns $u_{1,2}, u_{2,2}, u_{3,2}, ..., u_{n-1,2}$. Continue in the indicated fashion for each subsequent row j, $j=3,4,...,m$. The numerical method so obtained is called the *implicit central difference* method.

In general, the mildly nonlinear system (7.75) may not have a solution, or may have more than one solution. Note, however, that the linear part of system (7.75) is tridiagonal with $-(1+2\alpha P_{ij})$ on the *main* diagonal, $(\alpha P_{ij}-\beta Q_{ij})$ on the *subdiagonal*, and $(\alpha P_{ij}+\beta Q_{ij})$ on the *superdiagonal*. Thus, if h satisfies the inequality hM<2v, then system (7.75) satisfies all the assumptions of Theorem 1.4, and in fact,

$$-(1+2\alpha P_{ij}) \le -(1+2\alpha v) < 0$$

$$(\alpha P_{ij}-\beta Q_{ij}) \ge (\alpha v-\beta M) > 0$$

$$(\alpha P_{ij}+\beta Q_{ij}) \ge (\alpha v-\beta M) > 0 ,$$

and

$$|\alpha P_{ij}-\beta Q_{ij}|+|\alpha P_{ij}+\beta Q_{ij}| = |2\alpha P_{ij}| < |1+2\alpha P_{ij}| = |-(1+2\alpha P_{ij})| .$$

Hence the numerical solution u_{ij}, i=0,1,2,...,n, *exists* and is *unique* for each j=0,1,2,...,m. Moreover, the solution of these systems can be found by the generalized Newton's method which, in this case, will converge for all ω in a subrange of $0<\omega<2$ and for all initial guesses. For efficiency, however, since the numerical solution at time level t_j, in general, is expected to be close to the solution at previous time level t_{j-1}, it is convenient to use $u_{i,j-1}$ as an initial guess for $u_{i,j}$, i=1,2,...,n-1 in the Newtonian iteration formulas.

The implicit central difference method for linear initial-boundary problems can be derived from (7.73) as a particular case. In fact, for F(x,t,u)=R(x,t)u+S(x,t), (7.73) becomes

(7.76) $(\alpha P_{ij}-\beta Q_{ij})u_{i-1,j}-(1-kR_{ij}+2\alpha P_{ij})u_{i,j}+(\alpha P_{ij}+\beta Q_{ij})u_{i+1,j}+u_{i,j-1}+kS_{ij} = 0 .$

Equations (7.76), for each j=1,2,...,m, constitute a linear tridiagonal system of n−1 equations in the n−1 unknowns $u_{1,j}$, $u_{2,j}$, ..., $u_{n-1,j}$. Each of these systems, when Mh<2v, is diagonally dominant, with negative elements on the main diagonal and positive ones on the subdiagonal and on the superdiagonal. Thus, by Theorem 1.2, the numerical solution *exists* and is *unique* on the entire $R_{n+1,m+1}$ set. It is interesting to note that the condition Mh<2v can be extended to Mh≤2v. However, the proof must then invoke Theorem 1.3 rather than Theorem 1.2. Moreover, when Mh≤2v, the implicit finite difference method (7.76) also possesses the *max-min* property.

THEOREM 7.7. Let the initial-boundary problem (7.25)-(7.27) be defined on $0 \leq x \leq a$, $0 \leq t \leq T$, with $S(x,t) \equiv 0$, and with continuous $P(x,t)$, $Q(x,t)$ and $R(x,t)$. Assume that $P(x,t) \geq v > 0$, $|Q(x,t)| \leq M$ and $R(x,t) \leq 0$ on $0 \leq x \leq a$, $0 \leq t \leq T$. If $Mh \leq 2v$ then the numerical solution of (7.25)-(7.27), obtained with the implicit central difference method, possesses the max-min property.

PROOF. Let $u_{pj} = \max(u_{ij})$. If $p=0$, $p=n$, or $j=0$, the theorem is valid. Hence assume $j>0$ and $0<p<n$. Then, since $S_{ij}=0$, (7.76) yields

(7.77)
$$(1 - kR_{pj})u_{p,j} = u_{p,j-1} + (\alpha P_{pj} - \beta Q_{pj})u_{p-1,j} - 2\alpha P_{pj}u_{p,j} + (\alpha P_{pj} + \beta Q_{pj})u_{p+1,j}$$

$$\leq u_{p,j-1} + (\alpha P_{pj} - \beta Q_{pj})u_{p,j} - 2\alpha P_{pj}u_{p,j} + (\alpha P_{pj} + \beta Q_{pj})u_{p,j} = u_{p,j-1} .$$

from which follows, independently of the sign of u_{pj}, since $R_{pj} \leq 0$,

(7.78)
$$\max_i(u_{i,j}) \leq \max[0, u_{0,j}, \max_p(u_{p,j-1}), u_{n,j}] .$$

Now, by repeating the above argument, one finds

$$\max_i(u_{i,j}) \leq \max[0, u_{0,j}, \max_{0<p<n}(u_{p,j-1}), u_{n,j}]$$

$$\leq \max[0, u_{0,j}, u_{0,j-1}, \max_{0<p<n}(u_{p,j-2}), u_{n,j-1}, u_{n,j}]$$

$$\leq \max[0, u_{0,j}, u_{0,j-1}, u_{0,j-2}, \max_{0<p<n}(u_{p,j-3}), u_{n,j-2}, u_{n,j-1}, u_{n,j}]$$

$$\vdots$$

$$\leq \max[0, \max_{1 \leq q \leq j}(u_{0,q}), \max_{0<p<n}(u_{p,0}), \max_{1 \leq q \leq j}(u_{n,q})]$$

which implies (7.42). Inequality (7.41) can be proved in an entirely similar way.

If $R_{ij}=0$, $i=0,1,2,\ldots,n$, $j=0,1,2,\ldots,m$, inequality (7.77) implies

(7.79)
$$\max_i(u_{i,j}) \leq \max[u_{0,j}, \max_p(u_{p,j-1}), u_{n,j}] ,$$

which implies, as above,

$$\max_i(u_{i,j}) \leq \max[u_{0,j}, \max_{0<p<n}(u_{p,j-1}), u_{n,j}]$$

$$\leq \max[u_{0,j}, u_{0,j-1}, \max_{0<p<n}(u_{p,j-2}), u_{n,j-1}, u_{n,j}]$$

$$\leq \max[u_{0,j}, u_{0,j-1}, u_{0,j-2}, \max_{0<p<n}(u_{p,j-3}), u_{n,j-2}, u_{n,j-1}, u_{n,j}]$$

$$\cdot$$
$$\cdot$$
$$\cdot$$

$$\leq \max[\max_{1\leq q\leq j}(u_{0,q}), \max_{0<p<n}(u_{p,0}), \max_{1\leq q\leq j}(u_{n,q})] .$$

Hence, (7.44) is established. Inequality (7.43) can be proved in a similar fashion.

Note that under the hypotheses of Theorem 7.7, if $f(x)$, $g_1(t)$ and $g_2(t)$ are bounded, the numerical solution of initial-boundary problem (7.25)-(7.27), obtained with the implicit central difference method remains bounded for any fixed T. Hence, inequality (7.46) can also be regarded as a *sufficient stability condition* for the method.

EXAMPLE. Let us reconsider the initial-boundary problem (7.35)-(7.37), namely

$$u_t = u_{xx} + (x-2)u_x - 3u$$

$$u(x,0) = x^2 - 4x + 5 , \qquad 0 \leq x \leq 4$$

$$u(0,t) = u(4,t) = 5e^{-t}, \qquad t > 0 .$$

Let T=1, n=4, and, in order to take advantage of the less restrictive condition imposed by the implicit method, fix m=2, so that h=1 and k=0.5. Now, at the initial time $t_0=0$, the given initial condition implies

$$u_{00} = 5 , \qquad u_{10} = 2 , \qquad u_{20} = 1 , \qquad u_{30} = 2 , \qquad u_{40} = 5 ,$$

while, at the boundary grid points the given boundary conditions, to three decimal places, yield

$$u_{0,1} = 3.033 \qquad\qquad\qquad u_{4,1} = 3.033$$

$$u_{0,2} = 1.839 \qquad\qquad\qquad u_{4,2} = 1.839 .$$

At the interior grid points, since $\alpha=1/2$ and $\beta=1/4$, equation (7.76) becomes

$$\left[\tfrac{1}{2}-\tfrac{1}{4}(x_i-2)\right]u_{i-1,j}-\left[1+\tfrac{3}{2}+1\right]u_{i,j}+\left[\tfrac{1}{2}+\tfrac{1}{4}(x_i-2)\right]u_{i+1,j}+u_{i,j-1}=0\ .$$

Now, for $j=1$ and $i=1,2,3$, one has

$$\tfrac{3}{4}u_{01}-\tfrac{7}{2}u_{11}+\tfrac{1}{4}u_{21}+u_{10}=0$$

$$\tfrac{1}{2}u_{11}-\tfrac{7}{2}u_{21}+\tfrac{1}{2}u_{31}+u_{20}=0$$

$$\tfrac{1}{4}u_{21}-\tfrac{7}{2}u_{31}+\tfrac{3}{4}u_{41}+u_{30}=0$$

which, by using the known initial and boundary values yields

$$-14u_{11}\ +u_{21}\qquad\ =-17.099$$
$$2u_{11}-14u_{21}+2u_{31}=-4$$
$$u_{21}-14u_{31}=-17.099\ ,$$

the unique solution of which, to three decimal places, is

$$u_{11}=1.268,\qquad\quad u_{21}=0.648\ ,\qquad\quad u_{31}=1.268\ .$$

Next, for $j=2$, by using the above numerical results, and the required boundary conditions, one obtains

$$-14u_{12}\ +u_{22}\qquad\ =-10.589$$
$$2u_{12}-14u_{22}+2u_{32}=-2.592$$
$$u_{22}-14u_{32}=-10.589\ ,$$

which, to three decimal places, yields the following results at $t_2=1$:

$$u_{12} = 0.786 , \qquad u_{22} = 0.410 , \qquad u_{32} = 0.786 ,$$

and the example is complete.

7.8 IMPLICIT UPWIND METHOD

If one wishes to eliminate also the restriction (7.42) on the space step h, one can use an implicit upwind difference approximation to discretize (7.60). Specifically, the terms u_{xx}, u_x and u_t are approximated by (7.32), (7.55) and (7.72), respectively. The resulting approximation for (7.60) is then given by

$$\frac{u_{i,j}-u_{i,j-1}}{k} = P_{ij}\frac{u_{i+1,j}-2u_{i,j}+u_{i-1,j}}{h^2} + Q_{ij}\frac{u_{i+1,j}-u_{i,j}}{h} + F(x_i,t_j,u_{i,j}) , \quad \text{if } Q_{ij} \geq 0 ,$$

$$\frac{u_{i,j}-u_{i,j-1}}{k} = P_{ij}\frac{u_{i+1,j}-2u_{i,j}+u_{i-1,j}}{h^2} + Q_{ij}\frac{u_{i,j}-u_{i-1,j}}{h} + F(x_i,t_j,u_{i,j}) , \quad \text{if } Q_{ij} < 0 ,$$

or, equivalently, by

$$
\begin{align}
(7.80) \qquad & [\alpha P_{ij}+\beta(|Q_{ij}|-Q_{ij})]u_{i-1,j}-[1+2(\alpha P_{ij}+\beta|Q_{ij}|)]u_{i,j} \\
& +[\alpha P_{ij}+\beta(|Q_{ij}|+Q_{ij})]u_{i+1,j}+kF(x_i,t_j,u_{i,j})+u_{i,j-1} = 0 .
\end{align}
$$

The algorithm to be used to generate the numerical solution at each time level t_j, $j=1,2,\ldots,m$, is the same as described above, but with (7.80) replacing (7.73). Specifically, starting with $j=1$, if one writes (7.80) consecutively for fixed j and $i=1,2,\ldots,n-1$, *and* one also inserts the known numerical solution at time level t_{j-1} and the known boundary values, there results a mildly nonlinear linear system of $n-1$ equations in the $n-1$ unknowns $u_{1,j}$, $u_{2,j}, \ldots, u_{n-1,j}$. This system satisfies all the assumptions of Theorem 1.4 for *any* given h *and* k. Thus, the numerical solution u_{ij} *exists* and is *unique* on the entire $R_{n+1,m+1}$ set. The method so obtained is called the *implicit upwind* method.

The implicit upwind method for *linear* initial-boundary problems can be derived directly from (7.80). By setting $F(x,t,u)=R(x,t)u+S(x,t)$, (7.80) becomes

(7.81) $\qquad [\alpha P_{ij} + \beta(|Q_{ij}| - Q_{ij})]u_{i-1,j} - [1 - kR_{ij} + 2(\alpha P_{ij} + \beta|Q_{ij}|)]u_{i,j}$

$$+ [\alpha P_{ij} + \beta(|Q_{ij}| + Q_{ij})]u_{i+1,j} + u_{i,j-1} + kS_{ij} = 0 .$$

Equations (7.81), for each $j=1,2,\ldots,m$, constitute a linear tridiagonal system of $n-1$ equations in the $n-1$ unknowns $u_{1,j}$, $u_{2,j}$, ..., $u_{n-1,j}$. For *all* h and k, these systems are diagonally dominant, with negative elements on the main diagonal and positive ones on the subdiagonal and the superdiagonal. Thus, by Theorem 1.2, the numerical solution *exists* and is *unique* on the entire $R_{n+1,m+1}$ set.

Although the implicit upwind method is relatively less accurate, it yields a numerical solution which possesses the discrete max-min property for any choice of h and k. This is stated in the next theorem.

THEOREM 7.8. **Let the initial-boundary problem (7.25)-(7.27) be defined on $0 \le x \le a$, $0 \le t \le T$, with $S(x,t) \equiv 0$, and with continuous $P(x,t)$, $Q(x,t)$ and $R(x,t)$. Assume that $R(x,t) \le 0$, $|Q(x,t)| \le M$, $P(x,t) \ge \nu > 0$ on $0 \le x \le a$, $0 \le t \le T$. Then the numerical solution of (7.25)-(7.27), obtained with the implicit upwind method, possesses the max-min property.**

The proof of this theorem is entirely similar to that of Theorem 7.7.

Note that under the hypotheses of Theorem 7.8, if $f(x)$, $g_1(t)$ and $g_2(t)$ are bounded, the numerical solution of initial-boundary problem (7.25)-(7.27), obtained with the implicit upwind method remains bounded for any T. Hence, since no restrictions are imposed on h and k, this method is *unconditionally* stable.

EXAMPLE. Let us consider, once again, the initial-boundary problem (7.35)-(7.37), that is,

$$u_t = u_{xx} + (x-2)u_x - 3u$$

$$u(x,0) = x^2 - 4x + 5 , \qquad 0 \le x \le 4$$

$$u(0,t) = u(4,t) = 5e^{-t}, \qquad t > 0 .$$

For the implicit upwind method, set T=1, n=4 and m=2, so that h=1 and k=0.5, $\alpha = 1/2$, $\beta = 1/4$. The given initial and boundary conditions, to three decimal places, are

$u_{02} = 1.839$ $\qquad\qquad\qquad\qquad\qquad\qquad\qquad\qquad\qquad u_{42} = 1.839$

$u_{01} = 3.033$ $\qquad\qquad\qquad\qquad\qquad\qquad\qquad\qquad\qquad u_{41} = 3.033$

$u_{00} = 5.000 \qquad u_{10} = 2.000 \qquad u_{20} = 1.000 \qquad u_{30} = 2.000 \qquad u_{40} = 5.000 .$

Thus, for j=1 and i=1,2,3, and by use of the required initial and boundary conditions, equation (7.81) implies

$$-8u_{11} + u_{21} \qquad = -10.066$$

$$u_{11} - 7u_{21} + u_{31} = -2$$

$$u_{21} - 8u_{31} = -10.066 \ ,$$

the unique solution of which, to three decimal places, is

$$u_{11} = 1.342 \ , \qquad u_{21} = 0.669 \ , \qquad u_{31} = 1.342 \ .$$

Next, for j=2, and i=1,2,3, equation (7.81) implies

$$-8u_{12} + u_{22} \qquad = -6.362$$

$$u_{12} - 7u_{22} + u_{32} = -1.338$$

$$u_{22} - 8u_{32} = -6.362 \ ,$$

which, at $t_2=1$, yields

$$u_{12} = 0.849 \ , \qquad u_{22} = 0.434 \ , \qquad u_{32} = 0.849 \ .$$

Table 7.2 Implicit methods.

exact	$u_{1,2}=0.736$	$u_{2,2}=0.368$	$u_{3,2}=0.736$
central	$u_{1,2}=0.786$	$u_{2,2}=0.410$	$u_{3,2}=0.786$
upwind	$u_{1,2}=0.849$	$u_{2,2}=0.434$	$u_{3,2}=0.849$

Table 7.3 Grid size restrictions.

	Central difference	Upwind
Explicit	$hM \leq 2v$, $k \leq \dfrac{h^2}{Nh^2 + 2V}$	$k \leq \dfrac{h^2}{Nh^2 + Mh + 2V}$
Implicit	$hM \leq 2v$	*No restrictions*

For comparison, Table 7.2 shows the values of the exact solution at time T=1, the results from the implicit central difference method from the example in the previous section and the results just generated. As expected, both numerical approximations possess the discrete max-min property, and better accuracy is obtained when central differences are used. Finally, a comparison between Table 7.1 and Table 7.2 indicates that the numerical results listed in Table 7.1, obtained for the very same initial-boundary problem using the explicit upwind and the central difference method, have higher accuracy because a smaller time step has been used.

In summary, we observe that the simplest method for a general initial-boundary problem is the explicit central difference method. This method, however, as indicated in Table 7.3, requires limitations on h an k. If the limitation on h is too restrictive, one can use the explicit upwind method. When the limitation on k is too restrictive, one can use an implicit central difference method. If both limitations on h and k are too restrictive, then the implicit upwind method has to be used. Of course, upwind methods are only first order accurate in space, hence their use should be limited only to the cases when the limitation on h is relatively severe. On the other hand, implicit methods require one to solve a system of n–1 equations at each time step, so that the limitation on k is eliminated at the price of greater computational complexity. Note, finally, that while explicit methods apply to *both* initial value problems and initial-boundary problems, the implicit methods apply *only* to initial-boundary problems.

7.9 THE CRANK-NICOLSON METHOD

Since central difference methods have relatively high accuracy in space, the next problem is to improve on the time accuracy. For this purpose, note that the use of symmetry in the construction of difference equations can lead to better accuracy in the sense that the truncation error is of a higher order of magnitude than when symmetry is not used. Thus, for example, the error in approximation (3.30) is $O(h)$, while that of (3.37), which uses symmetry, is $O(h^2)$. With this observation as an intuitive guide, let us modify the previous methods to yield greater time accuracy in the following simple way. In place of the point patterns shown in Figures 7.4 and 7.6, consider the expanded point pattern shown in Figure 7.7. The center of symmetry of the six points shown there is the point $(x_i, t_{j-1/2}) = [ih, (j-1/2)k]$ which is *not* a grid point. If one wishes, now, to develop formulas *symmetrically* about $(x_i, t_{j-1/2})$, then note first that

$$(7.82) \qquad u_t(x_i, t_{j-1/2}) \approx \frac{u_{i,j} - u_{i,j-1}}{k},$$

does use points located symmetrically about $(x_i, t_{j-1/2})$. Further, since

$$u_{xx}(x_i, t_j) \approx \frac{u_{i+1,j} - 2u_{i,j} + u_{i-1,j}}{h^2}$$

and

$$u_{xx}(x_i, t_{j-1}) \approx \frac{u_{i+1,j-1} - 2u_{i,j-1} + u_{i-1,j-1}}{h^2},$$

it is reasonable to set

$$u_{xx}(x_i, t_{j-1/2}) \approx \frac{1}{2}[u_{xx}(x_i, t_j) + u_{xx}(x_i, t_{j-1})],$$

so that

$$(7.83) \qquad u_{xx}(x_i, t_{j-1/2}) \approx \frac{1}{2}\left[\frac{u_{i+1,j} - 2u_{i,j} + u_{i-1,j}}{h^2} + \frac{u_{i+1,j-1} - 2u_{i,j-1} + u_{i-1,j-1}}{h^2}\right].$$

Similarly, when centered finite differences are used to approximate u_x, we set

$$(7.84) \qquad u_x(x_i, t_{j-1/2}) \approx \frac{1}{2}\left[\frac{u_{i+1,j} - u_{i-1,j}}{2h} + \frac{u_{i+1,j-1} - u_{i-1,j-1}}{2h}\right].$$

Using (7.82)-(7.84) in (7.60) yields

$$(7.85) \qquad \frac{u_{i,j} - u_{i,j-1}}{k} = P_{i,j-1/2}\left[\frac{u_{i+1,j} - 2u_{i,j} + u_{i-1,j}}{2h^2} + \frac{u_{i+1,j-1} - 2u_{i,j-1} + u_{i-1,j-1}}{2h^2}\right]$$

$$+ Q_{i,j-1/2}\left[\frac{u_{i+1,j} - u_{i-1,j}}{4h} + \frac{u_{i+1,j-1} - u_{i-1,j-1}}{4h}\right] + F(x_i, t_{j-1/2}, \frac{u_{i,j} + u_{i,j-1}}{2}),$$

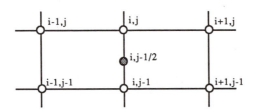

Figure 7.7

which, in each of its parts, is symmetrical about $(x_i, t_{j-1/2})$. Formula (7.85) is called the *Crank-Nicolson* formula, or the central difference Crank-Nicolson formula. It has greater accuracy than both the fully explicit and the fully implicit formulas. When this formula is used in place of (7.73) in the implicit method, the resulting method is called the Crank-Nicolson method, or the central difference Crank-Nicolson method. It does lead, at each time step, to a mildly nonlinear system which, for $hM<2v$, satisfies all the assumptions of Theorem 1.4. Moreover, for linear initial-boundary problems, when $hM \leq 2v$, the Crank-Nicolson method at each time step yields a system which satisfies all the assumptions of Theorem 1.3. Thus, it has a *unique* solution for all k, but, when $S(x,t) \equiv 0$, in order to be assured that this method possesses the *max-min* property, the time step k must satisfy the condition $k \leq 2h^2/(Nh^2+2V)$.

EXAMPLE. For completness, calculating to three decimal places, let us solve the initial boundary problem (7.35)-(7.37) with a Crank-Nicolson method. In this case, for T=1, n=4 and m=10, one has h=1, k=0.1, and (7.85) reduces to

$$\frac{u_{i,j}-u_{i,j-1}}{0.1} = \frac{u_{i+1,j}-2u_{i,j}+u_{i-1,j}}{2} + \frac{u_{i+1,j-1}-2u_{i,j-1}+u_{i-1,j-1}}{2}$$

$$+ (x_i-2)\left[\frac{u_{i+1,j}-u_{i-1,j}}{4} + \frac{u_{i+1,j-1}-u_{i-1,j-1}}{4}\right] - 3\frac{u_{i,j}+u_{i,j-1}}{2},$$

or, equivalently,

$$0.1(4-x_i)u_{i-1,j}-5u_{i,j}+0.1x_iu_{i+1,j} = -0.1(4-x_i)u_{i-1,j-1}-3u_{i,j-1}-0.1x_iu_{i+1,j-1},$$

which, by using the initial conditions (7.38) and the boundary conditions (7.39), for j=1, and i=1,2,3 yields the system

$$-5u_{11}+0.1u_{21} \qquad\qquad = -8.957$$
$$0.2u_{11} \quad - 5u_{21}+0.2u_{31} = -3.800$$
$$0.1u_{21} \quad - 5\,u_{31} = -8.957 \ .$$

The solution of this system is $u_{11}=1.810$, $u_{21}=0.905$, $u_{31}=1.810$. Next, by using the boundary conditions (3.39) and the results just obtained, one has for j=2

$$-5u_{12}+0.1u_{22} \qquad\qquad = -8.104$$
$$0.2u_{12} \quad - 5u_{22}+0.2u_{32} = -3.438$$
$$0.1u_{22} \quad - 5u_{32} = -8.104$$

whose solution is $u_{12}=1.637$, $u_{22}=0.819$, $u_{32}=1.637$. Continuing in the indicated fashion for j=3,4,...,10, one finds at $t_{10}=1$

$$u_{1,10} = 0.736 \ , \qquad u_{2,10} = 0.368 \ , \qquad u_{3,10} = 0.736 \ ,$$

which, by comparison with the results shown in Tables 7.1 and 7.2, indicates clearly the superiority in accuracy of the Crank-Nicolson method.

As usual, if one is willing to sacrifice a degree of accuracy in space, the restriction on h can be eliminated by using, in place of (7.85), an *upwind* Crank-Nicolson formula which is given by

(7.86)
$$\frac{u_{i,j}-u_{i,j-1}}{k} = P_{i,j-1/2}\Big[\frac{u_{i+1,j}-2u_{i,j}+u_{i-1,j}}{2h^2} + \frac{u_{i+1,j-1}-2u_{i,j-1}+u_{i-1,j-1}}{2h^2}\Big]$$

$$+\frac{(|Q_{i,j-1/2}|+Q_{i,j-1/2})u_{i+1,j}-2|Q_{i,j-1/2}|u_{i,j}+(|Q_{i,j-1/2}|-Q_{i,j-1/2})u_{i-1,j}}{4h}$$

$$+\frac{(|Q_{i,j-1/2}|+Q_{i,j-1/2})u_{i+1,j-1}-2|Q_{i,j-1/2}|u_{i,j-1}+(|Q_{i,j-1/2}|-Q_{i,j-1/2})u_{i-1,j-1}}{4h}$$

$$+ F(x_i,t_{j-1/2}, \frac{u_{i,j}+u_{i,j-1}}{2}) \ .$$

The mildly nonlinear system that one obtains at each time step from (7.86) satisfies the conditions of Theorem 1.4 for any h and k. Thus the numerical solution always exists, is unique, and also possesses the max-min property for linear equations with $S(x,t)\equiv0$ provided k satisfies the following inequality:

(7.87) $k \leq \dfrac{2h^2}{Nh^2+Mh+2V}$.

For programming purposes, the above results are often unified as follows. The fully explicit, fully implicit and Crank-Nicolson space centered formulas can be written compactly as

(7.88) $\dfrac{u_{i,j}-u_{i,j-1}}{k} = P_{i,j-\theta}\left[(1-\theta)\dfrac{u_{i+1,j}-2u_{i,j}+u_{i-1,j}}{h^2} + \theta\dfrac{u_{i+1,j-1}-2u_{i,j-1}+u_{i-1,j-1}}{h^2}\right]$

$+ Q_{i,j-\theta}\left[(1-\theta)\dfrac{u_{i+1,j}-u_{i-1,j}}{2h} + \theta\dfrac{u_{i+1,j-1}-u_{i-1,j-1}}{2h}\right] + F(x_i,t_{j-\theta},(1-\theta)u_{i,j}+\theta u_{i,j-1})$,

which, for $\theta=1$, is equivalent to the explicit formula (7.62), for $\theta=0$ yields the implicit formula (7.73), and for $\theta=1/2$ yields the Crank-Nicolson formula (7.85). In general, a method that results for $0 \leq \theta \leq 1$ is called a θ-method. The general conditions for a θ-method to possess the max-min property for linear equations with $S(x,t) \equiv 0$ are as follows:

(7.89) $Mh \leq 2v$

(7.90) $\theta k \leq \dfrac{h^2}{Nh^2+2V}$.

The fully explicit, fully implicit and Crank-Nicolson upwind formulas can be written, more compactly, as

(7.91) $\dfrac{u_{i,j}-u_{i,j-1}}{k} = P_{i,j-\theta}\left[(1-\theta)\dfrac{u_{i+1,j}-2u_{i,j}+u_{i-1,j}}{h^2} + \theta\dfrac{u_{i+1,j-1}-2u_{i,j-1}+u_{i-1,j-1}}{h^2}\right]$

$+ (1-\theta)\dfrac{(|Q_{i,j-\theta}|+Q_{i,j-\theta})u_{i+1,j}-2|Q_{i,j-\theta}|u_{i,j}+(|Q_{i,j-\theta}|-Q_{i,j-\theta})u_{i-1,j}}{2h}$

$+ \theta\dfrac{(|Q_{i,j-\theta}|+Q_{i,j-\theta})u_{i+1,j-1}-2|Q_{i,j-\theta}|u_{i,j-1}+(|Q_{i,j-\theta}|-Q_{i,j-\theta})u_{i-1,j-1}}{2h}$

$+ F(x_i,t_{j-\theta},(1-\theta)u_{i,j}+\theta u_{i,j-1})$, which, for $\theta=1$, is equivalent to the explicit upwind formula (7.63), for $\theta=0$ yields the implicit upwind formula (7.80), and for $\theta=1/2$ yields the upwind Crank-Nicolson formula (7.86). In general, an upwind θ-method that one obtains from (7.91) for $0\le\theta\le1$ yields a unique solution which possesses the max-min property for linear equations with $S(x,t)\equiv0$ under the following condition:

$$(7.92) \qquad \theta k \le \frac{h^2}{Nh^2+Mh+2V} .$$

It should be emphasized, finally, that the conditions (7.89), (7.90), (7.92) yield numerical solutions in every case which are unique and possess the max-min property for appropriate linear equations, just as their analytical counterparts do. The *stability* of the methods can be established, usually, with less restrictive assumptions [Casulli (1987)]. Thus, for example, the Crank-Nicolson method for the heat equation is stable for all h and k, but need not yield the physically significant max-min property.

EXERCISES

Basic Exercises

1. Given the initial-boundary problem for $u_t=u_{xx}$ with $a=1$, $g_1(t)=e^{-t}$, $g_2(t)=2e^{-t}$, and $f(x)=x^2+1$, find the numerical solution by the explicit central difference method up to t_{10} for each of the following choices:

   (a) $h = 1/4$, $k = 1/10$
   (b) $h = 1/4$, $k = 1/20$
   (c) $h = 1/4$, $k = 1/40$
   (d) $h = 1/4$, $k = 1/80$.

   Which of the above calculations are stable? Which will lead eventually to overflow? Which possess the max-min property?

2. Given the initial-boundary problem for $u_t=u_{xx}+xu_x-3u$ with $a=1$, $g_1(t)=e^{-t}$, $g_2(t)=2e^{-t}$, and $f(x)=x^2+1$, find the numerical solution by the explicit central difference method at $T=1$ for each of the following choices:

    (a)   h = 1/4 ,   k = 1/10

• (b)   h = 1/4 ,   k = 1/20

    (c)   h = 1/4 ,   k = 1/40

    (d)   h = 1/4   k = 1/80 .

Which of the above calculations are stable? Which will lead eventually to overflow? Which possess the max-min property? Compare your results with the exact solution $u=(x^2+1)e^{-t}$.

3. Repeat Exercise 2 by using the explicit upwind method. Then compare the results of the two methods with the exact solution.

4. Complete the proof of Theorem 7.2.

5. By using an explicit method, find the numerical solution at T=3 for the initial-boundary problem defined by

$$u_t = u_{xx} - \arctan(u) , \quad 0<x<1 , \quad t>0$$

$$f(x) = x , \quad\quad\quad\quad 0 \le x \le 1 ,$$

$$g_1(t) = 0 , \quad\quad\quad\quad g_2(t) = e^{-t}, \quad t \ge 0 .$$

6. Give a proof to Theorem 7.5.

7. Repeat Exercise 1 by using the implicit method with h=1/5 and k=1.

8. Given the initial-boundary problem for $u_t=u_{xx}+xu_x-3u$ with a=1, $g_1(t)=e^{-t}$, $g_2(t)=2e^{-t}$, and $f(x)=x^2+1$, find the numerical solution by the implicit central difference method at T=1 for each of the following choices:

    (a)   h = 1/4 ,   k = 1/10

    (b)   h = 1/4 ,   k = 1/5

    (c)   h = 1/4 ,   k = 1/2

    (d)   h = 1/4 ,   k = 1 .

Which of the above calculations are stable? Which possess the max-min property? Compare your results with the exact solution $u=(x^2+1)e^{-t}$.

9. Give an example of an initial-boundary problem for which, when Mh>2v, the implicit central difference method, when applied to linear equations with S(x,t)≡0, does not possess the max-min property.

10. Repeat Exercise 8 but use the implicit upwind method. Then compare the numerical results with those of Exercise 8.

11. Give a proof to Theorem 7.8.

12. Repeat Exercise 1 but use the Crank-Nicolson method with h=1/5 and k=1.

13. Using the Crank-Nicolson method, repeat Exercise 8 and compare the numerical results obtained.

14. Prove that when Mh≤2v and k≤h²/(Nh²+V), the central difference Crank-Nicolson method possesses the max-min property when applied to linear equations with S(x,t)≡0.

15. Give an example of an initial-boundary problem for which, when Mh>2v, the central difference Crank-Nicolson method does not possess the max-min property.

16. Give an example of an initial-boundary problem for which, when k>h²/(Nh²+V), the central difference Crank-Nicolson method does not possess the max-min property.

17. Prove that when k≤h²/(Nh²+Mh+V), the upwind Crank-Nicolson method, when applied to linear equations with S(x,t)≡0, possesses the max-min property.

18. Give an example of an initial-boundary problem for which, when k>h²/(Nh²+Mh+V), the upwind Crank-Nicolson method, when applied to linear equations with S(x,t)≡0, does not possess the max-min property.

Supplementary Exercises

19. Extend the explicit methods developed for initial-boundary problems to initial value problems.

20. Consider the initial and boundary conditions

$$u(x,0) = x \ , \qquad\qquad 0 \le x \le 1$$
$$u(0,t) = 0 \ , \qquad\qquad t \ge 0$$
$$u(1,t) = e^{-t}, \qquad\qquad t \ge 0 \ .$$

Find an approximate solution of the associated initial-boundary problem for

$$u_t = u_{xx} - x u_x$$

at $(0.5, 5)$ and compare your result with the exact solution $u = x e^{-t}$.

21. Repeat Exercise 20 but replace the parabolic equation by

$$u_t = u_{xx} - u - u^3 - x^3 e^{-3t}.$$

22. Consider the initial and boundary conditions

$$u(x,0) = 0 \ , \qquad\qquad 0 \le x \le \pi$$
$$u(0,t) = 0 \ , \qquad\qquad t \ge 0$$
$$u(\pi,t) = 0 \ , \qquad\qquad t \ge 0 \ .$$

Find an approximate solution of the associated initial-boundary problem for

$$u_t = u_{xx} + \sin(x)[\sin(t) + \cos(t)]$$

when $T = 4\pi$. Compare your result with the exact solution

$$u = \sin(x)\sin(t) \ .$$

8

Hyperbolic Equations

8.1 INTRODUCTION

One can study hyperbolic equations either as second-order partial differential equations or as equivalent systems of two first-order equations. Because of the exceptional difficulties encountered for this class of equations, in this chapter we will consider **both** approaches. For simplicity, let us consider first the wave equation

(8.1) $u_{xx} - u_{yy} = 0$.

In (8.1), the variable y, again, represents time. However, the literature is sufficiently divided between the use of (8.1) and $u_{xx} - u_{tt} = 0$, that we will continue this time to use (8.1).

Two kinds of problems, similar to those described for the heat equation, are of fundamental interest both mathematically and physically with regard to (8.1). These are the *initial value problem*, which, for the wave equation is called, more commonly, the *Cauchy problem,* and the *initial-boundary problem*. A Cauchy problem for (8.1) is an initial value problem in which one must find a function u(x,y) which is defined and continuous for $-\infty<x<\infty$, $y\geq0$; which satisfies (8.1) for $-\infty<x<\infty$, $y>0$; and which satisfies the initial conditions

(8.2) $u(x,0) = f_1(x)$, $-\infty<x<\infty$

(8.3) $u_y(x,0) = f_2(x)$, $-\infty<x<\infty$,

where f_1 and f_2 are given functions of x. As shown in Figure 8.1, the Cauchy problem is defined on the upper half of the XY plane.

An initial-boundary problem for the wave equation is one in which one is given a constant a>0 and four continuous functions

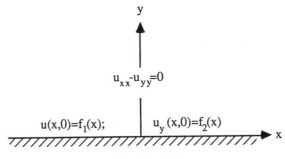

Figure 8.1

$$g_1(y), \quad y \geq 0; \qquad g_2(y), \quad y \geq 0; \qquad f_1(x), \quad 0 \leq x \leq a; \qquad f_2(x), \quad 0 < x < a,$$

and one is asked to find a function $u(x,y)$ which is continuous for $y \geq 0$, $0 \leq x \leq a$; satisfies (8.1) for $0 < x < a$, $y > 0$; and which satisfies the initial and boundary conditions

(8.4) $u(x,0) = f_1(x)$, $0 \leq x \leq a$,

                                                           initial conditions,

(8.5) $u_y(x,0) = f_2(x)$, $0 < x < a$,

(8.6) $u(0,y) = g_1(y)$, $y \geq 0$,

                                                          boundary conditions.

(8.7) $u(a,y) = g_2(y)$, $y \geq 0$,

As shown in Figure 8.2, initial-boundary problems are defined on a semi-infinite strip. Note also that we will avoid corner discontinuities by assuming throughout that $g_1(0) = f_1(0)$ and $g_2(0) = f_1(a)$.

The solution of the Cauchy problem can be given by means of the formula of D'Alembert, while that of an initial-boundary problem can be given in terms of series. However, the methods for generating such solutions do not extend to nonlinear problems, and analytical solutions so given are not usually evaluated easily at particular points of interest. We will turn then to numerical methods, the development of which will be facilitated by first developing the D'Alembert formula.

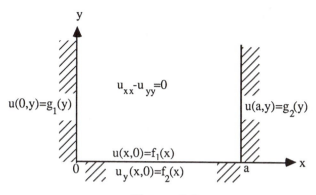

Figure 8.2

8.2 THE CAUCHY PROBLEM

For practical reasons which will become apparent later, let us determine first the analytical solution of the Cauchy problem. Consider the wave equation (8.1). Under the rotation of axes given by

(8.8) $\xi = x+y$, $\psi = x-y$,

equation (8.1) has the equivalent form

(8.9) $u_{\xi\psi} = 0$.

Integrating (8.9) with respect to ψ yields

(8.10) $u_\xi = F_1(\xi)$,

where F_1 is an arbitrary differentiable function. (This can be verified easily by differentiating both sides of (8.10) with respect to ψ.) Next, integrating (8.10) with respect to ξ yields

$$u = \int_0^{\xi} F_1(z)dz + G_2(\psi) \, ,$$

where G_2 is an arbitrary differentiable function. Setting

$$G_1(\xi) = \int_0^{\xi} F_1(z)dz$$

then implies

$$u = G_1(\xi) + G_2(\psi) \, ,$$

or, finally,

(8.11) $u(x,y) = G_1(x+y) + G_2(x-y) \, .$

The function $u(x,y)$ in (8.11) is called the general solution of the wave equation. In it, G_1 and G_2 are arbitrary differentiable functions. From (8.11), one has

(8.12) $u_y = \dfrac{\partial G_1}{\partial \xi} \dfrac{\partial \xi}{\partial y} + \dfrac{\partial G_2}{\partial \psi} \dfrac{\partial \psi}{\partial y} = \dfrac{\partial G_1(\xi)}{\partial \xi} - \dfrac{\partial G_2(\psi)}{\partial \psi} \, .$

From (8.11) and (8.12), it follows, with the aid of (8.2), (8.3) and (8.8), that

(8.13) $u(x,0) = G_1(x) + G_2(x) = f_1(x)$

(8.14) $u_y(x,0) = G_1'(x) - G_2'(x) = f_2(x) \, .$

From (8.13), then

(8.15) $G_1'(x) + G_2'(x) = f_1'(x) \, ,$

so that, from (8.14) and (8.15),

(8.16) $G'_1(x) = \frac{1}{2}[f'_1(x) + f_2(x)], \qquad G'_2(x) = \frac{1}{2}[f'_1(x) - f_2(x)]$.

By integration, (8.16) implies

(8.17) $G_1(x) = \frac{1}{2}\left[f_1(x) + \int\limits_0^x f_2(z)dz\right], \qquad G_2(x) = \frac{1}{2}\left[f_1(x) - \int\limits_0^x f_2(z)dz\right]$.

Thus, from (8.11) and (8.17),

$$u(x,y) = \frac{1}{2}\left[f_1(x+y) + \int\limits_0^{x+y} f_2(z)dz\right] + \frac{1}{2}\left[f_1(x-y) - \int\limits_0^{x-y} f_2(z)dz\right] ,$$

or, equivalently,

(8.18) $$u(x,y) = \frac{1}{2}\left[f_1(x+y) + f_1(x-y) + \int\limits_{x-y}^{x+y} f_2(z)dz\right] ,$$

which is the formula of D'Alembert.

EXAMPLE. The solution of the Cauchy problem with $f_1(x)=x^2$, $f_2(x)=e^{-x^2}$, is given by

$$u(x,y) = \frac{1}{2}\left[(x+y)^2 + (x-y)^2 + \int\limits_{x-y}^{x+y} e^{-z^2}dz\right] .$$

    If one wishes to evaluate the solution of a Cauchy problem at a particular point, then, at worst, one may have to apply a numerical integration technique in the formula of D'Alembert. Because this simple state of affairs exists, we will not pursue the Cauchy problem further, but, instead, will make some important observations from (8.18) which will be fundamental for the study of initial-boundary problems.

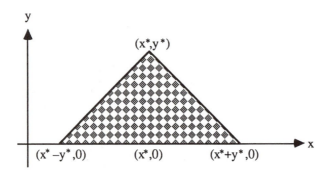

Figure 8.3

Suppose one is given a Cauchy problem and wishes to know the solution at a point (x^*,y^*), as shown in Figure 8.3. From (8.18),

(8.19) $$u(x^*,y^*) = \frac{1}{2}\left[f_1(x^*+y^*)+f_1(x^*-y^*) + \int_{x^*-y^*}^{x^*+y^*} f_2(z)dz\right].$$

From (8.19), it follows that $u(x^*,y^*)$ is determined completely by a knowledge of f_1 and f_2 only between the two points $(x^*-y^*,0)$ and $(x^*+y^*,0)$ on the X axis, as shown in Figure 8.3. The interval $x^*-y^*\leq x\leq x^*+y^*$ is therefore called the *interval of dependence* of the point (x^*,y^*). The region interior to the triangle with vertices (x^*,y^*), $(x^*-y^*,0)$, $(x^*+y^*,0)$ is called the *region of dependence*. The line through (x^*,y^*) and $(x^*-y^*,0)$ and the line through (x^*,y^*) and $(x^*+y^*,0)$ are called the *characteristics* of the wave equation through (x^*,y^*). The equations of the characteristics are

$$y-y^* = x-x^*, \qquad y-y^* = -(x-x^*).$$

Finally, it is important to note that, as a consequence of the above discussion, if one has to solve a Cauchy problem but has been given the initial conditions only on $0\leq x\leq a$, then one can only find the solution in the region of dependence determined by the point $(x^*,y^*)=(a/2,a/2)$, that is, in the triangle whose vertices are $(a/2,a/2)$, $(0,0)$, $(a,0)$.

8.3. AN EXPLICIT METHOD FOR INITIAL-BOUNDARY PROBLEMS

Let us begin the study of initial-boundary problems by developing some intuition with regard to difference approximations for the wave equation. Fix b>0. Then, as for the parabolic problems in Section 7.2, construct a covering $R_{n+1,m+1}$ for the rectangle $0 \leq x \leq a$, $0 \leq y \leq b$ with a set of planar grid points, where h=a/n, k=b/m. At present, n and m are arbitrary.

By means of (3.40) and for the point arrangement shown in Figure 8.4, consider first the elementary difference approximations

$$u_{xx}(x_i, y_j) \approx \frac{u_{i+1,j} - 2u_{i,j} + u_{i-1,j}}{h^2}, \qquad u_{yy}(x_i, y_j) \approx \frac{u_{i,j+1} - 2u_{i,j} + u_{i,j-1}}{k^2},$$

so that the wave equation (8.1) is approximated by

$$(8.20) \qquad \frac{u_{i+1,j} - 2u_{i,j} + u_{i-1,j}}{h^2} - \frac{u_{i,j+1} - 2u_{i,j} + u_{i,j-1}}{k^2} = 0,$$

or, equivalently,

$$(8.21) \qquad u_{i,j+1} = 2u_{i,j} - u_{i,j-1} + \frac{k^2}{h^2} (u_{i+1,j} - 2u_{i,j} + u_{i-1,j}).$$

Now, (8.21) is an explicit formula for $u_{i,j+1}$ in terms of u on $y=y_j$ and $y=y_{j-1}$, that is, (8.21) is an explicit formula for generating u at any point of a row in terms of values of u on the *two* previous rows. Thus, it does not appear that (8.21) can be applied on the *first* row of $R_{n+1,m+1}$, but only on the second and higher rows. For this reason, u is approximated on the first row by using the following truncated Taylor series expansion:

$$u_{i,1} \approx u_{i,0} + k u_y(x_i, 0) + \frac{k^2}{2} u_{yy}(x_i, 0).$$

Next, let us assume that, for the present considerations, the wave equation is valid also on the point set (x,0), 0<x<a, so that $u_{i,1}$ can be rewritten as

$$u_{i,1} \approx u_{i,0} + k u_y(x_i, 0) + \frac{k^2}{2} u_{xx}(x_i, 0),$$

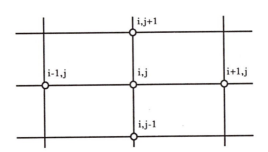

Figure 8.4

which, by using (8.4), (8.5) and the corresponding finite difference approximation for $u_{xx}(x_i,0)$, yields

$$(8.22) \qquad u_{i,1} \approx f_1(x_i)+kf_2(x_i)+ \frac{k^2}{2} \frac{f_1(x_{i+1})-2f_1(x_i)+f_1(x_{i-1})}{h^2}.$$

Consider now a simple illustrative example which will lead directly to a stability condition for (8.21). Consider the initial-boundary problem defined by (8.1) and

$$(8.23) \qquad u(x,0) = x , \qquad 0 \le x \le 1$$
$$(8.24) \qquad u_y(x,0) = 1 , \qquad 0 < x < 1$$
$$(8.25) \qquad u(0,y) = 0 , \qquad y \ge 0$$
$$(8.26) \qquad u(1,y) = 1 , \qquad y \ge 0 ,$$

and let us examine the consequences of setting $h=1/6$, $k=1/2$, and of *neglecting* boundary conditions (8.25)-(8.26). The grid points $R_{7,4}$ are shown in Figure 8.5.

From (8.22) and (8.23), one has the first row approximation

$$(8.27) \qquad u_{11} = 2/3 , \qquad u_{21} = 5/6 , \qquad u_{31} = 1 , \qquad u_{41} = 7/6 , \qquad u_{51} = 4/3 .$$

Next, u_{22}, u_{32} and u_{42} can be generated explicitly from (8.21) to yield

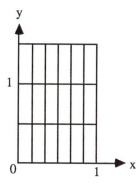

Figure 8.5

$$u_{22} = 2u_{21} - u(1/3,0) + 9(u_{11} - 2u_{21} + u_{31}) = 4/3$$

$$u_{32} = 2u_{31} - u(1/2,0) + 9(u_{21} - 2u_{31} + u_{41}) = 3/2$$

$$u_{42} = 2u_{41} - u(2/3,0) + 9(u_{31} - 2u_{41} + u_{51}) = 5/3 .$$

Note that u_{12} and u_{52} *cannot* be so approximated, because we are deliberately neglecting the given boundary data. From u_{22}, u_{32}, u_{42}, then, one can generate, by (8.21),

$$u_{33} = 2u_{32} - u_{31} + 9(u_{22} - 2u_{32} + u_{42}) = 2 .$$

But, now, observe that since we considered only initial conditions (8.23) and (8.24), u can be determined *only in the region of dependence* determined by the point (1/2,1/2), that is, u can be determined only in the triangle whose vertices are (0,0), (1,0) and (1/2,1/2). Since the point (1/2,1/2) is (x_3,y_1) in Figure 8.5, it is somewhat unreasonable that any numerical method should yield an approximation for u at (x_3,y_3). But this inconsistency is rectified easily by insisting that

(8.28) $k \leq h$,

which will yield approximations only within the domain of dependence. Inequality (8.28) is also the *Courant, Friedrichs and Lewy* stability *condition*, which is derivable mathematically by more rigorous arguments [Forsythe and Wasow (1960)].

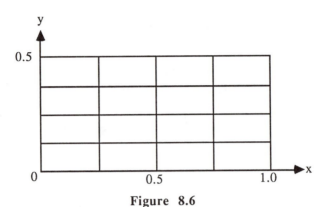

Figure 8.6

From the above discussion, an algorithm for approximating the solution of initial-boundary problem (8.1), (8.4)-(8.7), can be summarized as follows. Fix b>0, choose n and m so that $k=(b/m)\leq(a/n)=h$. Then, apply (8.22) to approximate $u_{i,1}$, i=1,2,...,n−1. Next, using (8.6) and (8.7), for each j=1,2,...,m−1, apply (8.21) to approximate explicitly $u_{i,j+1}$, i=1,2,...,n−1.

EXAMPLE. Consider the initial-boundary problem defined by (8.1) and

(8.29) $u(x,0) = x(1-x)$, $0\leq x\leq 1$

(8.30) $u_y(x,0) = 1$, $0<x<1$

(8.31) $u(0,y) = y(1-y)$, $y\geq 0$

(8.32) $u(1,y) = y(1-y)$, $y\geq 0$.

For b=0.5, n=4 and m=4, one has h=1/4, k=1/8, and the $R_{5,5}$ set of grid points is shown in Figure 8.6. From (8.22) one has

$$u_{11} = u(1/4,0)+(1/8)(1)+(1/8)[u(1/2,0)-2u(1/4,0)+u(0,0)] = 19/64$$

$$u_{21} = u(1/2,0)+(1/8)(1)+(1/8)[u(3/4,0)-2u(1/2,0)+u(1/4,0)] = 23/64$$

$$u_{31} = u(3/4,0)+(1/8)(1)+(1/8)[u(1,0)-2u(3/4,0)+u(1/2,0)] = 19/64 .$$

Next, since (8.21) has the form

$$u_{i,j+1} = 2u_{i,j} - u_{i,j-1} + \frac{1}{4}\left(u_{i+1,j} - 2u_{i,j} + u_{i-1,j}\right),$$

or, equivalently,

$$u_{i,j+1} = \frac{3}{2}u_{i,j} - u_{i,j-1} + \frac{1}{4}\left(u_{i-1,j} + u_{i+1,j}\right),$$

it follows that

$$u_{12} = (3/2)u_{11} - u(1/4,0) + (1/4)[u(0,1/8) + u_{21}] = 3/8$$

$$u_{22} = (3/2)u_{21} - u(1/2,0) + (1/4)[u_{11} + u_{31}] = 7/16$$

$$u_{32} = (3/2)u_{31} - u(3/4,0) + (1/4)[u_{21} + u(1,1/8)] = 3/8 .$$

Moreover,

$$u_{13} = (3/2)u_{12} - u_{11} + (1/4)[u(0,1/4) + u_{22}] = 27/64$$

$$u_{23} = (3/2)u_{22} - u_{21} + (1/4)[u_{12} + u_{32}] = 31/64$$

$$u_{33} = (3/2)u_{32} - u_{31} + (1/4)[u_{22} + u(1,1/4)] = 27/64 ,$$

and, finally,

$$u_{14} = (3/2)u_{13} - u_{12} + (1/4)[u(0,3/8) + u_{23}] = 7/16$$

$$u_{24} = (3/2)u_{23} - u_{22} + (1/4)[u_{13} + u_{33}] = 1/2$$

$$u_{34} = (3/2)u_{33} - u_{32} + (1/4)[u_{23} + u(1,3/8)] = 7/16 .$$

Note that, in this simple example, the numerical solution agrees exactly with the analytical solution $u = x(1-x) + y(1-y)$.

If the physics of the problem does not allow the assumption that $u_{xx} - u_{yy} = 0$ is valid on the X-axis, then a simple forward difference procedure, like,

$$u_{i,1} = f_1(x_i) + k f_2(x_i) ,$$

must be employed in place of (8.22), with the accompanying loss of accuracy, and with the Courant, Friedrichs and Lewy condition still being required.

8.4 AN IMPLICIT METHOD FOR INITIAL-BOUNDARY PROBLEMS

Now that a first method has been constructed for the solution of initial-boundary problems, one can proceed to try to improve on it. A simple implicit method, which requires the solution of a tridiagonal system on each row of grid points, but which is stable for all choices of h and k, can be developed as follows. In the notation of Figure 8.7, use the point (x_i, y_j) as a center of symmetry and substitute

$$u_{xx} \approx \frac{1}{2} \left[\frac{u_{i+1,j+1} - 2u_{i,j+1} + u_{i-1,j+1}}{h^2} + \frac{u_{i+1,j-1} - 2u_{i,j-1} + u_{i-1,j-1}}{h^2} \right]$$

$$u_{yy} \approx \frac{u_{i,j+1} - 2u_{i,j} + u_{i,j-1}}{k^2}$$

into (8.1) to yield

$$(8.33) \quad u_{i-1,j+1} - 2\left[1 + \left(\tfrac{h}{k}\right)^2\right]u_{i,j+1} + u_{i+1,j+1} = -u_{i-1,j-1} - u_{i+1,j-1} - 4\left(\tfrac{h}{k}\right)^2 u_{i,j} + 2\left[1 + \left(\tfrac{h}{k}\right)^2\right]u_{i,j-1}.$$

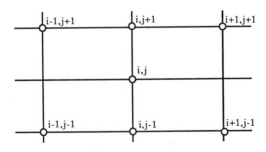

Figure 8.7

If the numerical solution is known at levels y_j and y_{j-1}, then, with the aid of boundary conditions (8.6) and (8.7), equations (8.33) constitute a linear system of $n-1$ equations in the $n-1$ unknowns $u_{i,j+1}$, $i=1,2,\ldots,n-1$. This system is tridiagonal and diagonally dominant, with negative elements on the main diagonal and nonnegative ones elsewhere. Thus, by Theorem 1.2, it has a unique solution for all j.

The general algorithm for the implicit method is, then, as follows. Fix $b>0$. Choose n and m and define $h=a/n$, $k=b/m$. At the initial and at the boundary grid points, (8.4) and (8.6)-(8.7) yield

$$
\begin{array}{lll}
& u_{i0} = f_1(x_i) \,, & i=0,1,2,\ldots,n \\[6pt]
(8.34) & u_{0j} = g_1(y_j) \,, & j=1,2,\ldots,m \\[6pt]
& u_{nj} = g_2(y_j) \,, & j=1,2,\ldots,m.
\end{array}
$$

At $y_1=k$, compute u_{i1}, $i=1,2,\ldots,n-1$, from (8.22). Then at each level y_j, $j=1,2,\ldots,m-1$, apply (8.33) to generate a tridiagonal system and then solve for $u_{i,j+1}$, $i=1,2,\ldots,n-1$.

EXAMPLE. Consider the initial-boundary problem defined by (8.1) and (8.29)-(8.32). For $b=0.5$, $n=4$ and $m=4$, one has $h=1/4$, $k=1/8$, so that the $R_{5,5}$ set of grid points is as shown in Figure 8.6. From (8.22) one has, as in the previous example,

$$
u_{11} = 19/64 \,, \qquad u_{21} = 23/64 \,, \qquad u_{31} = 19/64 \,.
$$

For the given parameter choices, (8.33) becomes

$$
(8.35) \qquad u_{i-1,j+1} - 10u_{i,j+1} + u_{i+1,j+1} = -u_{i-1,j-1} + 10u_{i,j-1} - u_{i+1,j-1} - 16u_{i,j} \,,
$$

which, for $j=1$ and $i=1,2,3$, yields

$$
\begin{array}{l}
u(0,1/4) - 10u_{12} + u_{22} = -u(0,0) + 10u(1/4,0) - u(1/2,0) - 16u_{11} \\[6pt]
u_{12} - 10u_{22} + u_{32} = -u(1/4,0) + 10u(1/2,0) - u(3/4,0) - 16u_{21} \\[6pt]
u_{22} - 10u_{32} + u(1,1/4) = -u(1/2,0) + 10u(3/4,0) - u(1,0) - 16u_{31} \,,
\end{array}
$$

or, equivalently,

$$-10u_{12} + u_{22} \qquad = -53/16$$

$$u_{12} - 10u_{22} + u_{32} = -29/8$$

$$u_{22} - 10u_{32} = -53/16$$

the solution of which is

$$u_{12} = 3/8 , \qquad\qquad u_{22} = 7/16 , \qquad\qquad u_{32} = 3/8 .$$

One then proceeds to generate a solution at level y_3 using (8.35), the given boundary conditions, and the approximations at levels y_1 and y_2, and so on. The results are

$$u_{13} = 27/64 , \qquad u_{23} = 31/64 , \qquad u_{33} = 27/64 ,$$

$$u_{14} = 7/16 , \qquad u_{24} = 1/2 , \qquad u_{34} = 7/16 ,$$

which, again, are identical with the analytical solution.

8.5 MILDLY NONLINEAR PROBLEMS

The explicit and the implicit methods extend in a natural way to mildly nonlinear problems defined by (8.4)-(8.7) and

$$(8.36) \qquad u_{xx} - u_{yy} = f(x,y,u) .$$

The only modifications necessary occur when (8.36) replaces (8.1), and then one need only replace the difference equations accordingly. In the explicit method, then, approximate (8.36) by

$$(8.37) \qquad \frac{u_{i+1,j} - 2u_{i,j} + u_{i-1,j}}{h^2} - \frac{u_{i,j+1} - 2u_{i,j} + u_{i,j-1}}{k^2} = f(x_i, y_j, u_{i,j}) ,$$

while, in the implicit method, approximate (8.36) by

(8.38)
$$\frac{1}{2}\left[\frac{u_{i+1,j+1}-2u_{i,j+1}+u_{i-1,j+1}}{h^2}+\frac{u_{i+1,j-1}-2u_{i,j-1}+u_{i-1,j-1}}{h^2}\right]$$

$$-\frac{u_{i,j+1}-2u_{i,j}+u_{i,j-1}}{k^2}=f(x_i,y_j,u_{i,j}).$$

Unfortunately, stability conditions now vary from problem to problem and no meaningful, comprehensive results are available.

8.6 HYPERBOLIC SYSTEMS

A second approach to the study of hyperbolic equations is by means of systems. This approach seems to have special value for equations which are more complex than the wave equation and is developed as follows.

The general, *quasilinear* system of two first order partial differential equations in the two unknowns $v(x,y)$, $w(x,y)$ is defined to be

$$p_{11}v_x+p_{12}w_x+q_{11}v_y+q_{12}w_y+r_1 = 0$$

(8.39)

$$p_{21}v_x+p_{22}w_x+q_{21}v_y+q_{22}w_y+r_2 = 0,$$

where p_{ij}, q_{ij}, r_i can be functions of x, y, v and w. If, however, p_{ij}, q_{ij}, r_i depend only on x and y, then (8.39) is called *linear*. Introducing the matrices

$$P=\begin{bmatrix} p_{11} & p_{12} \\ p_{21} & p_{22} \end{bmatrix}, \qquad Q=\begin{bmatrix} q_{11} & q_{12} \\ q_{21} & q_{22} \end{bmatrix}, \qquad R=\begin{bmatrix} r_1 \\ r_2 \end{bmatrix}, \qquad U=\begin{bmatrix} v \\ w \end{bmatrix}$$

enables one to write (8.39) in the equivalent form

(8.40) $$PU_x+QU_y+R = 0.$$

Here, we will restrict our attention to the case that $|Q|\neq 0$, so that (8.40) can also be written in the form

(8.41) $U_y + AU_x + C = 0$,

where $A = Q^{-1}P$ and $C = Q^{-1}R$. System (8.40), is said to be hyperbolic *if and only if* matrix A has two eigenvalues λ_1, λ_2 which are real and distinct. When (8.40) is hyperbolic, matrix A has two eigenvectors which are linearly independent.

Now let H be a matrix whose column vectors are eigenvectors of A in such a fashion that the first column of H corresponds to λ_1, while the second column corresponds to λ_2. Of course, for a hyperbolic system, matrix H is nonsingular. Denote by T the inverse of H. Throughout, we will assume that (8.40) is hyperbolic, so that T exists and is nonsingular. Now, by setting

$$\Lambda = \begin{bmatrix} \lambda_1 & 0 \\ 0 & \lambda_2 \end{bmatrix},$$

one has, by definition, $AH = H\Lambda$, which, since $T = H^{-1}$, implies

$$TA = \Lambda T.$$

Thus, by multiplying (8.41) by T, there results the following equivalent system:

$$TU_y + TAU_x + D = 0,$$

in which $D = TC$, and which is the same as

(8.42) $TU_y + \Lambda TU_x + D = 0$.

System (8.42) is said to be a *normal form* of (8.40). Note that, since H is nonsingular, a normal form of a hyperbolic system always exists.

The *characteristics* of (8.40) are defined to be the solutions of

$$\frac{dx}{dy} = \lambda_1, \qquad\qquad \frac{dx}{dy} = \lambda_2$$

which are given in this particular form because, in general, one wants to allow one of λ_1, λ_2 to be zero. If neither of λ_1, λ_2 is zero, then the directions of the characteristics are

$$\frac{dy}{dx} = \frac{1}{\lambda_1}, \qquad \frac{dy}{dx} = \frac{1}{\lambda_2}.$$

EXAMPLE 1. Consider the wave equation

(8.43) $u_{xx} - u_{yy} = 0$.

If one sets $v = u_x$ and $w = u_y$, then

(8.44) $v_y - w_x = 0$

$w_y - v_x = 0$,

which is a linear system of the form (8.40) with $\mathbf{Q} = \mathbf{I}$ and

$$\mathbf{P} = \mathbf{A} = \begin{bmatrix} 0 & -1 \\ -1 & 0 \end{bmatrix}, \qquad \mathbf{R} = \mathbf{C} = \begin{bmatrix} 0 \\ 0 \end{bmatrix}.$$

Next, since

$$|\mathbf{A} - \lambda \mathbf{I}| = \lambda^2 - 1 ,$$

it follows that the eigenvalues of $\mathbf{A}$ are

$$\lambda_1 = 1 , \qquad \lambda_2 = -1 .$$

Thus, system (8.43) is hyperbolic, which is consistent with calling (8.42) a hyperbolic second-order equation. To get a normal form of (8.44), let

$$T = \begin{bmatrix} t_{11} & t_{12} \\ t_{21} & t_{22} \end{bmatrix}, \qquad \Lambda = \begin{bmatrix} 1 & 0 \\ 0 & -1 \end{bmatrix}.$$

Then we wish to have

$$TA = \Lambda T.$$

But

$$TA = \begin{bmatrix} -t_{12} & -t_{11} \\ -t_{22} & -t_{21} \end{bmatrix}, \qquad \Lambda T = \begin{bmatrix} t_{11} & t_{12} \\ -t_{21} & -t_{22} \end{bmatrix},$$

so that one wishes to have

$$t_{11} = -t_{12}$$
$$t_{21} = t_{22},$$

one solution of which is

$$t_{11} = 1, \qquad t_{12} = -1, \qquad t_{21} = t_{22} = 1.$$

Thus, one possible choice of T is

$$T = \begin{bmatrix} 1 & -1 \\ 1 & 1 \end{bmatrix}.$$

But, then,

$$TA = \Lambda T = \begin{bmatrix} 1 & -1 \\ -1 & -1 \end{bmatrix} ,$$

and system (8.44), in normal form, is

$$\begin{bmatrix} 1 & -1 \\ 1 & 1 \end{bmatrix} \begin{bmatrix} v_y \\ w_y \end{bmatrix} + \begin{bmatrix} 1 & -1 \\ -1 & -1 \end{bmatrix} \begin{bmatrix} v_x \\ w_x \end{bmatrix} = 0$$

or, equivalently,

(8.45)
$$(v_y + v_x) - (w_y + w_x) = 0$$
$$(v_y - v_x) + (w_y - w_x) = 0 ,$$

As indicated above, we emphasize that a normal form need not be unique. Finally, the characteristics of the system are the solutions of

$$\frac{dx}{dy} = 1 , \qquad \frac{dx}{dy} = -1 ,$$

or

$$x - y = c_1 , \qquad x + y = c_2 ,$$

which is consistent with the definition of characteristics given in Section 8.2.

EXAMPLE 2. The one-dimensional equations of isentropic flow are

(8.46)
$$\rho(u_t + u u_x) + c^2 \rho_x = 0$$

$$\rho_t + u \rho_x + \rho u_x = 0 ,$$

where u is the velocity, ρ is a positive density function, and $c=c(\rho)$ is the speed of sound. With the change of variables

$$v = u , \qquad w = \rho , \qquad y = t ,$$

system (8.46) can be rewritten as

(8.47)
$$v_y + vv_x + \frac{c^2}{w} w_x = 0$$

$$w_y + vw_x + wv_x = 0 .$$

This is a quasilinear system of form (8.40) with $\mathbf{Q}=\mathbf{I}$ and

$$\mathbf{P} = \mathbf{A} = \begin{bmatrix} v & c^2/w \\ w & v \end{bmatrix} , \qquad \mathbf{R} = \mathbf{C} = \begin{bmatrix} 0 \\ 0 \end{bmatrix} .$$

Hence, the eigenvalues of $\mathbf{A}$ are solutions of

$$(v-\lambda)^2 - c^2 = 0 .$$

Thus,

$$\lambda = v \pm c .$$

Since c is the speed of sound, assume c>0, so that

$$\lambda_1 = v+c , \qquad \lambda_2 = v-c .$$

Assuming that v exists and is real, then λ_1 and λ_2 are real and distinct so that the system is hyperbolic.

To get a normal form of (8.47), let

$$
T = \begin{bmatrix} t_{11} & t_{12} \\ t_{21} & t_{22} \end{bmatrix}, \qquad \Lambda = \begin{bmatrix} v+c & 0 \\ 0 & v-c \end{bmatrix}.
$$

Then

$$
TA = \begin{bmatrix} t_{11} & t_{12} \\ t_{21} & t_{22} \end{bmatrix} \begin{bmatrix} v & c^2/w \\ w & v \end{bmatrix} = \begin{bmatrix} t_{11}v+t_{12}w & t_{11}(c^2/w)+t_{12}v \\ t_{21}v+t_{22}w & t_{21}(c^2/w)+t_{22}v \end{bmatrix},
$$

$$
\Lambda T = \begin{bmatrix} v+c & 0 \\ 0 & v-c \end{bmatrix} \begin{bmatrix} t_{11} & t_{12} \\ t_{21} & t_{22} \end{bmatrix} = \begin{bmatrix} (v+c)t_{11} & (v+c)t_{12} \\ (v-c)t_{21} & (v-c)t_{22} \end{bmatrix}.
$$

Thus, $TA = \Lambda T$ implies

$$
t_{11}c = t_{12}w
$$

$$
t_{21}c = -t_{22}w .
$$

A solution of the latter system is $t_{11}=w$, $t_{12}=c$, $t_{21}=w$, $t_{22}=-c$, so that one can choose

$$
T = \begin{bmatrix} w & c \\ w & -c \end{bmatrix}.
$$

Then

$$
TA = \Lambda T = \begin{bmatrix} w(v+c) & c(v+c) \\ w(v-c) & -c(v-c) \end{bmatrix},
$$

and a normal form is

$$\begin{bmatrix} w & c \\ w & -c \end{bmatrix} \begin{bmatrix} v_y \\ w_y \end{bmatrix} + \begin{bmatrix} w(v+c) & c(v+c) \\ w(v-c) & -c(v-c) \end{bmatrix} \begin{bmatrix} v_x \\ w_x \end{bmatrix} = 0 ,$$

or, equivalently,

(8.48)

$$w[v_y+(v+c)v_x]+c[w_y+(v+c)w_x] = 0$$

$$w[v_y+(v-c)v_x]-c[w_y+(v-c)w_x] = 0 .$$

The characteristics of system (8.47) are the solutions of

$$\frac{dx}{dy} = v+c , \qquad \frac{dx}{dy} = v-c ,$$

which *cannot* be solved because v is unknown. Thus, most interestingly, even though the characteristics of system (8.47) cannot be found, the system *can* be put into normal form.

8.7 THE METHOD OF CHARACTERISTICS FOR INITIAL VALUE PROBLEMS

In formulating initial value problems for hyperbolic systems, assume that the given system is already in normal form. The normal form may be quasilinear. Also, assume that v and w are given on a curve, which, for simplicity, will be taken to be the X axis. We will wish to find solutions v, w of the system for y>0 which take on given initial values on the X axis. Precisely, an initial value problem is defined as follows. Given

(8.49)

$$v(x,0) = f_1(x) ,$$

$$-\infty < x < \infty,$$

$$w(x,0) = f_2(x) ,$$

and given two matrices **T** and Λ such that $|T| \neq 0$ and Λ is a diagonal matrix whose diagonal elements λ_1 and λ_2 are real and unequal, then, find v(x,y), w(x,y) which satisfy initial conditions (8.49) and which, for y>0, are solutions of

$$TU_y + \Lambda TU_x + D = 0 ,$$

or, equivalently,

(8.50) $t_{11}(v_y + \lambda_1 v_x) + t_{12}(w_y + \lambda_1 w_x) + d_1 = 0$

(8.51) $t_{21}(v_y + \lambda_2 v_x) + t_{22}(w_y + \lambda_2 w_x) + d_2 = 0 .$

Throughout, without loss of generality, we will assume that, *at each point of definition*, one has $\lambda_1 > \lambda_2$.

In general, existence and uniqueness of solutions of initial value problem (8.49)-(8.51) are known only "in the small", that is, in a neighborhood close to the X axis. Since no general analytical methodology is available for solving (8.49)-(8.51), let us turn next to a very general, useful numerical method.

The numerical method to be derived is based on the following observations. At a point (x,y), let v(x,y) be differentiable and let λ_1 be a given *nonzero* constant. Let the angle ϕ at (x,y) be defined as that angle, measured counterclockwise from the positive X direction, which satisfies $\tan(\phi) = 1/\lambda_1$. Then the directional derivative of v, denoted by D_1, in the direction defined by ϕ is given by

$$D_1 v = v_y \sin(\phi) + v_x \cos(\phi) = \frac{1}{\sqrt{1+\lambda_1^2}} v_y + \frac{\lambda_1}{\sqrt{1+\lambda_1^2}} v_x ,$$

so that

$$\sqrt{1+\lambda_1^2} \, D_1 v = v_y + \lambda_1 v_x .$$

For a second differentiable function w(x,y), one has, similarly,

$$\sqrt{1+\lambda_1^2} \, D_1 w = w_y + \lambda_1 w_x .$$

Now, if a characteristic at (x,y) is given by $(dx/dy)=\lambda_1$, then the direction of the characteristic is $(dy/dx)=1/\lambda_1$, so that $(dy/dx)=\tan(\phi)$. Thus, equation (8.50) can be written as

$$t_{11}\sqrt{1+\lambda_1^2}\ D_1v+t_{12}\sqrt{1+\lambda_1^2}\ D_1w+d_1 = 0\ ,$$

which reveals that the coefficients of t_{11} and t_{12} are, in fact, proportional to the directional derivatives of v and w in the direction of the given characteristic. Similarly, for nonzero λ_2, the terms $v_y+\lambda_2v_x$, $w_y+\lambda_2w_x$, in (8.51), are proportional to the directional derivative, denoted by D_2, of v and w in the direction of the characteristic whose equation is $(dx/dy)=\lambda_2$. Hence, a normal form has the important property that each equation contains derivatives in *one direction only*. Each such direction is a characteristic direction. Note that, in the particular case when λ_1 or λ_2 is zero, the corresponding characteristic direction is that of the Y axis and, accordingly, D_1 or D_2 simply reduces to $(\partial/\partial y)$. Equations (8.50)-(8.51), for any λ_1 and λ_2, can be written then as follows:

(8.52) $$\sqrt{1+\lambda_1^2}\ (t_{11}D_1v+t_{12}D_1w)+d_1 = 0$$

(8.53) $$\sqrt{1+\lambda_2^2}\ (t_{21}D_2v+t_{22}D_2w)+d_2 = 0\ .$$

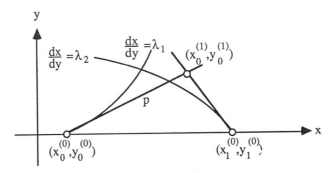

Figure 8.8

Now, in order to solve the initial value problem for equations (8.52)-(8.53), consider first on the X axis an R_{n+1} set of points whose coordinates are $(x_i, 0)$, i=0,1,2,…,n, which, for later convenience, we shall denote at present by $(x_i^{(0)}, y_i^{(0)})$. From (8.49) one has that the solution at each point $(x_i^{(0)}, y_i^{(0)})$ is

$$v_i^{(0)} = f_1(x_i^{(0)})$$

$$w_i^{(0)} = f_2(x_i^{(0)}) .$$

On the X axis, let us begin by considering two consecutive grid points, say $(x_0^{(0)}, y_0^{(0)})$ and $(x_1^{(0)}, y_1^{(0)})$, so that the λ_1-characteristic through $(x_0^{(0)}, y_0^{(0)})$ satisfies the equation

$$\frac{dx}{dy} = \lambda_1 , \qquad x(y_0^{(0)}) = x_0^{(0)} ,$$

while the λ_2-characteristic through $(x_1^{(0)}, y_1^{(0)})$ satisfies the equation

$$\frac{dx}{dy} = \lambda_2 , \qquad x(y_1^{(0)}) = x_1^{(0)} .$$

Since, in general, the characteristics are curved lines which depend upon x, y, v and w, as indicated in Figure 8.8, we will approximate these lines with the straight lines which, at $(x_0^{(0)}, y_0^{(0)})$ and at $(x_1^{(0)}, y_1^{(0)})$ have the same slope as the characteristics. Thus, the approximated characteristics, have the following equations:

$$x - x_0^{(0)} = (\lambda_1)_0^{(0)} [y - y_0^{(0)}]$$

$$x - x_1^{(0)} = (\lambda_2)_1^{(0)} [y - y_1^{(0)}] .$$

The point of intersection between the λ_1-characteristic through $(x_0^{(0)}, y_0^{(0)})$ with the λ_2-characteristic through $(x_1^{(0)}, y_1^{(0)})$ has coordinates $(x_0^{(1)}, y_0^{(1)})$ which satisfy

$$x_0^{(1)} - x_0^{(0)} = (\lambda_1)_0^{(0)} [y_0^{(1)} - y_0^{(0)}]$$

(8.54)

$$x_0^{(1)} - x_1^{(0)} = (\lambda_2)_1^{(0)} [y_0^{(1)} - y_1^{(0)}] .$$

Now, if $(\lambda_1)_0^{(0)} > (\lambda_2)_1^{(0)}$, system (8.54) yields, uniquely, the coordinates of a new grid point, namely $x_0^{(1)}$ and $y_0^{(1)}$. In order to approximate v and w at $(x_0^{(1)}, y_0^{(1)})$, observe that since $(x_0^{(1)}, y_0^{(1)})$ lies on the λ_1-characteristic through $(x_0^{(0)}, y_0^{(0)})$, it is reasonable to approximate equation (8.52), at the point $(x_0^{(0)}, y_0^{(0)})$ with a finite difference equation which is obtained from (8.52). This is done by replacing the derivative D_1 with the corresponding finite difference between $(x_0^{(1)}, y_0^{(1)})$ and $(x_0^{(0)}, y_0^{(0)})$, that is, by

(8.55) $$\sqrt{1+\lambda_1^2} \left[t_{11} \frac{v_0^{(1)} - v_0^{(0)}}{p} + t_{12} \frac{w_0^{(1)} - w_0^{(0)}}{p} \right] + d_1 = 0 ,$$

in which all the coefficients are evaluated at $(x_0^{(0)}, y_0^{(0)})$, and p, for the case shown in Figure 8.8, is given by

(8.56) $$p = \sqrt{\left[x_0^{(1)} - x_0^{(0)} \right]^2 + \left[y_0^{(1)} - y_0^{(0)} \right]^2} = \left[y_0^{(1)} - y_0^{(0)} \right] \sqrt{1+\lambda_1^2} .$$

Substitution of (8.56) into (8.55) yields

(8.57) $$(t_{11})_0^{(0)} \left[v_0^{(1)} - v_0^{(0)} \right] + (t_{12})_0^{(0)} \left[w_0^{(1)} - w_0^{(0)} \right] + (d_1)_0^{(0)} \left[y_0^{(1)} - y_0^{(0)} \right] = 0 .$$

Similarly, since $(x_0^{(1)}, y_0^{(1)})$ also lies on the λ_2-characteristic through $(x_1^{(0)}, y_1^{(0)})$, if one approximates equation (8.53) at $(x_1^{(0)}, y_1^{(0)})$ with a finite difference equation obtained from (8.53) by replacing the derivative D_2 with the corresponding finite difference between $(x_0^{(1)}, y_0^{(1)})$ and $(x_1^{(0)}, y_1^{(0)})$, one finds

(8.58) $$(t_{21})_1^{(0)} \left[v_0^{(1)} - v_1^{(0)} \right] + (t_{22})_1^{(0)} \left[w_0^{(1)} - w_1^{(0)} \right] + (d_2)_1^{(0)} \left[y_0^{(1)} - y_1^{(0)} \right] = 0 .$$

Equations (8.57) and (8.58) can be written, equivalently, as

$$(t_{11})_0^{(0)} v_0^{(1)} + (t_{12})_0^{(0)} w_0^{(1)} = (t_{11})_0^{(0)} v_0^{(0)} + (t_{12})_0^{(0)} w_0^{(0)} - (d_1)_0^{(0)} \left[y_0^{(1)} - y_0^{(0)} \right]$$

(8.59)

$$(t_{21})_1^{(0)} v_0^{(1)} + (t_{22})_1^{(0)} w_0^{(1)} = (t_{21})_1^{(0)} v_1^{(0)} + (t_{22})_1^{(0)} w_1^{(0)} - (d_2)_1^{(0)} \left[y_0^{(1)} - y_1^{(0)} \right] ,$$

which is a system of two linear equations in the two unknowns $v_0^{(1)}$ and $w_0^{(1)}$. Thus, an approximate solution $v_0^{(1)}$ and $w_0^{(1)}$ at $(x_0^{(1)}, y_0^{(1)})$ is determined by solving (8.59).

Formulas (8.54) and (8.59) are the basic ones and can be applied to approximate v and w in the entire region of dependence defined by $[x_0, x_n]$. In particular, from (8.54), the equations which determine the coordinates of a new grid point are

$$x_i^{(k+1)} - (\lambda_1)_i^{(k)} y_i^{(k+1)} = x_i^{(k)} - (\lambda_1)_i^{(k)} y_i^{(k)}$$

(8.60)

$$x_i^{(k+1)} - (\lambda_2)_{i+1}^{(k)} y_i^{(k+1)} = x_{i+1}^{(k)} - (\lambda_2)_{i+1}^{(k)} y_{i+1}^{(k)} ,$$

while, from (8.59), the approximate solutions for v and w at $(x_i^{(k+1)}, y_i^{(k+1)})$ are determined from

$$(t_{11})_i^{(k)} v_i^{(k+1)} + (t_{12})_i^{(k)} w_i^{(k+1)} = (t_{11})_i^{(k)} v_i^{(k)} + (t_{12})_i^{(k)} w_i^{(k)} - (d_1)_i^{(k)} \left(y_i^{(k+1)} - y_i^{(k)} \right)$$

(8.61)

$$(t_{21})_{i+1}^{(k)} v_i^{(k+1)} + (t_{22})_{i+1}^{(k)} w_i^{(k+1)} = (t_{21})_{i+1}^{(k)} v_{i+1}^{(k)} + (t_{22})_{i+1}^{(k)} w_{i+1}^{(k)} - (d_2)_{i+1}^{(k)} \left(y_i^{(k+1)} - y_{i+1}^{(k)} \right) .$$

The above method is called the **method of characteristics**, and the general algorithm for the method of characteristics is as follows. Given, on the X axis, an R_{n+1} set, define $(x_i^{(0)}, y_i^{(0)}) = (x_i, 0)$. Then, starting with k=0, and by using the known initial conditions, apply (8.60) to determine the new grid points $(x_i^{(1)}, y_i^{(1)})$, i=0,1,2,...,n−1. Then, solve the linear system (8.61) to determine $v_i^{(1)}$ and $w_i^{(1)}$, i=0,1,2,...,n−1. Consider, next, k=1 and calculate the points $(x_i^{(2)}, y_i^{(2)})$ and the function values $v_i^{(2)}$, $w_i^{(2)}$, for i=0,1,2,...,n−2. Continue in the indicated fashion until, finally, $v_0^{(n)}$ and $w_0^{(n)}$ have been computed.

Note that, for the general case when λ_1 and λ_2 are functions of x, y, v and w, the various possible grid point structures may render the method of characteristics to be impractical. However, for constant λ_1 and λ_2, from (8.60) one obtains a uniform mesh distribution even for nonlinear systems.

EXAMPLE 1. Consider the wave equation (8.44) in the normal form (8.45), that is,

$$(v_y+v_x)-(w_y+w_x) = 0$$

$$(v_y-v_x)+(w_y-w_x) = 0 \, ,$$

so that $t_{11}=t_{21}=t_{22}=1$, $t_{12}=-1$, $\lambda_1=1$, $\lambda_2=-1$, $d_1=d_2=0$. Let

$$v(x,0) = 1-2x \, , \qquad w(x,0) = 1 \, .$$

With $n=2$, consider the R_3 set $x_0=-1$, $x_1=0$, $x_2=1$. Then, $(x_0^{(0)},y_0^{(0)})=(-1,0)$, $(x_1^{(0)},y_1^{(0)})=(0,0)$, $(x_2^{(0)},y_2^{(0)})=(1,0)$, and

$$v_0^{(0)} = 3 \, , \qquad v_1^{(0)} = 1 \, , \qquad v_2^{(0)} = -1 \, ,$$

$$w_0^{(0)} = 1 \, , \qquad w_1^{(0)} = 1 \, , \qquad w_2^{(0)} = 1 \, .$$

Now, for $i=0$, and $k=0$, application of (8.60) yields

$$x_0^{(1)} - y_0^{(1)} = -1$$

$$x_0^{(1)} + y_0^{(1)} = 0 \, ,$$

the solution of which is $x_0^{(1)}=-1/2$, $y_0^{(1)}=1/2$, while, (8.61) yields

$$v_0^{(1)} - w_0^{(1)} = 2$$

$$v_0^{(1)} + w_0^{(1)} = 2 \, ,$$

the solution of which is $v_0^{(1)}=2$, $w_0^{(1)}=0$. Next, for $i=1$, and $k=0$, application of (8.60) yields

$$x_1^{(1)} - y_1^{(1)} = 0$$

$$x_1^{(1)} + y_1^{(1)} = 1 \, ,$$

whose solution is $x_1^{(1)}=1/2$, $y_1^{(1)}=1/2$, while equations (8.61) yield

$$v_1^{(1)} - w_1^{(1)} = 0$$

$$v_1^{(1)} + w_1^{(1)} = 0 \, ,$$

the solution of which is $v_1^{(1)}=0$, $w_1^{(1)}=0$. Finally, for k=1 and i=0, equations (8.60) yield

$$x_0^{(2)} - y_0^{(2)} = -1$$

$$x_0^{(2)} + y_0^{(2)} = 1 \, ,$$

whose solution is $x_0^{(2)}=0$, $y_0^{(2)}=1$, while (8.61) yield

$$v_0^{(2)} - w_0^{(2)} = 2$$

$$v_0^{(2)} + w_0^{(2)} = 0 \, ,$$

the solution of which is $v_0^{(2)}=1$, $w_0^{(2)}=-1$. Note that all the numerical results are in complete agreement with the analytical solution v=1–2x, w=1–2y.

EXAMPLE 2. Consider the isentropic flow equations (8.47) in the normal form (8.48) with c=1, that is,

$$w[v_y+(v+1)v_x]+[w_y+(v+1)w_x] = 0$$

$$w[v_y+(v-1)v_x]-[w_y+(v-1)w_x] = 0 \, ,$$

so that $t_{11}=t_{21}=w$, $t_{12}=1$, $t_{22}=-1$, $\lambda_1=v+1$, $\lambda_2=v-1$, $d_1=d_2=0$. Let

$$v(x,0) = 0, \qquad w(x,0) = x+1.$$

With n=2, consider the R_3 set $x_0=0$, $x_1=1$, $x_2=2$. Then, $(x_0^{(0)}, y_0^{(0)})=(0,0)$, $(x_1^{(0)}, y_1^{(0)})=(1,0)$, $(x_2^{(0)}, y_2^{(0)})=(2,0)$, and

$$v_0^{(0)} = 0, \qquad v_1^{(0)} = 0, \qquad v_2^{(0)} = 0,$$

$$w_0^{(0)} = 1, \qquad w_1^{(0)} = 2, \qquad w_2^{(0)} = 3.$$

Now, for i=0, and k=0, application of (8.60) with $(\lambda_1)_0^{(0)}=v_0^{(0)}+1$, $(\lambda_2)_1^{(0)}=v_1^{(0)}-1$, yields

$$x_0^{(1)} - y_0^{(1)} = 0$$

$$x_0^{(1)} + y_0^{(1)} = 1,$$

the solution of which is $x_0^{(1)}=1/2$, $y_0^{(1)}=1/2$, while, (8.61) yields

$$v_0^{(1)} + w_0^{(1)} = 1$$

$$2v_0^{(1)} - w_0^{(1)} = -2,$$

the solution of which is $v_0^{(1)}=-1/3$, $w_0^{(1)}=4/3$. Next, for i=1, and k=0, application of (8.60) with $(\lambda_1)_1^{(0)}=v_1^{(0)}+1$, $(\lambda_2)_2^{(0)}=v_2^{(0)}-1$, yields

$$x_1^{(1)} - y_1^{(1)} = 1$$

$$x_1^{(1)} + y_1^{(1)} = 2,$$

whose solution is $x_1^{(1)}=3/2$, $y_1^{(1)}=1/2$, while equations (8.61) yield

$$2v_1^{(1)} + w_1^{(1)} = 2$$

$$3v_1^{(1)} - w_1^{(1)} = -3 ,$$

the solution of which is $v_1^{(1)} = -1/5$, $w_1^{(1)} = 12/5$. Finally, for k=1 and i=0, equations (8.60) with $(\lambda_1)_0^{(1)} = v_0^{(1)} + 1$, $(\lambda_2)_1^{(1)} = v_1^{(1)} - 1$, yield

$$x_0^{(2)} - (2/3)y_0^{(2)} = 1/6$$

$$x_0^{(2)} + (6/5)y_0^{(2)} = 21/10 ,$$

whose solution is $x_0^{(2)} = 6/7$, $y_0^{(2)} = 29/28$, while (8.61) yield

$$(4/3)v_0^{(2)} + w_0^{(2)} = 8/9$$

$$(12/5)v_0^{(2)} - w_0^{(2)} = -72/25 ,$$

the solution of which is $v_0^{(2)} = -8/15$, $w_0^{(2)} = 8/5$, and the example is complete.

8.8 THE METHOD OF COURANT, ISAACSON AND REES

The method of characteristics, when applicable, is a very accurate one. The main disadvantage of this method arises when it is applied to systems where λ_1 and λ_2 are not constants. In such cases, for example, the location of the grid points cannot be controlled by the user, a distorted grid will be generated, and accuracy decreases very quickly. For this reason one would like to choose a uniform finite difference grid on which to discretize the differential equations. For this purpose, reconsider system (8.50)-(8.51), which is in the normal form

(8.62) $t_{11}(v_y + \lambda_1 v_x) + t_{12}(w_y + \lambda_1 w_x) + d_1 = 0$

(8.63) $t_{21}(v_y + \lambda_2 v_x) + t_{22}(w_y + \lambda_2 w_x) + d_2 = 0 .$

Also, recall that this system can be written, equivalently, as

(8.64) $\sqrt{1+\lambda_1^2}\,(t_{11}D_1v+t_{12}D_1w)+d_1 = 0$

(8.65) $\sqrt{1+\lambda_2^2}\,(t_{21}D_2v+t_{22}D_2w)+d_2 = 0$,

where D_1 and D_2 denote the directional derivatives in the directions of the λ_1- and λ_2-characteristics, respectively.

In order to solve the initial value problem for equations (8.62)-(8.63), consider, first, an interval $[-a,a]$ on the X axis. Divide $[-a,a]$ into $2n$ equal parts, each of length h, by the R_{2n+1} set $x_i=-a+ih$, $i=0,1,2,\ldots,2n$. Then, choose a grid size k and define $y_j=jk$, $j=0,1,2,\ldots,n$. In order to develop the general finite difference formulas, assume, as shown in Figure 8.9, that (x_i,y_j) is a typical interior grid point, and $v_{i,j}$, $v_{i-1,j}$, $v_{i+1,j}$, $w_{i,j}$, $w_{i-1,j}$, $w_{i+1,j}$ are known, or have been computed previously. One proceeds to approximate $v_{i,j+1}$ and $w_{i,j+1}$ as follows.

Let the λ_1-characteristic through (x_i,y_{j+1}) be approximated by a straight line whose equation is

$$x-x_i = (\lambda_1)_{i,j}(y-y_{j+1}) \ .$$

The characteristic so approximated intersects the line $y=y_j$ at a point whose abscissa is

$$x_{i-\alpha} = x_i-(\lambda_1)_{i,j}(y_{j+1}-y_j) = -a+\left[i - \frac{k}{h}(\lambda_1)_{i,j}\right]h \ ,$$

where $\alpha=(k/h)(\lambda_1)_{i,j}$. Thus, the point of intersection of the λ_1-characteristic with the line $y=y_j$ is $(x_{i-\alpha},y_j)$. The distance between $(x_{i-\alpha},y_j)$ and (x_i,y_{j+1}) is

(8.66) $p = \sqrt{(x_i-x_{i-\alpha})^2+(y_{j+1}-y_j)^2} = k\sqrt{1+\lambda_1^2}$,

where $\lambda_1=(\lambda_1)_{i,j}$. Now, equation (8.64), which contains derivatives along the λ_1-characteristic only, is discretized as follows:

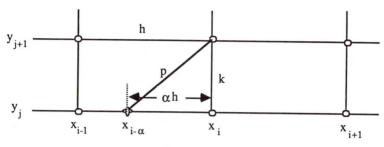

Figure 8.9

(8.67) $$\sqrt{1+\lambda_1^2}\left[t_{11}\frac{v_{i,j+1}-v_{i-\alpha,j}}{p} + t_{12}\frac{w_{i,j+1}-w_{i-\alpha,j}}{p}\right]+d_1 = 0 \, ,$$

where all the coefficients are evaluated at (x_i, y_j). By use of (8.66), (8.67) yields

(8.68) $$t_{11}\frac{v_{i,j+1}-v_{i-\alpha,j}}{k} + t_{12}\frac{w_{i,j+1}-w_{i-\alpha,j}}{k}+d_1 = 0 \, .$$

Note, however, that since α, in general, is not an integer, $(x_{i-\alpha}, y_j)$ is not a mesh point. Thus $v_{i-\alpha,j}$ and $w_{i-\alpha,j}$ are not known and need to be approximated. For this purpose, choose a linear approximation formula, so that, if, for example, $\alpha \geq 0$, $x_{i-\alpha}$ is located to the left of x_i. Hence $v_{i-\alpha,j}$ and $w_{i-\alpha,j}$ will be determined from the values of v and w at the mesh points (x_i, y_j) and (x_{i-1}, y_j) as follows:

(8.69) $$v_{i-\alpha,j} = \alpha v_{i-1,j}+(1-\alpha)v_{i,j}, \quad w_{i-\alpha,j} = \alpha w_{i-1,j}+(1-\alpha)w_{i,j} \, .$$

Substitution of (8.69) into (8.68) yields, then,

$$t_{11}\left[\frac{v_{i,j+1}-v_{i,j}}{k} + \alpha\frac{v_{i,j}-v_{i-1,j}}{k}\right]+t_{12}\left[\frac{w_{i,j+1}-w_{i,j}}{k} + \alpha\frac{w_{i,j}-w_{i-1,j}}{k}\right]+d_1 = 0 \, ,$$

or, since $\alpha = \dfrac{k}{h}\lambda_1$,

$$(8.70) \qquad t_{11}\left[\frac{v_{i,j+1}-v_{i,j}}{k} + \lambda_1\frac{v_{i,j}-v_{i-1,j}}{h}\right] + t_{12}\left[\frac{w_{i,j+1}-w_{i,j}}{k} + \lambda_1\frac{w_{i,j}-w_{i-1,j}}{h}\right] + d_1 = 0 .$$

If, instead, $\alpha<0$, then $x_{i-\alpha}$ is located to the right of x_i. Hence $v_{i-\alpha,j}$ and $w_{i-\alpha,j}$ will be determined from the values of v and w at the mesh points (x_i,y_j) and (x_{i+1},y_j) as follows:

$$(8.71) \qquad v_{i-\alpha,j} = -\alpha v_{i+1,j} + (1+\alpha)v_{i,j} , \qquad\qquad w_{i-\alpha,j} = -\alpha w_{i+1,j} + (1+\alpha)w_{i,j} ,$$

so that substitution of (8.71) into (8.68) yields

$$(8.72) \qquad t_{11}\left[\frac{v_{i,j+1}-v_{i,j}}{k} + \lambda_1\frac{v_{i+1,j}-v_{i,j}}{h}\right] + t_{12}\left[\frac{w_{i,j+1}-w_{i,j}}{k} + \lambda_1\frac{w_{i+1,j}-w_{i,j}}{h}\right] + d_1 = 0 .$$

Note now that the finite difference equations (8.70) and (8.72), at (x_i,y_j), could have been derived immediately from (8.62) simply by approximating the derivatives in y with a forward finite difference between (x_i,y_{j+1}) and (x_i,y_j), and by approximating the derivatives in x by a backward finite difference when $\lambda_1\geq 0$ (equation (8.70)), or by a forward finite difference when $\lambda_1<0$ (equation (8.72)).

By a similar argument, one can now derive a finite difference approximation for equation (8.63), which, for $\lambda_2\geq 0$, is given by

$$(8.73) \qquad t_{21}\left[\frac{v_{i,j+1}-v_{i,j}}{k} + \lambda_2\frac{v_{i,j}-v_{i-1,j}}{h}\right] + t_{22}\left[\frac{w_{i,j+1}-w_{i,j}}{k} + \lambda_2\frac{w_{i,j}-w_{i-1,j}}{h}\right] + d_2 = 0 ,$$

and which, for $\lambda_2<0$, is given by

$$(8.74) \qquad t_{21}\left[\frac{v_{i,j+1}-v_{i,j}}{k} + \lambda_2\frac{v_{i+1,j}-v_{i,j}}{h}\right] + t_{22}\left[\frac{w_{i,j+1}-w_{i,j}}{k} + \lambda_2\frac{w_{i+1,j}-w_{i,j}}{h}\right] + d_2 = 0 .$$

In (8.70), (8.72)-(8.74) the coefficients t, λ, d are evaluated at (x_i,y_j). Equation (8.70), or (8.72), together with equation (8.73), or (8.74), constitute a linear system of two equations in two unknowns whose associated matrix is T, which is nonsingular. The solution of this system is taken to be the required approximation $v_{i,j+1}$ and $w_{i,j+1}$ at (x_i, y_{j+1}).

The above method, called the *method of Courant, Isaacson and Rees*, can be summarized as follows. Assume, for example, that one is interested in knowing the solution of an inital value problem at a particular point, which, without loss of generality, one can assume to have coordinates $(0,b)$, $b>0$. Consider an interval $[-a,a]$ on the X axis, as indicated in Figure 8.10. Subdivide this interval into $2n$ equal parts, each of length $h=a/n$, by the R_{2n+1} set $x_i=-a+ih$, $i=0,1,2,\ldots,2n$. Subdivide the interval $[0,b]$ into n equal parts, each of length $k=b/n$, by the R_{n+1} set $y_j=jk$, $j=0,1,2,\ldots,n$. At each grid point (x_i, y_0) on the X axis define, from (8.49), $v_{i,0}$ and $w_{i,0}$ as follows

$$v_{i,0} = f_1(x_i), \qquad w_{i,0} = f_2(x_i), \qquad i=0,1,2,\ldots,2n.$$

Then, for $j=0$ and for each $i=1,2,\ldots,2n-1$, apply and solve the linear system whose first equation is either (8.70) if $\lambda_1 \geq 0$, or (8.72) if $\lambda_1 < 0$, and whose second equation is either (8.73) if $\lambda_2 \geq 0$, or (8.74) if $\lambda_2 < 0$. Once $v_{i,1}$ and $w_{i,1}$, $i=1,2,\ldots,2n-1$, have been computed, set $j=1$, and, for each $i=2,3,\ldots,2n-2$, solve a linear system to compute $v_{i,2}$ and $w_{i,2}$ for each $i=3,4,\ldots,2n-3$. Continue in the indicated fashion with $j=2,3,\ldots,n-1$, until $v_{n,n}$ and $w_{n,n}$ have been computed.

For a related grid condition for the Courant, Isaacson and Rees method, v and w can be determined only in the region of dependence of (x_n, y_n). Thus, one needs to choose the number a large enough so that the interval whose end points are $\pm b \cdot \max(|\lambda_1|, |\lambda_2|)$ is contained in $[-a,a]$, that is,

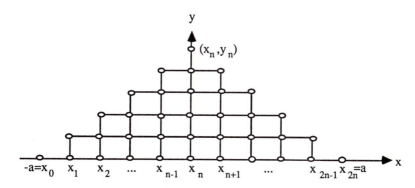

Figure 8.10

$$b \max(|\lambda_1|,|\lambda_2|) \le a \ ,$$

or, equivalently,

$$\max(|\lambda_1|,|\lambda_2|) \le \frac{nh}{nk} \ ,$$

so that

(8.75) $\qquad k \le \dfrac{h}{\max(|\lambda_1|,|\lambda_2|)} \ .$

Of course, inequality (8.75) has to be valid over the entire region of interest.

EXAMPLE. Reconsider equations (8.44) in the normal form (8.45), that is,

$$(v_y+v_x)-(w_y+w_x) = 0$$

$$(v_y-v_x)+(w_y-w_x) = 0 \ ,$$

so that $t_{11}=t_{21}=t_{22}=1$, $t_{12}=-1$, $\lambda_1=1$, $\lambda_2=-1$, $d_1=d_2=0$. Let

$$v(x,0) = 1-2x \ , \qquad w(x,0) = 1 \ .$$

Suppose one wishes to approximate the solution at (0,1). Then, let a=2, n=2, so that h=1, k=0.5, and inequality (8.75) is satisfied. Moreover, once the interval [–2,2] is subdivided into four equal parts by the R_5 set $x_i=i-2$, i=0,1,2,3,4, the given initial conditions imply

$$v_{00} = 5 \ , \qquad v_{10} = 3 \ , \qquad v_{20} = 1 \ , \qquad v_{30} = -1 \ , \qquad v_{40} = -3 \ ,$$
$$w_{00} = 1 \ , \qquad w_{10} = 1 \ , \qquad w_{20} = 1 \ , \qquad w_{30} = 1 \ , \qquad w_{40} = 1 \ .$$

Now, since $\lambda_1>0$ and $\lambda_2<0$, the differential system is discretized by using (8.70) and (8.74), which, in the present case are

$$\left[\frac{v_{i,j+1}-v_{i,j}}{0.5}+\frac{v_{i,j}-v_{i-1,j}}{1}\right]-\left[\frac{w_{i,j+1}-w_{i,j}}{0.5}+\frac{w_{i,j}-w_{i-1,j}}{1}\right]=0$$

$$\left[\frac{v_{i,j+1}-v_{i,j}}{0.5}-\frac{v_{i+1,j}-v_{i,j}}{1}\right]+\left[\frac{w_{i,j+1}-w_{i,j}}{0.5}-\frac{w_{i+1,j}-w_{i,j}}{1}\right]=0,$$

or, equivalently,

$$2v_{i,j+1}-2w_{i,j+1}=v_{i,j}+v_{i-1,j}-(w_{i,j}+w_{i-1,j})$$

$$2v_{i,j+1}+2w_{i,j+1}=v_{i,j}+v_{i+1,j}+(w_{i,j}+w_{i+1,j}).$$

Thus, for i=1 and j=0, one has

$$2v_{11}-2w_{11}=6$$

$$2v_{11}+2w_{11}=6,$$

which yields v_{11}=3, w_{11}=0. Next, for i=2 and j=0, one has

$$2v_{21}-2w_{21}=2$$

$$2v_{21}+2w_{21}=2,$$

which yields v_{21}=1, w_{21}=0. Similarly, for i=3 and j=0, one has

$$2v_{31}-2w_{31}=-2$$

$$2v_{31}+2w_{31}=-2,$$

which yields v_{31}=-1, w_{31}=0. Finally, by using the results just generated, for i=2 and j=1, one has

$$2v_{22} - 2w_{22} = 4$$

$$2v_{22} + 2w_{22} = 0 ,$$

the solution of which is $v_{22}=1$, $w_{22}=-1$, which, again, is in complete agreement with the analytical solution $v=1-2x$, $w=1-2y$.

8.9 THE LAX-WENDROFF METHOD

In gas dynamical problems, one often encounters a hyperbolic system of equations in the special form

(8.76)

$$\frac{\partial v}{\partial y} + \frac{\partial}{\partial x}[F_1(v,w)] = 0$$

$$\frac{\partial w}{\partial y} + \frac{\partial}{\partial x}[F_2(v,w)] = 0 .$$

Such a system is said to be in conservative form because F_1 and F_2 often represent such conserved quantities as mass, energy, or momentum. The Lax-Wendroff method attempts to conserve these very same quantities by using difference approximations of equations (8.76) *as they stand*, that is, *without* carrying out the indicated differentiations with respect to x. The method proceeds as follows.

If one has determined v and w at grid points (x_i, y_j) then one proceeds to the grid points at level y_{j+1} in two steps. First one determines v and w at the points whose coordinates have the form $(x_{i+1/2}, y_{j+1/2})$ by

(8.77)

$$v_{i+1/2,j+1/2} = \frac{v_{i+1,j} + v_{i,j}}{2} - \frac{k}{2h}\left[F_1(v_{i+1,j}, w_{i+1,j}) - F_1(v_{i,j}, w_{i,j})\right]$$

$$w_{i+1/2,j+1/2} = \frac{w_{i+1,j} + w_{i,j}}{2} - \frac{k}{2h}\left[F_2(v_{i+1,j}, w_{i+1,j}) - F_2(v_{i,j}, w_{i,j})\right] .$$

Then, one determines v and w at mesh points of the form (x_i, y_{j+1}) from

$$v_{i,j+1} = v_{i,j} - \frac{k}{h}\left[F_1(v_{i+1/2,j+1/2}, w_{i+1/2,j+1/2}) - F_1(v_{i-1/2,j+1/2}, w_{i-1/2,j+1/2})\right]$$

(8.78)

$$w_{i,j+1} = w_{i,j} - \frac{k}{h}\left[F_2(v_{i+1/2,j+1/2}, w_{i+1/2,j+1/2}) - F_2(v_{i-1/2,j+1/2}, w_{i-1/2,j+1/2})\right].$$

EXAMPLE. Let us consider equations (8.44), which are of the form (8.76) with $F_1(v,w)=-w$ and $F_2(v,w)=-v$. In this case (8.77) imply

(8.79)
$$v_{i+1/2,j+1/2} = \frac{v_{i+1,j}+v_{i,j}}{2} + \frac{k}{2h}(w_{i+1,j}-w_{i,j})$$

(8.80)
$$w_{i+1/2,j+1/2} = \frac{w_{i+1,j}+w_{i,j}}{2} + \frac{k}{2h}(v_{i+1,j}-v_{i,j}).$$

Similarly, at $(x_{i-1/2}, y_{j+1/2})$ one has

(8.81)
$$v_{i-1/2,j+1/2} = \frac{v_{i,j}+v_{i-1,j}}{2} + \frac{k}{2h}(w_{i,j}-w_{i-1,j})$$

(8.82)
$$w_{i-1/2,j+1/2} = \frac{w_{i,j}+w_{i-1,j}}{2} + \frac{k}{2h}(v_{i,j}-v_{i-1,j}).$$

Formulas (8.78) become

(8.83)
$$v_{i,j+1} = v_{i,j} + \frac{k}{h}(w_{i+1/2,j+1/2}-w_{i-1/2,j+1/2})$$

(8.84)
$$w_{i,j+1} = w_{i,j} + \frac{k}{h}(v_{i+1/2,j+1/2}-v_{i-1/2,j+1/2}).$$

Now, substitution of (8.80) and (8.82) into (8.83) yields

(8.85)
$$v_{i,j+1} = v_{i,j} + \frac{k}{2h}(w_{i+1,j}-w_{i-1,j}) + \frac{k^2}{2h^2}(v_{i+1,j}-2v_{i,j}+v_{i-1,j}).$$

Similarly, substitution of (8.79) and (8.81) into (8.84) yields

(8.86) $w_{i,j+1} = w_{i,j} + \dfrac{k}{2h}(v_{i+1,j} - v_{i-1,j}) + \dfrac{k^2}{2h^2}(w_{i+1,j} - 2w_{i,j} + w_{i-1,j})$,

so that the Lax-Wendroff formulas (8.77)-(8.78), in this case, have been reduced to (8.85)-(8.86). Now, let the initial conditions be

$$v(x,0) = 1 - 2x , \qquad\qquad w(x,0) = 1 .$$

In order to approximate the solution of equations (8.44) at $(0,1)$, let $n=2$, $h=1$, and consider the R_5 set $x_i = i-2$, $i=0,1,2,3,4$. On this set the given initial conditions imply

$$v_{00} = 5 , \qquad v_{10} = 3 , \qquad v_{20} = 1 , \qquad v_{30} = -1 , \qquad v_{40} = -3 ,$$
$$w_{00} = 1 , \qquad w_{10} = 1 , \qquad w_{20} = 1 , \qquad w_{30} = 1 , \qquad w_{40} = 1 .$$

Next, let $k=0.5$, so that use of (8.85) with $j=0$ and $i=1,2,3$, yields $v_{11}=3$, $v_{21}=1$, $v_{31}=-1$, while (8.86) yields $w_{11}=0$, $w_{21}=0$, $w_{31}=0$. Finally, by using the results just generated, for $j=1$ and $i=2$, (8.85) yields $v_{22}=1$, while (8.86) yields $w_{22}=-1$, in agreement with the analytical solution $v=1-2x$, $w=1-2y$.

EXERCISES

Basic Exercises

1. Show that the change of variables $\xi=x+y$, $\psi=x-y$ transforms the wave equation into

$$4u_{\xi\psi} = 0 ,$$

which is equivalent to $u_{\xi\psi}=0$.

2. Find the solution of the Cauchy problem for each of the following.

(a) $f_1 = 1$, $f_2 = -1$

(b) $f_1 = x$, $f_2 = x^2$

(c) $f_1 = x^2$, $f_2 = e^{-x^2}$.

3. Using the explicit method with h=2 and k=1, determine u(0,10) for each of the Cauchy problems in Exercise 2. Compare your numerical results with the exact solution.

4. Find the interval of dependence for each of the following points: (0,3), (1,3), (–3,3), (7,8), (–7,–1), (–3,6).

5. Find the equations of the characteristics through each point of Exercise 4.

6. Given the initial-boundary problem for $u_{xx}-u_{yy}=0$, with

$$a = 1 , \quad g_1(y) = e^{-y}, \quad g_2(y) = 2+e^{1-y}, \quad f_1(x) = 2x+e^x, \quad f_2(x) = -e^x,$$

find the numerical solution at y=5, by the explicit method. Compare your results with the exact solution $u=2x+e^{x-y}$. (Notice that the choice of the grid sizes is yours.)

7. Repeat Exercise 6 by using the implicit method.

8. Given the initial-boundary problem for $u_{xx}-u_{yy}=0$, with

$$a = 1 , \quad g_1(y) = 0 , \quad g_2(y) = e^{-y}, \quad f_1(x) = x , \quad f_2(x) = x^2,$$

find the numerical solution at y=5. (Notice that the choice of the method is yours.)

9. Determine which of the following systems are hyperbolic. For those which are, find the characteristics and a normal form.

(a) $v_x+w_x+v_y-w_y = 0$

         $v_x+w_x-v_y-w_y = 0$

(b) $v_x - w_x + v_y - w_y = 0$

       $v_x - w_x - v_y - w_y = 0$

(c) $v_x - w_x + v_y - w_y = 0$

       $v_x + w_x - v_y - w_y = 0$

(d) $v_x + w_y = 0$

       $v_y + w_x = 0$.

10. Find a normal form, different from the one given in Section 8.6, for the one dimensional, isentropic flow equations (8.47).

11. Derive the particular formulas for the method of characteristics for each of the following hyperbolic systems.

(a) $v_x - w_x + v_y - w_y = 0$

       $v_x + w_x - v_y - w_y = 0$

(b) $v_x + v_y = w$

       $w_x - w_y = v$

(c) $v_x - w_x + 10(v_y - w_y) = 0$

       $v_x + w_x - v_y - w_y = 0$

(d) $v_x - v_y = w$

       $w_x - 2w_y = v$.

12. Derive the particular formulas for the method of Courant, Isaacson and Rees, and the corresponding grid condition for each system of Exercise 11.

13. For the one dimensional, isentropic flow equations (8.47) with c=1100, let v(x,0)=x, w(x,0)=x². Approximate v(1,1) and w(1,1) using h=0.5, k=0.25.

14. Given $F_1=v^2+w^2$; $F_2=1$; $f_1(x)=0$, $x<0$; $f_1(x)=1$, $x\geq0$; $f_2(x)=0$, approximate $v(0,1)$, $w(0,1)$ by the Lax-Wendroff method with $h=k=0.5$.

Supplementary Exercises

15. Consider the initial and boundary conditions

$$u(x,0) = x , \qquad 0\leq x\leq1$$
$$u_y(x,0) = -x , \qquad 0\leq x\leq1$$
$$u(0,y) = 0 , \qquad y\geq0$$
$$u(1,y) = e^{-y}, \qquad y\geq0 .$$

Find a numerical solution of the initial-boundary problem with b=10 for

$$u_{xx}-u_{yy}-u_y = 0$$

and compare your results with the exact solution $u=xe^{-y}$.

16. Consider the initial and boundary conditions

$$u(x,0) = \frac{1}{1+x}$$
$$u_y(x,0) = -\frac{1}{1+x}$$
$$u(0,y) = \frac{1}{1+y}$$
$$u(1,y) = \frac{1}{2(1+y)} .$$

Find a numerical solution of the initial-boundary problem with b=10 for

$$u_{xx}-u_{yy} = 2(x-y)(x+y+2)u^3$$

and compare your results with the exact solution $u=1/[(1+x)(1+y)]$.

17. Consider the isentropic flow equations (8.48) with $c=1$. Let

$$v(x,0) = 0 , \qquad w(x,0) = x+1 .$$

Find the numerical solution by the method of characteristics with $n=4$ on the R_5 set $x_0=0$, $x_1=1$, $x_2=2$, $x_3=3$, $x_4=4$.

18. Consider the isentropic flow equations (8.48) with $c=1$. Let

$$v(x,0) = 0 , \qquad w(x,0) = x+1 .$$

Find the numerical solution at $(5,10)$ by the method of Courant, Isaacson and Rees.

19. For $F_1(v,w)=-w$, $F_2(v,w)=-v$, find an approximate solution at $(0,10)$ by the Lax-Wendroff method for system (8.76) when the initial conditions are $v(x,0)=1-2x$, $w(x,0)=1$.

*9

The Navier-Stokes Equations

9.1 INTRODUCTION

When dealing with applied problems, sometimes theory and practice are often worlds apart. In this final chapter we will confront an initial-boundary problem for the fundamental continuous equations of all of fluid dynamics, the Navier-Stokes equations. The nonlinear behavior is of such complexity that classical existence and uniqueness theory is not available. And yet, the equations are of such fundamental importance that related numerical methods are being developed constantly and then evaluated by comparing computer results with laboratory experimental results. The niceties developed in Chapters 1-8 will be of value in our study, but will not "solve" our problem without the employment of some additional problematic devices. The discussion which follows will provide a view of an area of numerical analysis which is quasi-mathematical and quasi-experimental.

9.2 THE GOVERNING DYNAMICAL EQUATIONS

Dynamical equations of fluid flow can be derived from the physical principles of conservation of linear momentum and mass. We will consider here, in detail, only fluid flow in a closed, two-dimensional domain Ω whose boundary Γ is piecewise regular, though the ideas and techniques extend dimensionwise in a natural way.

For an incompressible and homogeneous fluid at constant temperature, the principle of conservation of *linear momentum* in the X and Y directions, respectively, yields

$$(9.1) \qquad \frac{\partial u}{\partial t} + u\frac{\partial u}{\partial x} + v\frac{\partial u}{\partial y} = -\frac{\partial p}{\partial x} + v\left(\frac{\partial^2 u}{\partial x^2} + \frac{\partial^2 u}{\partial y^2}\right)$$

$$(9.2) \qquad \frac{\partial v}{\partial t} + u\frac{\partial v}{\partial x} + v\frac{\partial v}{\partial y} = -\frac{\partial p}{\partial y} + v\left(\frac{\partial^2 v}{\partial x^2} + \frac{\partial^2 v}{\partial y^2}\right),$$

where t denotes the time; u(x,y,t) and v(x,y,t) are the velocity components in the X and Y directions, respectively; p(x,y,t) is the pressure; and v is a nonnegative viscosity coefficient, which is assumed to be constant.

The principle of conservation of *mass* for an incompressible fluid reduces to the following incompressibility condition:

(9.3) $$\frac{\partial u}{\partial x} + \frac{\partial v}{\partial y} = 0 \ .$$

Equations (9.1)-(9.3) will be called the *Navier-Stokes* equations, though they are, more precisely, the two-dimensional Navier-Stokes equations. They form a system of three differential equations in the three unknown functions $u(x,y,t)$, $v(x,y,t)$ and $p(x,y,t)$.

An *initial-boundary* problem for the Navier-Stokes equations is one in which one is given four continuous functions $u_0(x,y)$, $v_0(x,y)$, $u_\Gamma(x,y,t)$ and $v_\Gamma(x,y,t)$ and one is asked to find three functions $u(x,y,t)$, $v(x,y,t)$ and $p(x,y,t)$ which are continuous for $(x,y)\in(\Omega\cup\Gamma)$, $t\geq0$; satisfy (9.1)-(9.3) for $(x,y)\in\Omega$, $t>0$; and which satisfy the initial and boundary conditions

(9.4) $u(x,y,0) = u_0(x,y)$,

             $v(x,y,0) = v_0(x,y)$, $(x,y)\in\Omega$ (initial conditions),

(9.5) $u(x,y,t) = u_\Gamma(x,y,t)$,

             $v(x,y,t) = v_\Gamma(x,y,t)$, $(x,y)\in\Gamma$, $t\geq0$ (boundary conditions).

Since the Navier-Stokes equations cannot be classified as being elliptic, parabolic or hyperbolic, none of the numerical methods developed thus far applies directly to determine an approximate solution of the given initial-boundary problem. However, the development of a special method can and will be facilitated by using concepts from the last three chapters.

9.3 THE FINITE DIFFERENCE EQUATIONS

In order to solve the initial-boundary problem for the Navier-Stokes equations numerically, assume, for illustrative simplicity, that Γ is a rectangle with vertices $(0,0)$, $(a,0)$, (a,b), $(0,b)$, with $a>0$, $b>0$. Let Ω is the interior of Γ. As indicated in Figure 9.1, the interval $[0,a]$ is now subdivided into n equal parts, each of length $\Delta x=a/n$, by an R_{n+1} set $x_i=i\Delta x$, $i=0,1,2,\ldots,n$, and the interval $[0,b]$ is subdivided into m equal parts, each of length $\Delta y=b/m$, by an R_{m+1} set $y_j=j\Delta y$, $j=0,1,2,\ldots,m$. Thereby, the region Ω is subdivided naturally into a set of nm rectangular *cells* of width Δx and height Δy, as shown in Figure 9.1. Each cell is labelled at its center by the index pair (i,j), where $i=1,2,\ldots,n$, and

$j=1,2,\ldots,m$, and (x_i,y_j) is the upper-right vertex of the cell. The discrete field variables u, v and p are defined at the locations of cell (i,j) which are shown in Figure 9.2: the horizontal velocity u is defined at the center of each vertical side of the cell; the vertical velocity v is defined at the center of each horizontal side; and the pressure p is defined at the cell center.

Denote the time step by Δt. For consistency with current journal literature, we will use the notation shown in Figure 9.2 and will also denote u, v, p at time $t_k=k\Delta t$, by u^k, v^k, p^k, respectively. Hence, at the initial time $t_0=0$, the initial conditions (9.4), imply

(9.6) $$u^0_{i+1/2,j} = u_0(x_i,y_{j-1/2}) , \qquad i=0,1,2,\ldots,n, \qquad j=1,2,\ldots,m,$$

(9.7) $$v^0_{i,j+1/2} = v_0(x_{i-1/2},y_j) , \qquad i=1,2,\ldots,n, \qquad j=0,1,2,\ldots,m.$$

Since no initial conditions have been given for the pressure, $p^0_{i,j}$ must be approximated from the differential equations. We will show how to do this presently.

If the discrete velocity field is known at time step t_k, from (9.5) one can set tangential velocity boundary conditions at time t_k and normal velocity boundary conditions at time t_{k+1}. This will be done as follows. As indicated in Figure 9.3, since the discrete u velocities are not located at the top and at the bottom boundaries, boundary conditions on the tangential velocity are taken to be

$$\frac{u^k_{i+1/2,0}+u^k_{i+1/2,1}}{2} = u_\Gamma(x_i,0,t_k) , \qquad i=1,2,\ldots,n-1,$$

$$\frac{u^k_{i+1/2,m}+u^k_{i+1/2,m+1}}{2} = u_\Gamma(x_i,b,t_k) , \qquad i=1,2,\ldots,n-1,$$

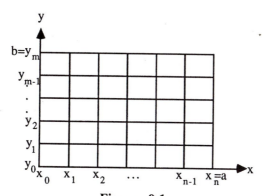

Figure 9.1

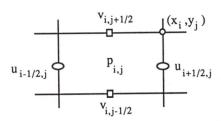

Figure 9.2

from which one has

$$u^k_{i+1/2,0} = 2u_\Gamma(x_i,0,t_k) - u^k_{i+1/2,1} \, ,$$

(9.8) $i=1,2,\ldots,n-1.$

$$u^k_{i+1/2,m+1} = 2u_\Gamma(x_i,b,t_k) - u^k_{i+1/2,m} \, ,$$

Note, immediately, that $u_{i+1/2,0}$ and $u_{i+1/2,m+1}$ have not been defined by (9.6), so that, in effect, these are now being defined by (9.8). Similarly, for the tangential velocity boundary conditions at the left and at the right boundaries, set

$$v^k_{0,j+1/2} = 2v_\Gamma(0,y_j,t_k) - v^k_{1,j+1/2} \, ,$$

(9.9) $j=1,2,\ldots,m-1.$

$$v^k_{n+1,j+1/2} = 2v_\Gamma(a,y_j,t_k) - v^k_{n,j+1/2} \, ,$$

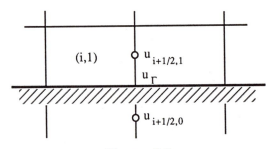

Figure 9.3

The normal velocity boundary conditions at time t_{k+1} are taken to be

(9.10) $\qquad u_{1/2,j}^{k+1} = u_\Gamma(0,y_{j-1/2},t_{k+1})\,, \qquad u_{n+1/2,j}^{k+1} = u_\Gamma(a,y_{j-1/2},t_{k+1})\,, \qquad j=1,2,\ldots,m,$

(9.11) $\qquad v_{i,1/2}^{k+1} = v_\Gamma(x_{i-1/2},0,t_{k+1})\,, \qquad v_{i,m+1/2}^{k+1} = v_\Gamma(x_{i-1/2},b,t_{k+1})\,, \qquad i=1,2,\ldots,n.$

Let us show now how one can approximate the pressure at the discrete times t_k, and the velocities at the discrete times t_{k+1}, $k=0,1,2,\ldots$. At the center of the right-vertical side of a cell, approximate the differential equation (9.1) with a corresponding explicit, space centered, finite difference equation as follows:

$$\frac{u_{i+1/2,j}^{k+1}-u_{i+1/2,j}^{k}}{\Delta t} + u_{i+1/2,j}^{k}\frac{u_{i+3/2,j}^{k}-u_{i-1/2,j}^{k}}{2\Delta x} + \bar{v}_{i+1/2,j}^{k}\frac{u_{i+1/2,j+1}^{k}-u_{i+1/2,j-1}^{k}}{2\Delta y}$$

$$= -\frac{p_{i+1,j}^{k}-p_{i,j}^{k}}{\Delta x} + \nu\Big(\frac{u_{i+3/2,j}^{k}-2u_{i+1/2,j}^{k}+u_{i-1/2,j}^{k}}{(\Delta x)^2} + \frac{u_{i+1/2,j+1}^{k}-2u_{i+1/2,j}^{k}+u_{i+1/2,j-1}^{k}}{(\Delta y)^2}\Big)\,,$$

so that

(9.12) $\qquad u_{i+1/2,j}^{k+1} = Fu_{i+1/2,j}^{k} - \Delta t\frac{p_{i+1,j}^{k}-p_{i,j}^{k}}{\Delta x}\,, \qquad i=1,2,\ldots,n-1, \qquad j=1,2,\ldots,m,$

where the finite difference operator F is defined by

$$Fu_{i+1/2,j}^{k} = u_{i+1/2,j}^{k} - \Delta t\left[u_{i+1/2,j}^{k}\frac{u_{i+3/2,j}^{k}-u_{i-1/2,j}^{k}}{2\Delta x} + \bar{v}_{i+1/2,j}^{k}\frac{u_{i+1/2,j+1}^{k}-u_{i+1/2,j-1}^{k}}{2\Delta y}\right]$$

$$+ \nu\,\Delta t\Big(\frac{u_{i+3/2,j}^{k}-2u_{i+1/2,j}^{k}+u_{i-1/2,j}^{k}}{(\Delta x)^2} + \frac{u_{i+1/2,j+1}^{k}-2u_{i+1/2,j}^{k}+u_{i+1/2,j-1}^{k}}{(\Delta y)^2}\Big)\,.$$

Similarly, at the center of the upper-horizontal side of a cell, a finite difference approximation to the differential equation (9.2) yields

(9.13) $v_{i,j+1/2}^{k+1} = Gv_{i,j+1/2}^{k} - \Delta t \dfrac{p_{i,j+1}^{k} - p_{i,j}^{k}}{\Delta y}$, $i=1,2,\ldots,n$, $j=1,2,\ldots,m-1$,

where the finite difference operator G is

$$Gv_{i,j+1/2}^{k} = v_{i,j+1/2}^{k} - \Delta t \left[\bar{u}_{i,j+1/2}^{k} \frac{v_{i+1,j+1/2}^{k} - v_{i-1,j+1/2}^{k}}{2\Delta x} + v_{i,j+1/2}^{k} \frac{v_{i,j+3/2}^{k} - v_{i,j-1/2}^{k}}{2\Delta y} \right]$$

$$+ v \Delta t \left(\frac{v_{i+1,j+1/2}^{k} - 2v_{i,j+1/2}^{k} + v_{i-1,j+1/2}^{k}}{(\Delta x)^2} + \frac{v_{i,j+3/2}^{k} - 2v_{i,j+1/2}^{k} + v_{i,j-1/2}^{k}}{(\Delta y)^2} \right).$$

The special terms $\bar{u}$ and $\bar{v}$, above, are defined as follows. In the formula for $Fu_{i+1/2,j}^{k}$, the discrete v velocity is not defined at the center of a vertical side of the computational cell. A value for $v_{i+1/2,j}^{k}$ is then taken to be the average at four neighboring grid points at which v^{k} is known, that is,

$$\bar{v}_{i+1/2,j}^{k} = \frac{v_{i,j+1/2}^{k} + v_{i,j-1/2}^{k} + v_{i+1,j+1/2}^{k} + v_{i+1,j-1/2}^{k}}{4}.$$

Similarly, in the formula for $Gv_{i,j+1/2}^{k}$, a value for $u_{i,j+1/2}^{k}$ is defined as follows:

$$\bar{u}_{i,j+1/2}^{k} = \frac{u_{i+1/2,j}^{k} + u_{i-1/2,j}^{k} + u_{i+1/2,j+1}^{k} + u_{i-1/2,j+1}^{k}}{4}.$$

Note now that, although an explicit difference approximation has been used to discretize equations (9.1) and (9.2), the finite difference formulas (9.12), (9.13) cannot, as yet, be used to determine the new velocity at time t_{k+1} because the pressure field at time t_{k} is still unknown. In order to determine $p_{i,j}^{k}$ we require that the new velocity field satisfies, in each cell, the finite difference analogue of the incompressibility condition (9.3), that is,

(9.14) $\dfrac{u_{i+1/2,j}^{k+1} - u_{i-1/2,j}^{k+1}}{\Delta x} + \dfrac{v_{i,j+1/2}^{k+1} - v_{i,j-1/2}^{k+1}}{\Delta y} = 0$, $i=1,2,\ldots,n$, $j=1,2,\ldots,m$.

By inserting the known boundary conditions (9.10)-(9.11) into (9.14), the set of equations (9.12), (9.13) and (9.14) yields a linear algebraic system of (3nm–n–m) equations with nm unknown pressure values at time t_k, $m(n-1)$ unknown u-velocities at time t_{k+1}, and $n(m-1)$ unknown v-velocities at time t_{k+1}. A solution of this system can then be used to generate a numerical solution at the next time step.

9.4 PRESSURE APPROXIMATION

The finite difference method discussed in the previous section is a very general one. However, an efficient computer algorithm must be devised for solving the large linear system of $(3nm-n-m)$ equations. It is this problem that we begin to consider now.

Let (i,j) be a typical computational cell, so that the incompressibility condition (9.14) at (i,j) is

$$(9.15) \qquad \frac{u_{i+1/2,j}^{k+1} - u_{i-1/2,j}^{k+1}}{\Delta x} + \frac{v_{i,j+1/2}^{k+1} - v_{i,j-1/2}^{k+1}}{\Delta y} = 0 \, .$$

Now, if (i,j) is an interior cell, that is, a cell which has no sides in common with the boundary, equation (9.12) implies

$$(9.16) \qquad u_{i+1/2,j}^{k+1} = Fu_{i+1/2,j}^{k} - \Delta t \frac{p_{i+1,j}^{k} - p_{i,j}^{k}}{\Delta x} \, , \qquad u_{i-1/2,j}^{k+1} = Fu_{i-1/2,j}^{k} - \Delta t \frac{p_{i,j}^{k} - p_{i-1,j}^{k}}{\Delta x} \, .$$

Similarly, since the top and the bottom sides of cell (i,j) are not on the boundary, equation (9.13) implies

$$(9.17) \qquad v_{i,j+1/2}^{k+1} = Gv_{i,j+1/2}^{k} - \Delta t \frac{p_{i,j+1}^{k} - p_{i,j}^{k}}{\Delta y} \, , \qquad v_{i,j-1/2}^{k+1} = Gv_{i,j-1/2}^{k} - \Delta t \frac{p_{i,j}^{k} - p_{i,j-1}^{k}}{\Delta y} \, .$$

Now, substitution of (9.16) and (9.17) into (9.15), yields

$$(9.18) \qquad \Delta t \Big[\frac{p_{i+1,j}^{k} - 2p_{i,j}^{k} + p_{i-1,j}^{k}}{(\Delta x)^2} + \frac{p_{i,j+1}^{k} - 2p_{i,j}^{k} + p_{i,j-1}^{k}}{(\Delta y)^2} \Big] = \frac{Fu_{i+1/2,j}^{k} - Fu_{i-1/2,j}^{k}}{\Delta x}$$

$$+ \frac{Gv_{i,j+1/2}^{k} - Gv_{i,j-1/2}^{k}}{\Delta y} \, , \qquad i=2,3,\dots,n-1, \qquad j=2,3,\dots,m-1.$$

For the rows of cells adjacent to the bottom boundary, but not the corner cells, substitution of (9.16) and the first of (9.17) into (9.15) yields

$$(9.19) \quad \Delta t\left[\frac{p^k_{i+1,1}-2p^k_{i,1}+p^k_{i-1,1}}{(\Delta x)^2}+\frac{p^k_{i,2}-p^k_{i,1}}{(\Delta y)^2}\right]=\frac{Fu^k_{i+1/2,1}-Fu^k_{i-1/2,1}}{\Delta x}+\frac{Gv^k_{i,3/2}-v^{k+1}_{i,1/2}}{\Delta y},$$

$$i=2,3,\ldots,n-1.$$

For the row of cells adjacent to the top boundary, but not the corner cells, substitution of (9.16) and the second of (9.17) into (9.15) yields

$$(9.20) \quad \Delta t\left[\frac{p^k_{i+1,m}-2p^k_{i,m}+p^k_{i-1,m}}{(\Delta x)^2}+\frac{p^k_{i,m-1}-p^k_{i,m}}{(\Delta y)^2}\right]=\frac{Fu^k_{i+1/2,m}-Fu^k_{i-1/2,m}}{\Delta x}+\frac{v^{k+1}_{i,m+1/2}-Gv^k_{i,m-1/2}}{\Delta y},$$

$$i=2,3,\ldots,n-1.$$

Similarly, at the columns of cells adjacent to the left and adjacent to the right boundaries, but not the corner cells, one obtains, respectively,

$$(9.21) \quad \Delta t\left[\frac{p^k_{2,j}-p^k_{1,j}}{(\Delta x)^2}+\frac{p^k_{1,j+1}-2p^k_{1,j}+p^k_{1,j-1}}{(\Delta y)^2}\right]=\frac{Fu^k_{3/2,j}-u^{k+1}_{1/2,j}}{\Delta x}+\frac{Gv^k_{1,j+1/2}-Gv^k_{1,j-1/2}}{\Delta y},$$

$$(9.22) \quad \Delta t\left[\frac{p^k_{n-1,j}-p^k_{n,j}}{(\Delta x)^2}+\frac{p^k_{n,j+1}-2p^k_{n,j}+p^k_{n,j-1}}{(\Delta y)^2}\right]=\frac{u^{k+1}_{n+1/2,j}-Fu^k_{n-1/2,j}}{\Delta x}+\frac{Gv^k_{n,j+1/2}-Gv^k_{n,j-1/2}}{\Delta y},$$

$$j=2,3,\ldots,m-1.$$

Finally, at the four corner cells, one gets

$$(9.23) \quad \Delta t\left[\frac{p^k_{2,1}-p^k_{1,1}}{(\Delta x)^2}+\frac{p^k_{1,2}-p^k_{1,1}}{(\Delta y)^2}\right]=\frac{Fu^k_{3/2,1}-u^{k+1}_{1/2,1}}{\Delta x}+\frac{Gv^k_{1,3/2}-v^{k+1}_{1,1/2}}{\Delta y},$$

$$(9.24) \quad \Delta t\left[\frac{p^k_{2,m}-p^k_{1,m}}{(\Delta x)^2}+\frac{p^k_{1,m-1}-p^k_{1,m}}{(\Delta y)^2}\right]=\frac{Fu^k_{3/2,m}-u^{k+1}_{1/2,m}}{\Delta x}+\frac{v^{k+1}_{1,m+1/2}-Gv^k_{1,m-1/2}}{\Delta y},$$

$$(9.25) \quad \Delta t\left[\frac{p_{n-1,1}^{k}-p_{n,1}^{k}}{(\Delta x)^2}+\frac{p_{n,2}^{k}-p_{n,1}^{k}}{(\Delta y)^2}\right]=\frac{u_{n+1/2,1}^{k+1}-Fu_{n-1/2,1}^{k}}{\Delta x}+\frac{Gv_{n,3/2}^{k}-v_{n,1/2}^{k+1}}{\Delta y},$$

$$(9.26) \quad \Delta t\left[\frac{p_{n-1,m}^{k}-p_{n,m}^{k}}{(\Delta x)^2}+\frac{p_{n,m-1}^{k}-p_{n,m}^{k}}{(\Delta y)^2}\right]=\frac{u_{n+1/2,m}^{k+1}-Fu_{n-1/2,m}^{k}}{\Delta x}+\frac{v_{n,m+1/2}^{k+1}-Gv_{n,m-1/2}^{k}}{\Delta y}.$$

Now, by using the known values for the velocity at time t_k and the known normal velocity boundary values at t_{k+1}, the right-hand sides of equations (9.18)-(9.26) *are all known*. These equations constitute, then, a linear algebraic system of nm equations in the nm unknown pressure values p_{ij}^{k}, i=1,2,...,n, j=1,2,...,m. Note, however, that the coefficient matrix of system (9.18)-(9.26) is *singular*. In fact, as in the continuous differential problem, if p_{ij}^{k}, i=1,2,...,n, j=1,2,...,m, is a solution of (9.18)-(9.26) and c is an arbitrary constant, then $p_{ij}^{k}+c$ is also a solution of (9.18)-(9.26). Numerically, the approximated pressure will be determined up to an arbitrary additive constant which *does not affect the velocity field* because this constant cancels in the finite difference equations (9.12)-(9.13). Thus, once a value for the pressure is found for each cell, the corresponding velocity field is determined *uniquely* from (9.12)-(9.13). We will then next organize all the above formulas and procedures into a convenient, efficient, and compact algorithm.

9.5 SOLUTION ALGORITHM

Consider first the way one determines an approximate pressure field. Interestingly enough, as will be shown, the velocity field will be determined simultaneously.

Let us solve system (9.18)-(9.26) with a special form of the generalized Newton's method, which, even in this singular case, can be shown to be convergent [Forsythe and Wasow (1960)] for any initial guess and for ω in the range $0<\omega<2$. Specifically, in sweeping the cells consecutively from left to right and from bottom to top, at any *interior* computational cell the generalized Newton's formula for p_{ij}^{k} is, from (9.18),

$$(9.27) \quad (p_{i,j}^{k})^{(r+1)}=(p_{i,j}^{k})^{(r)}-\frac{\omega}{-2\left[\frac{1}{(\Delta x)^2}+\frac{1}{(\Delta y)^2}\right]}\left[\frac{(p_{i+1,j}^{k})^{(r)}-2(p_{i,j}^{k})^{(r)}+(p_{i-1,j}^{k})^{(r+1)}}{(\Delta x)^2}\right.$$

$$\left.+\frac{(p_{i,j+1}^{k})^{(r)}-2(p_{i,j}^{k})^{(r)}+(p_{i,j-1}^{k})^{(r+1)}}{(\Delta y)^2}-\frac{Fu_{i+1/2,j}^{k}-Fu_{i-1/2,j}^{k}}{(\Delta t)\Delta x}-\frac{Gv_{i,j+1/2}^{k}-Gv_{i,j-1/2}^{k}}{(\Delta t)\Delta y}\right],$$

or, equivalently,

(9.28) $(p_{i,j}^k)^{(r+1)} = (p_{i,j}^k)^{(r)} - \dfrac{\omega}{2\Delta t\left[\dfrac{1}{(\Delta x)^2} + \dfrac{1}{(\Delta y)^2}\right]} \times$

$$\times \left\{ \dfrac{\left[Fu_{i+1/2,j}^k - \Delta t\, \dfrac{(p_{i+1,j}^k)^{(r)} - (p_{i,j}^k)^{(r)}}{\Delta x}\right] - \left[Fu_{i-1/2,j}^k - \Delta t\, \dfrac{(p_{i,j}^k)^{(r)} - (p_{i-1,j}^k)^{(r+1)}}{\Delta x}\right]}{\Delta x} \right.$$

$$\left. + \dfrac{\left[Gv_{i,j+1/2}^k - \Delta t\, \dfrac{(p_{i,j+1}^k)^{(r)} - (p_{i,j}^k)^{(r)}}{\Delta y}\right] - \left[Gv_{i,j-1/2}^k - \Delta t\, \dfrac{(p_{i,j}^k)^{(r)} - (p_{i,j-1}^k)^{(r+1)}}{\Delta y}\right]}{\Delta y} \right\} .$$

Now, in cell (i,j), define the following iterative formulas for equations (9.16)-(9.17):

(9.29) $(u_{i+1/2,j}^{k+1})^{(r)} = Fu_{i+1/2,j}^k - \Delta t\, \dfrac{(p_{i+1,j}^k)^{(r)} - (p_{i,j}^k)^{(r)}}{\Delta x}$

(9.30) $(u_{i-1/2,j}^{k+1})^{(r+1/2)} = Fu_{i-1/2,j}^k - \Delta t\, \dfrac{(p_{i,j}^k)^{(r)} - (p_{i-1,j}^k)^{(r+1)}}{\Delta x}$

(9.31) $(v_{i,j+1/2}^{k+1})^{(r)} = Gv_{i,j+1/2}^k - \Delta t\, \dfrac{(p_{i,j+1}^k)^{(r)} - (p_{i,j}^k)^{(r)}}{\Delta y}$

(9.32) $(v_{i,j-1/2}^{k+1})^{(r+1/2)} = Gv_{i,j-1/2}^k - \Delta t\, \dfrac{(p_{i,j}^k)^{(r)} - (p_{i,j-1}^k)^{(r+1)}}{\Delta y} .$

Substitution of (9.29)-(9.32) into (9.28) yields

$$(p_{i,j}^k)^{(r+1)} = (p_{i,j}^k)^{(r)} - \omega \dfrac{\dfrac{(u_{i+1/2,j}^{k+1})^{(r)} - (u_{i-1/2,j}^{k+1})^{(r+1/2)}}{\Delta x} + \dfrac{(v_{i,j+1/2}^{k+1})^{(r)} - (v_{i,j-1/2}^{k+1})^{(r+1/2)}}{\Delta y}}{2\Delta t\left[\dfrac{1}{(\Delta x)^2} + \dfrac{1}{(\Delta y)^2}\right]} ,$$

or, equivalently,

(9.33) $(p_{i,j}^k)^{(r+1)} = (p_{i,j}^k)^{(r)} + (\delta p_{i,j}^k)^{(r)},$

where

(9.34) $(\delta p_{i,j}^k)^{(r)} = -\omega \dfrac{\dfrac{(u_{i+1/2,j}^{k+1})^{(r)}-(u_{i-1/2,j}^{k+1})^{(r+1/2)}}{\Delta x} + \dfrac{(v_{i,j+1/2}^{k+1})^{(r)}-(v_{i,j-1/2}^{k+1})^{(r+1/2)}}{\Delta y}}{2\Delta t\left[\dfrac{1}{(\Delta x)^2} + \dfrac{1}{(\Delta y)^2}\right]}.$

After $(p_{i,j}^k)^{(r+1)}$ has been calculated, (9.29)-(9.32) can be updated by including this new result to yield

(9.35) $(u_{i+1/2,j}^{k+1})^{(r+1/2)} = Fu_{i+1/2,j}^k - \Delta t \dfrac{(p_{i+1,j}^k)^{(r)}-(p_{i,j}^k)^{(r+1)}}{\Delta x}$

(9.36) $(u_{i-1/2,j}^{k+1})^{(r+1)} = Fu_{i-1/2,j}^k - \Delta t \dfrac{(p_{i,j}^k)^{(r+1)}-(p_{i-1,j}^k)^{(r+1)}}{\Delta x}$

(9.37) $(v_{i,j+1/2}^{k+1})^{(r+1/2)} = Gv_{i,j+1/2}^k - \Delta t \dfrac{(p_{i,j+1}^k)^{(r)}-(p_{i,j}^k)^{(r+1)}}{\Delta y}$

(9.38) $(v_{i,j-1/2}^{k+1})^{(r+1)} = Gv_{i,j-1/2}^k - \Delta t \dfrac{(p_{i,j}^k)^{(r+1)}-(p_{i,j-1}^k)^{(r+1)}}{\Delta y}.$

Elimination of $Fu_{i+1/2,j}^k$, $Fu_{i-1/2,j}^k$, $Gv_{i,j+1/2}^k$ and $Gv_{i,j-1/2}^k$ between (9.29)-(9.32) and (9.35)-(9.38) then yields

(9.39) $(u_{i+1/2,j}^{k+1})^{(r+1/2)} = (u_{i+1/2,j}^{k+1})^{(r)} + \dfrac{\Delta t}{\Delta x}(\delta p_{i,j}^k)^{(r)}$

(9.40) $(u_{i-1/2,j}^{k+1})^{(r+1)} = (u_{i-1/2,j}^{k+1})^{(r+1/2)} - \dfrac{\Delta t}{\Delta x}(\delta p_{i,j}^k)^{(r)}$

$$(9.41) \qquad (v_{i,j+1/2}^{k+1})^{(r+1/2)} = (v_{i,j+1/2}^{k+1})^{(r)} + \frac{\Delta t}{\Delta y}(\delta p_{i,j}^{k})^{(r)}$$

$$(9.42) \qquad (v_{i,j-1/2}^{k+1})^{(r+1)} = (v_{i,j-1/2}^{k+1})^{(r+1/2)} - \frac{\Delta t}{\Delta y}(\delta p_{i,j}^{k})^{(r)}.$$

The formulas (9.33), (9.34) and (9.39)-(9.42) are the formulas for p^k, u^{k+1} and v^{k+1} on the interior cells. We will show now how to continue so that these are *also* the formulas on the boundary cells.

At any noncorner cell, for example, which is adjacent to the bottom boundary, let us use the overrelaxation factor ω^* given by

$$(9.43) \qquad \omega^* = \omega \left[\frac{2}{(\Delta x)^2} + \frac{2}{(\Delta y)^2}\right]^{-1} \left[\frac{2}{(\Delta x)^2} + \frac{1}{(\Delta y)^2}\right].$$

Since $0<\omega<2$, one has $0<\omega^*<\omega<2$. Thus, from (9.19),

$$(9.44) \qquad (p_{i,1}^{k})^{(r+1)} = (p_{i,1}^{k})^{(r)} - \frac{\omega^*}{-\left[\frac{2}{(\Delta x)^2} + \frac{1}{(\Delta y)^2}\right]} \left[\frac{(p_{i+1,1}^{k})^{(r)} - 2(p_{i,1}^{k})^{(r)} + (p_{i-1,1}^{k})^{(r+1)}}{(\Delta x)^2}\right.$$
$$\left. + \frac{(p_{i,2}^{k})^{(r)} - (p_{i,1}^{k})^{(r)}}{(\Delta y)^2} - \frac{Fu_{i+1/2,1}^{k} - Fu_{i-1/2,1}^{k}}{(\Delta t)\Delta x} - \frac{Gv_{i,3/2}^{k} - v_{i,1/2}^{k+1}}{(\Delta t)\Delta y}\right],$$

or, equivalently,

$$(9.45) \qquad (p_{i,1}^{k})^{(r+1)} = (p_{i,1}^{k})^{(r)} - \frac{\omega^*}{\Delta t\left[\frac{2}{(\Delta x)^2} + \frac{1}{(\Delta y)^2}\right]} \left\{\frac{\left[Gv_{i,3/2}^{k} - \Delta t\frac{(p_{i,2}^{k})^{(r)} - (p_{i,1}^{k})^{(r)}}{\Delta y}\right] - v_{i,1/2}^{k+1}}{\Delta y}\right.$$
$$\left. + \frac{\left[Fu_{i+1/2,1}^{k} - \Delta t\frac{(p_{i+1,1}^{k})^{(r)} - (p_{i,1}^{k})^{(r)}}{\Delta x}\right] - \left[Fu_{i-1/2,1}^{k} - \Delta t\frac{(p_{i,1}^{k})^{(r)} - (p_{i-1,1}^{k})^{(r+1)}}{\Delta x}\right]}{\Delta x}\right\}.$$

But, substitution of (9.29)-(9.31) into (9.45) and use of (9.43) yields

$$(p_{i,1}^k)^{(r+1)} = (p_{i,1}^k)^{(r)} - \omega \frac{\frac{(u_{i+1/2,1}^{k+1})^{(r)} - (u_{i-1/2,1}^{k+1})^{(r+1/2)}}{\Delta x} + \frac{(v_{i,3/2}^{k+1})^{(r)} - v_{i,1/2}^{k+1}}{\Delta y}}{2\Delta t \left[\frac{1}{(\Delta x)^2} + \frac{1}{(\Delta y)^2} \right]},$$

or, equivalently,

(9.46) $(p_{i,1}^k)^{(r+1)} = (p_{i,1}^k)^{(r)} + (\delta p_{i,1}^k)^{(r)},$

where

(9.47) $$(\delta p_{i,1}^k)^{(r)} = - \omega \frac{\frac{(u_{i+1/2,1}^{k+1})^{(r)} - (u_{i-1/2,1}^{k+1})^{(r+1/2)}}{\Delta x} + \frac{(v_{i,3/2}^{k+1})^{(r)} - v_{i,1/2}^{k+1}}{\Delta y}}{2\Delta t \left[\frac{1}{(\Delta x)^2} + \frac{1}{(\Delta y)^2} \right]}.$$

Then, if (9.29)-(9.31) are updated by including (9.46) there result, again, (9.35)-(9.37), or, equivalently, (9.39)-(9.41), in which $(v_{i,1/2}^{k+1})^{(r)} = v_{i,1/2}^{k+1}$ for all r, where $v_{i,1/2}^{k+1}$ is given by (9.11). In a similar fashion it follows that the iterative formulas (9.33) and (9.39)-(9.42) are the general ones which apply to *both* the interior and the boundary cells.

   The computer algorithm for the initial-boundary, Navier-Stokes problem is now given as follows.

i.   At the initial time $t_0=0$, for $k=0$, set the initial fluid velocities $u_{i+1/2,j}^0$ and $v_{i,j+1/2}^0$ from the given initial conditions (9.6)-(9.7), and set the initial pressure iterates $(p_{i,j}^0)^{(0)}$ equal to zero.

ii. Set the velocity boundary conditions (9.8)-(9.11).

iii. Set $r=0$ and determine, explicitly, the initial velocity iterates $(u_{i+1/2,j}^{k+1})^{(0)}$ and $(v_{i,j+1/2}^{k+1})^{(0)}$ by using equations (9.29) and (9.31), respectively.

iv. For each cell (i,j), $i=1,2,...,n$, $j=1,2,...,m$, calculate the pressure change $(\delta p_{i,j}^k)^{(r)}$ with (9.34). Then, update the pressure at the center and the velocities on the sides of the cell (i,j), by using (9.33) and (9.39)-(9.42), respectively. The boundary cells require no velocity adjustment on the boundary sides since the exact value for the velocity has been given in Step ii as a boundary condition.

v.   Repeat Step iv for $r=1,2,...$, until convergence is achieved, that is, until, for a fixed positive tolerance ε,

$$(9.48) \qquad |(\delta p_{i,j}^k)^{(r)}| < \varepsilon\,, \qquad i=1,2,\ldots,n, \quad j=1,2,\ldots,m.$$

Note from (9.36) and (9.38) that once the convergence for the pressure has been obtained, the corresponding velocity iterates have converged *simultaneously* to the desired velocity field u^{k+1}, v^{k+1}.

vi. Proceed to the next time step by setting the pressure guess $(p_{i,j}^{k+1})^{(0)}$ equal to the pressure at the previous time level, that is,

$$(9.49) \qquad (p_{i,j}^{k+1})^{(0)} = p_{i,j}^k\,, \qquad i=1,2,\ldots,n, \qquad j=1,2,\ldots,m.$$

Then, repeat Steps **ii-vi** for k=1,2,... .

vii. Terminate the computation when so desired.

9.6 THE CAVITY FLOW PROBLEM

For illustrative purposes, let us examine a widely studied cavity flow problem. Consider a viscous, incompressible fluid enclosed in a square cavity whose sides have length 1 (see Figure 9.4). Assume that the fluid has viscosity $v=0.1$ and that at the initial time $t_0=0$ the initial velocities are

$$(9.50) \qquad u_0(x,y) = v_0(x,y) = 0\,.$$

In addition, it is assumed that the fluid at the boundary has the same velocity as the boundary itself. Assume that the bottom and the side boundaries are fixed, while the top side moves with the given, uniform horizontal velocity $u_T=1$. Thus, the tangential velocity boundary conditions are

$$(9.51) \qquad u(x,0,t) = v(0,y,t) = v(1,y,t) = 0\,, \qquad u(x,1,t) = u_T = 1\,,$$

while the normal velocity boundary conditions are

$$(9.52) \qquad u(0,y,t) = u(1,y,t) = v(x,0,t) = v(x,1,t) = 0\,.$$

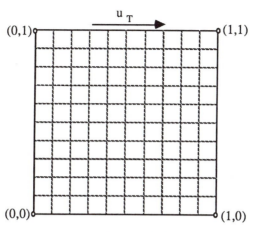

Figure 9.4

Let $n=m=10$ so that $\Delta x=\Delta y=0.1$, and let $\Delta t=0.01$, $\omega=1.0$, and $\varepsilon=0.001$. The calculations then proceed as follows.

i. At the initial time $t_0=0$, the initial conditions (9.50) imply

$$u^0_{i+1/2,j} = 0 , \qquad i=0,1,2,\ldots,10, \qquad j=1,2,\ldots,10,$$

$$v^0_{i,j+1/2} = 0 , \qquad i=1,2,\ldots,10, \qquad j=0,1,2,\ldots,10.$$

The initial guess for the pressure field is taken to be

$$(p^0_{i,j})^{(0)} = 0 , \qquad i=1,2,\ldots,10, \qquad j=1,2,\ldots,10.$$

ii. For $k=0$, the tangential velocity boundary conditions (9.51) imply

$$u^0_{i+1/2,0}= 0, \qquad u^0_{i+1/2,11} = 2u_T-u^0_{i+1/2,10} = 2 , \qquad i=1,2,\ldots,9,$$

$$v^0_{0,j+1/2} = 0, \qquad v^0_{11,j+1/2} = 0 , \qquad j=1,2,\ldots,9.$$

The normal velocity boundary conditions (9.52) imply

$$u^1_{1/2,j} = 0 \; , \qquad u^1_{10+1/2,j} = 0 \; , \qquad j=1,2,\ldots,10,$$

$$v^1_{i,1/2} = 0 \; , \qquad v^1_{i,10+1/2} = 0 \; , \qquad i=1,2,\ldots,10.$$

iii. For r=0, the initial velocity iterates are calculated from (9.29) and (9.31) by using the initial conditions, the initial pressure iterates from Step i, and the tangential velocity boundary conditions from Step ii. This yields

$$(u^1_{i+1/2,j})^{(0)} = 0 \; , \qquad i=1,2,\ldots,9, \qquad j=1,2,\ldots,9,$$

$$(u^1_{i+1/2,10})^{(0)} = 0.2 \; , \qquad i=1,2,\ldots,9,$$

$$(v^1_{i,j+1/2})^{(0)} = 0 \; , \qquad i=1,2,\ldots,10, \qquad j=1,2,\ldots,9.$$

iv. Now, at each cell (i,j), i=1,2,...,10, j=1,2,...,9, by using the these initial velocity iterates and the normal velocity boundary conditions from Step ii, equation (9.34) yields a zero pressure change $(\delta p^0_{i,j})^{(0)}$. Hence, use of (9.33) and (9.39)-(9.42) for i=1,2,...,10, and for j=1,2,...,9, yields

$$(p^0_{i,j})^{(1)} = 0 \; , \qquad i=1,2,\ldots,10, \qquad j=1,2,\ldots,9,$$

$$(u^1_{i+1/2,j})^{(1)} = 0 \; , \qquad i=1,2,\ldots,9, \qquad j=1,2,\ldots,9,$$

$$(v^1_{i,j+1/2})^{(1)} = 0 \; , \qquad i=1,2,\ldots,10, \qquad j=1,2,\ldots,8,$$

$$(v^1_{i,9+1/2})^{(1/2)} = 0 \; , \qquad i=1,2,\ldots,10.$$

However, when j=10, equation (9.34), for i=1, yields

$$(\delta p^0_{1,10})^{(0)} = -0.5 \; ,$$

so that from (9.33), (9.39) and (9.42) one has

$$(p^0_{1,10})^{(1)} = -0.5 \, ,$$

$$(u^1_{3/2,10})^{(1/2)} = 0.15 \, , \qquad (v^1_{1,10-1/2})^{(1)} = 0.05 \, .$$

Next, for j=10 and i=2, (9.34) yields

$$(\delta p^0_{2,10})^{(0)} = -0.125 \, ,$$

so that from (9.33), (9.39), (9.40) and (9.42) one has

$$(p^0_{2,10})^{(1)} = -0.125 \, ,$$

$$(u^1_{3/2,10})^{(1)} = 0.1625 \, , \qquad (u^1_{5/2,10})^{(1/2)} = 0.1875 \, , \qquad (v^1_{2,10-1/2})^{(1)} = 0.0125 \, .$$

Continuing in the indicated fashion with j=10, and for i=3,4,...,10, one finds, to five decimal places,

$$(p^0_{3,10})^{(1)} = -0.03125 \, , \qquad (u^1_{5/2,10})^{(1)} = 0.19062 \, , \qquad (v^1_{3,10-1/2})^{(1)} = 0.00312 \, ,$$

$$(p^0_{4,10})^{(1)} = -0.00781 \, , \qquad (u^1_{7/2,10})^{(1)} = 0.19765 \, , \qquad (v^1_{4,10-1/2})^{(1)} = 0.00078 \, ,$$

$$(p^0_{5,10})^{(1)} = -0.00195 \, , \qquad (u^1_{9/2,10})^{(1)} = 0.19941 \, , \qquad (v^1_{5,10-1/2})^{(1)} = 0.00019 \, ,$$

$$(p^0_{6,10})^{(1)} = -0.00048 \, , \qquad (u^1_{11/2,10})^{(1)} = 0.19985 \, , \qquad (v^1_{6,10-1/2})^{(1)} = 0.00004 \, ,$$

$$(p^0_{7,10})^{(1)} = -0.00012 \, , \qquad (u^1_{13/2,10})^{(1)} = 0.19996 \, , \qquad (v^1_{7,10-1/2})^{(1)} = 0.00001 \, ,$$

$$(p^0_{8,10})^{(1)} = -0.00003 \, , \qquad (u^1_{15/2,10})^{(1)} = 0.19999 \, , \qquad (v^1_{8,10-1/2})^{(1)} = 0.00000 \, ,$$

$$(p^0_{9,10})^{(1)} = -0.00000 \, , \qquad (u^1_{17/2,10})^{(1)} = 0.20000 \, , \qquad (v^1_{9,10-1/2})^{(1)} = 0.00000 \, ,$$

$$(p^0_{10,10})^{(1)} = 0.50000 \, , \qquad (u^1_{19/2,10})^{(1)} = 0.15000 \, , \qquad (v^1_{10,10-1/2})^{(1)} = -0.05000 \, .$$

v. By repeating Step iv for r=1,2,...,126, after 126 iterations, inequality (9.48) is satisfied throughout and the resulting pressure at time t_0 is given, to three decimal places, in Table 9.1, while the corresponding u and v velocity components at time t_1 are given in Tables 9.2 and 9.3, respectively.

Table 9.1: $(p_{i,j}^{0})^{(126)}$

	i=1	i=2	i=3	i=4	i=5	i=6	i=7	i=8	i=9	i=10
j=10	-2.824	-1.802	-1.152	-0.662	-0.241	0.159	0.580	1.070	1.720	2.742
j=9	-1.846	-1.430	-0.993	-0.593	-0.222	0.139	0.511	0.911	1.348	1.764
j=8	-1.284	-1.080	-0.798	-0.495	-0.102	0.110	0.413	0.715	0.998	1.202
j=7	-0.926	-0.810	-0.622	-0.399	-0.161	0.079	0.317	0.540	0.728	0.845
j=6	-0.685	-0.611	-0.481	-0.316	-0.134	0.052	0.235	0.399	0.529	0.604
j=5	-0.519	-0.468	-0.375	-0.252	-0.113	0.031	0.170	0.293	0.387	0.438
j=4	-0.405	-0.368	-0.298	-0.204	-0.097	0.015	0.123	0.217	0.287	0.324
j=3	-0.329	-0.300	-0.245	-0.171	-0.085	0.004	0.090	0.164	0.219	0.248
j=2	-0.282	-0.258	-0.213	-0.151	-0.078	-0.002	0.070	0.132	0.177	0.202
j=1	-0.259	-0.238	-0.197	-0.141	-0.075	-0.005	0.060	0.116	0.158	0.179

Table 9.2: $(u_{i+1/2,j}^{1})^{(126)}$

	i=1	i=2	i=3	i=4	i=5	i=6	i=7	i=8	i=9
j=10	0.097	0.135	0.151	0.158	0.160	0.158	0.151	0.135	0.097
j=9	-0.041	-0.043	-0.040	-0.037	-0.036	-0.037	-0.040	-0.043	-0.041
j=8	-0.020	-0.028	-0.030	-0.030	-0.030	-0.030	-0.030	-0.028	-0.020
j=7	-0.011	-0.018	-0.022	-0.024	-0.024	-0.024	-0.022	-0.018	-0.011
j=6	-0.007	-0.013	-0.016	-0.018	-0.019	-0.018	-0.016	-0.013	-0.007
j=5	-0.005	-0.009	-0.012	-0.014	-0.014	-0.014	-0.012	-0.009	-0.005
j=4	-0.003	-0.007	-0.009	-0.010	-0.012	-0.010	-0.009	-0.007	-0.003
j=3	-0.003	-0.005	-0.007	-0.008	-0.009	-0.008	-0.007	-0.005	-0.003
j=2	-0.002	-0.004	-0.006	-0.007	-0.007	-0.007	-0.006	-0.004	-0.002
j=1	-0.002	-0.004	-0.005	-0.006	-0.007	-0.006	-0.005	-0.004	-0.002

Table 9.3: $(v_{i,j+1/2}^{1})^{(126)}$

	i=1	i=2	i=3	i=4	i=5	i=6	i=7	i=8	i=9	i=10
j=9	0.097	0.037	0.015	0.007	0.002	-0.002	-0.007	-0.015	-0.037	-0.097
j=8	0.056	0.035	0.019	0.010	0.003	-0.003	-0.010	-0.019	-0.035	-0.056
j=7	0.035	0.027	0.017	0.009	0.003	-0.003	-0.009	-0.017	-0.027	-0.035
j=6	0.024	0.020	0.014	0.008	0.002	-0.002	-0.008	-0.014	-0.020	-0.024
j=5	0.016	0.014	0.010	0.006	0.002	-0.002	-0.006	-0.010	-0.014	-0.016
j=4	0.011	0.010	0.007	0.004	0.001	-0.001	-0.004	-0.007	-0.010	-0.011
j=3	0.007	0.006	0.005	0.003	0.001	-0.001	-0.003	-0.005	-0.006	-0.007
j=2	0.004	0.004	0.003	0.002	0.000	0.000	-0.002	-0.003	-0.002	-0.004
j=1	0.002	0.002	0.001	0.001	0.000	0.000	-0.001	-0.001	-0.002	-0.002

Table 9.4: $(p_{i,j}^{99})^{(1)}$

	i=1	i=2	i=3	i=4	i=5	i=6	i=7	i=8	i=9	i=10
j=10	-1.829	-0.833	-0.427	-0.243	-0.133	-0.029	0.120	0.392	0.949	2.158
j=9	-1.007	-0.697	-0.453	-0.288	-0.163	-0.037	0.126	0.370	0.730	1.156
j=8	-0.561	-0.486	-0.369	-0.259	-0.154	-0.040	0.098	0.273	0.466	0.593
j=7	-0.314	-0.313	-0.265	-0.198	-0.121	-0.032	0.071	0.182	0.274	0.290
j=6	-0.173	-0.194	-0.176	-0.137	-0.084	-0.020	0.049	0.114	0.152	0.132
j=5	-0.094	-0.118	-0.114	-0.092	-0.056	-0.012	0.032	0.068	0.080	0.052
j=4	-0.054	-0.075	-0.077	-0.063	-0.038	-0.008	0.020	0.040	0.041	0.016
j=3	-0.040	-0.056	-0.058	-0.048	-0.029	-0.006	0.014	0.027	0.025	0.006
j=2	-0.042	-0.054	-0.055	-0.045	-0.027	-0.005	0.014	0.025	0.023	0.010
j=1	-0.052	-0.062	-0.064	-0.052	-0.029	-0.002	0.021	0.033	0.032	0.021

Table 9.5: $(u_{i+1/2,j}^{100})^{(1)}$

	i=1	i=2	i=3	i=4	i=5	i=6	i=7	i=8	i=9
j=10	0.257	0.472	0.603	0.673	0.700	0.691	0.639	0.521	0.296
j=9	-0.024	0.053	0.148	0.219	0.250	0.237	0.177	0.076	-0.018
j=8	-0.058	-0.076	-0.057	-0.032	-0.020	-0.031	-0.060	-0.087	-0.072
j=7	-0.051	-0.104	-0.134	-0.146	-0.152	-0.155	-0.152	-0.127	-0.066
j=6	-0.040	-0.100	-0.149	-0.180	-0.194	-0.191	-0.167	-0.118	-0.050
j=5	-0.031	-0.085	-0.136	-0.171	-0.187	-0.180	-0.149	-0.096	-0.036
j=4	-0.023	-0.068	-0.112	-0.145	-0.158	-0.150	-0.119	-0.073	-0.025
j=3	-0.016	-0.050	-0.085	-0.111	-0.122	-0.114	-0.088	-0.051	-0.016
j=2	-0.009	-0.031	-0.055	-0.074	-0.082	-0.075	-0.057	-0.031	-0.009
j=1	-0.002	-0.010	-0.021	-0.030	-0.034	-0.031	-0.022	-0.010	-0.002

Table 9.6: $(v_{i,j+1/2}^{100})^{(1)}$

	i=1	i=2	i=3	i=4	i=5	i=6	i=7	i=8	i=9	i=10
j=9	0.257	0.215	0.131	0.070	0.027	-0.009	-0.052	-0.118	-0.224	-0.296
j=8	0.233	0.292	0.226	0.140	0.058	-0.022	-0.112	-0.219	-0.319	-0.278
j=7	0.175	0.274	0.244	0.165	0.070	-0.032	-0.141	-0.246	-0.304	-0.206
j=6	0.123	0.221	0.215	0.153	0.064	-0.036	-0.138	-0.220	-0.243	-0.140
j=5	0.083	0.162	0.166	0.122	0.050	-0.033	-0.114	-0.171	-0.174	-0.089
j=4	0.051	0.107	0.115	0.086	0.035	-0.026	-0.083	-0.119	-0.114	-0.053
j=3	0.028	0.063	0.071	0.054	0.021	-0.017	-0.053	-0.073	-0.066	-0.028
j=2	0.011	0.030	0.035	0.027	0.010	-0.009	-0.027	-0.036	-0.031	-0.011
j=1	0.002	0.008	0.001	0.008	0.003	-0.003	-0.008	-0.011	-0.008	-0.002

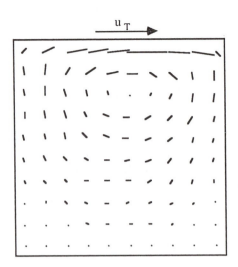

Figure 9.5

vi. One proceeds, then, to the next time step by setting the initial pressure iterates $(p_{i,j}^k)^{(0)} = p_{i,j}^{k-1}$ and by repeating steps ii-vi for k=1,2,.... The required iterations to achieve the pressure convergence at t_1, t_2, t_3, t_4, ..., are 75, 71, 60, 55, ..., respectively. At $t_{100}=1$, the numerical solution, which is listed in Tables 9.4, 9.5 and 9.6, required only 1 iteration.

vii. Since the numerical solution which can be obtained at later times shows no substantial changes, the numerical solution at t_{100} represents a steady state solution and hence the computation can be terminated at t_{100}.

Note that from the data in Tables 9.5 and 9.6 it is difficult to determine the corresponding flow pattern. For this purpose, we simply average the u and v velocity components on each cell and construct the velocity field at the center of the cell, as shown in Figure 9.5. As expected, a large, lid driven circulating eddy is clearly shown.

9.7 STABILITY OF THE METHOD

Because the variable t in the Navier-Stokes equations can vary in the unbounded range $0 \leq t < \infty$, it is important to study the possible instability of the method. It is also important to note that one wishes to allow great flexibility in the choice of the time step Δt. Unfortunately, however, due to the strong nonlinear structure of the governing equations, a precise stability condition is difficult to derive, and a stability analysis of the *linearized*

Navier-Stokes equations is usually studied. We will show next how this is done.

To develop a stability condition for the finite difference Navier-Stokes equations (9.12)-(9.14), consider the following heuristic argument. [A more rigorous stability analysis can be found in Casulli (1984)]. Assuming that p and all coefficients of partial derivatives are known in (9.1) and (9.2) implies that both (9.1) and (9.2) are of the form

$$(9.53) \qquad \frac{\partial w}{\partial t} = v\left(\frac{\partial^2 w}{\partial x^2} + \frac{\partial^2 w}{\partial y^2}\right) - u\frac{\partial w}{\partial x} - v\frac{\partial w}{\partial y} + f(x,y,t) \; ,$$

which is a general, linear parabolic equation in w.

Now, as was done for (9.1) and (9.2), equation (9.53) is discretized at the interior grid points of an $R_{n+1,m+1}$ set by using a forward finite difference for the time derivative, and central finite differences for the spatial derivatives. The following finite difference equation results:

$$(9.54) \qquad \frac{w_{i,j}^{k+1} - w_{i,j}^{k}}{\Delta t} = v\left(\frac{w_{i+1,j}^{k} - 2w_{i,j}^{k} + w_{i-1,j}^{k}}{(\Delta x)^2} + \frac{w_{i,j+1}^{k} - 2w_{i,j}^{k} + w_{i,j-1}^{k}}{(\Delta y)^2}\right)$$

$$- \left[u_{i,j}^{k}\frac{w_{i+1,j}^{k} - w_{i-1,j}^{k}}{2\Delta x} + v_{i,j}^{k}\frac{w_{i,j+1}^{k} - w_{i,j-1}^{k}}{2\Delta y}\right] + f(x_i,y_j,t_k) \; ,$$

or, equivalently,

$$(9.55) \qquad w_{i,j}^{k+1} = Fw_{i,j}^{k} + (\Delta t)f(x_i,y_j,t_k) \; ,$$

where, in analogy with (9.12) and (9.13), the finite difference operator F is defined by

$$(9.56) \qquad Fw_{i,j}^{k} = w_{i,j}^{k} - \Delta t\left[u_{i,j}^{k}\frac{w_{i+1,j}^{k} - w_{i-1,j}^{k}}{2\Delta x} + v_{i,j}^{k}\frac{w_{i,j+1}^{k} - w_{i,j-1}^{k}}{2\Delta y}\right]$$

$$+ v\Delta t\left(\frac{w_{i+1,j}^{k} - 2w_{i,j}^{k} + w_{i-1,j}^{k}}{(\Delta x)^2} + \frac{w_{i,j+1}^{k} - 2w_{i,j}^{k} + w_{i,j-1}^{k}}{(\Delta y)^2}\right) \; .$$

Under the simplifying assumptions made, the stability conditions for equation (9.55) apply to the Navier-Stokes difference equations (9.12)-(9.13).

By setting $\alpha = \Delta t/(\Delta x)^2$, $\beta = \Delta t/(\Delta y)^2$, $\gamma = \Delta t/(2\Delta x)$, $\delta = \Delta t/(2\Delta y)$, equation (9.56) can be written, equivalently, as

$$(9.57) \qquad Fw_{i,j}^k = (\beta v + \delta v_{i,j}^k) w_{i,j-1}^k + (\alpha v + \gamma u_{i,j}^k) w_{i-1,j}^k + [1 - 2v(\alpha + \beta)] w_{i,j}^k$$

$$+ (\alpha v - \gamma u_{i,j}^k) w_{i+1,j}^k + (\beta v - \delta v_{i,j}^k) w_{i,j+1}^k .$$

Note, now, that the summation of all the coefficients on the right-hand side of (9.57) is unity. Thus, if

$$(9.58) \qquad |\gamma u_{i,j}^k| \leq \alpha v , \qquad\qquad |\delta v_{i,j}^k| \leq \beta v ,$$

and

$$(9.59) \qquad 2v(\alpha + \beta) \leq 1 ,$$

then

$$\min(w_{i,j-1}^k, w_{i-1,j}^k, w_{i,j}^k, w_{i+1,j}^k, w_{i,j+1}^k) \leq Fw_{i,j}^k \leq \max(w_{i,j-1}^k, w_{i-1,j}^k, w_{i,j}^k, w_{i+1,j}^k, w_{i,j+1}^k) .$$

Hence, inequalities (9.58)-(9.59) are sufficient to assure that, when f(x,y,t)=0, any solution of (9.55) is bounded. Thus, sufficient (not necessary) stability conditions for method (9.12)-(9.14) are given by (9.58)-(9.59) or, equivalently, by

$$(9.60) \qquad |u_{i,j}^k|(\Delta x) \leq 2v , \qquad\qquad |v_{i,j}^k|(\Delta y) \leq 2v ,$$

$$(9.61) \qquad \Delta t \leq \frac{(\Delta x)^2 (\Delta y)^2}{2v[(\Delta x)^2 + (\Delta y)^2]} .$$

Inequalities (9.60) have to be valid over the entire flow field at each time step.

EXERCISES

Basic Exercises

1. Using $\Delta t=0.025$, compute to steady state the cavity flow problem of Section 9.6.

2. Using $v=0.01$, compute to steady state the cavity flow problem of Section 9.6. The choice of the time and space steps is yours. Assuming that the fluid velocity never exceeds that of the top boundary, verify that the stability conditions (9.60), (9.61) are satisfied.

3. Derive a finite difference method for the initial-boundary, Navier-Stokes problem which uses upwind, rather than central difference approximations.

4. Using $v=0.01$, compute to steady state the numerical solution of the cavity flow problem of Section 9.6. Use the upwind method with $\Delta x=\Delta y=0.1$ and $\Delta t=0.01$.

5. Derive, heuristically, sufficient stability conditions for the upwind method for the initial-boundary, Navier-Stokes problem.

Supplementary Exercises

6. Repeat Exercise 2 for the cavity flow problem with $v=10$.

7. Extend the method of this chapter to nonrectangular domains.

8. Extend the method of this chapter to free surface problems.

9. Extend the method of this chapter to three dimensions.

CHAPTER 1

1. $x_2=3$, $x_6=4$.

2. a, b, d.

3. b, e.

4. $x_1=x_2=1$, $x_3=x_4=-1$.

5. Exact solution is $x_1=2$, $x_2=3$, $x_3=5$, $x_4=7$, $x_5=-1$, $x_6=4$.

6.

(a) $L = \begin{bmatrix} -2 & 0 & 0 & 0 \\ 1 & -\frac{3}{2} & 0 & 0 \\ 0 & 1 & -\frac{4}{3} & 0 \\ 0 & 0 & 1 & -\frac{5}{4} \end{bmatrix}$, $U = \begin{bmatrix} 1 & -\frac{1}{2} & 0 & 0 \\ 0 & 1 & -\frac{2}{3} & 0 \\ 0 & 0 & 1 & -\frac{3}{4} \\ 0 & 0 & 0 & 1 \end{bmatrix}$

(b) $L = \begin{bmatrix} -3 & 0 & 0 & 0 \\ 2 & -\frac{7}{3} & 0 & 0 \\ 0 & 2 & -\frac{15}{7} & 0 \\ 0 & 0 & 2 & -\frac{31}{15} \end{bmatrix}$, $U = \begin{bmatrix} 1 & -\frac{1}{3} & 0 & 0 \\ 0 & 1 & -\frac{3}{7} & 0 \\ 0 & 0 & 1 & -\frac{7}{15} \\ 0 & 0 & 0 & 1 \end{bmatrix}$

(c) $L = \begin{bmatrix} 10 & 0 & 0 & 0 \\ 4 & \frac{56}{5} & 0 & 0 \\ 0 & 4 & \frac{155}{14} & 0 \\ 0 & 0 & 4 & \frac{1718}{155} \end{bmatrix}$, $U = \begin{bmatrix} 1 & -\frac{3}{10} & 0 & 0 \\ 0 & 1 & -\frac{15}{56} & 0 \\ 0 & 0 & 1 & -\frac{42}{155} \\ 0 & 0 & 0 & 1 \end{bmatrix}$

(d) $$L = \begin{bmatrix} 5 & 0 & 0 & 0 & 0 \\ -2 & \frac{21}{5} & 0 & 0 & 0 \\ 0 & -2 & \frac{85}{21} & 0 & 0 \\ 0 & 0 & -2 & \frac{341}{85} & 0 \\ 0 & 0 & 0 & -2 & \frac{1365}{341} \end{bmatrix}, \quad U = \begin{bmatrix} 1 & -\frac{2}{5} & 0 & 0 & 0 \\ 0 & 1 & -\frac{10}{21} & 0 & 0 \\ 0 & 0 & 1 & -\frac{42}{85} & 0 \\ 0 & 0 & 0 & 1 & -\frac{170}{341} \\ 0 & 0 & 0 & 0 & 1 \end{bmatrix}$$

(e) $$L = \begin{bmatrix} -4 & 0 & 0 & 0 & 0 \\ 1 & -\frac{13}{4} & 0 & 0 & 0 \\ 0 & 1 & -\frac{40}{13} & 0 & 0 \\ 0 & 0 & 1 & -\frac{121}{40} & 0 \\ 0 & 0 & 0 & 1 & -\frac{364}{121} \end{bmatrix}, \quad U = \begin{bmatrix} 1 & -\frac{3}{4} & 0 & 0 & 0 \\ 0 & 1 & -\frac{12}{13} & 0 & 0 \\ 0 & 0 & 1 & -\frac{39}{40} & 0 \\ 0 & 0 & 0 & 1 & -\frac{120}{121} \\ 0 & 0 & 0 & 0 & 1 \end{bmatrix}$$

(f) $$L = \begin{bmatrix} -4 & 0 & 0 & 0 & 0 \\ 2 & -\frac{7}{2} & 0 & 0 & 0 \\ 0 & 3 & -\frac{24}{7} & 0 & 0 \\ 0 & 0 & 4 & -\frac{7}{2} & 0 \\ 0 & 0 & 0 & 5 & -\frac{26}{7} \end{bmatrix}, \quad U = \begin{bmatrix} 1 & -\frac{3}{4} & 0 & 0 & 0 \\ 0 & 1 & -\frac{6}{7} & 0 & 0 \\ 0 & 0 & 1 & -\frac{7}{8} & 0 \\ 0 & 0 & 0 & 1 & -\frac{6}{7} \\ 0 & 0 & 0 & 0 & 1 \end{bmatrix}.$$

8. (a) $x_1 = x_2 = x_3 = 0$, $x_4 = -7$

(b) $x_1 = 1$, $x_2 = 2$, $x_3 = 3$, $x_4 = 4$, $x_5 = 5$

(c) $x_1 = -1$, $x_2 = x_3 = \ldots = x_{99} = 3$, $x_{100} = -5$

(d) $x_1 = x_3 = x_5 = 1$, $x_2 = x_4 = x_6 = -1$.

9. -32, 0.5, 0.5, i, $-i$.

10. Exact solution is $x_1 = 2$, $x_2 = 3$, $x_3 = 5$, $x_4 = 7$, $x_5 = -1$, $x_6 = 4$.

12. (a) $y_1=y_4=-0.2828$, $y_2=y_3=-0.4148$

(b) $x_1=x_2=x_3=0.0000$.

13. (a) $\lambda=15.80$, $v = \begin{bmatrix} 0.10 \\ 1.00 \\ 0.50 \end{bmatrix}$; (b) $\lambda=30.29$, $v = \begin{bmatrix} 0.96 \\ 0.69 \\ 1.00 \\ 0.94 \end{bmatrix}$;

(c) $\lambda=2.35$.

15. System (a) has no solutions.

16. $x_1=9$, $x_2=-36$, $x_3=30$.

19. 1.3.

20. (a) 1.72, (b) 0.59.

21. -1.934.

22. 0.607.

24. $x_1=1.8836$, $x_2=2.7159$.

25. Exact solutions are (a) $1,0,-1$ (b) $1,-1,1,-1$.

26. $x_1=0.57$, $x_2=1.86$.

28. 98.5.

CHAPTER 2

3. (a) $L(x)=4x$, $0 \leq x \leq 2$; $28x-48$, $2 \leq x \leq 4$;
$76x-240$, $4 \leq x \leq 6$; $148x-672$, $6 \leq x \leq 8$.

    (b) $L(x)=0.01x$, $0.0 \le x \le 0.1$; $0.07x-0.006$, $0.1 \le x \le 0.2$;

            $0.19x-0.03$, $0.2 \le x \le 0.3$; $0.37x-0.084$, $0.3 \le x \le 0.4$;

            $0.61x-0.18$, $0.4 \le x \le 0.5$; $0.91x-0.33$, $0.5 \le x \le 0.6$.

    (d) $L(x)=-11x+8$, $0 \le x \le 1$; $5x-8$, $1 \le x \le 2$; $-2x+6$, $2 \le x \le 3$; 0, $3 \le x \le 4$.

4.   (a) $P(x)=6x^2-8x$, $0 \le x \le 4$; $18x^2-104x+192$, $4 \le x \le 8$.

    (b) $P(x)=0.3x^2-0.02x$, $0.0 \le x \le 0.2$; $0.9x^2-0.26x+0.024$, $0.2 \le x \le 0.4$;

            $1.5x^2-0.74x+0.12$, $0.4 \le x \le 0.6$.

    (d) $P(x)=8x^2-19x+8$, $0 \le x \le 2$; $x^2-7x+12$, $2 \le x \le 4$.

7. $y=\dfrac{1}{2}(x-1)(x-2)-2x(x-2)+\dfrac{5}{2}x(x-1)$.

8.
$$y=-\frac{x(x-2)(x-3)(x-4)(x-5)(x-6)}{120}+\frac{x(x-1)(x-3)(x-4)(x-5)(x-6)}{24}$$
$$-\frac{x(x-1)(x-2)(x-4)(x-5)(x-6)}{9}+\frac{x(x-1)(x-2)(x-3)(x-5)(x-6)}{16}$$
$$+\frac{x(x-1)(x-2)(x-3)(x-4)(x-5)}{720}.$$

9. $y=0.38+1.52x$.

11. $y=1-x^3$.

12. $y=1.4\sin(x)-1.6\cos(x)+2.25\sin(2x)-4.08\cos(2x)$.

16. $(5.83)10^{-5}Re^{-(6.51)10^{-4}R}$, $(3.65)10^{-5}Re^{-(1.81)10^{-7}R^2}$,

      $(3.05)10^{-2}e^{-(2.63)10^{-8}(R-2880)^2}$.

17. $y=0.176x^3+0.532x^2+0.998x+0.996$.

18. $y=1.10x-0.29$.

19. $y=-0.763x^2+0.443x+3.195$.

22. $y=4.48e^{0.45x}$.

23. $y=0.2(x^3-13x^2+69x-92)$.

24. $y=x^5-2x+1$.

25. $y=1.01x^2-4.04x+5.05$.

26. $y=91.9x^{-0.4}$.

CHAPTER 3

1. (a) 0.256944.

3. 0.745119.

4. (a) 0.3.

5. (a) 0.250000.

11. $y=\dfrac{71}{21}+\dfrac{116}{35}x;$ $\dfrac{175}{3}$.

15. (a) 2.2033,   (b) 0.9461,   (c) 3.1416,
    (d) 1.3506,   (e) 0.7468.

16. 0.90.

17. 0.133.

CHAPTER 4

2. Numerical: $y_0=0$, $y_1=0.000$, $y_2=0.020$, $y_3=0.060$, $y_4=0.120$,
$y_5=0.199$, $y_6=0.297$, $y_7=0.413$, $y_8=0.546$,
$y_9=0.695$, $y_{10}=0.858$.

3. (a) $y_0=4.000$, $y_1=4.000$, $y_2=4.080$, $y_3=4.246$, $y_4=4.516$, $y_5=4.919$
(b) $y_0=1.000$, $y_1=1.200$, $y_2=1.488$, $y_3=1.931$, $y_4=2.676$, $y_5=4.103$.

5. Numerical:

$y_{11}=0.989105$,	$y_{24}=0.947537$,	$y_{36}=0.887659$,
$y_{42}=0.852463$,	$y_{53}=0.782915$,	$y_{60}=0.737271$,
$y_{73}=0.653787$,	$y_{79}=0.616891$,	$y_{90}=0.553198$,
$y_{98}=0.510522$,	$y_{109}=0.457108$,	$y_{116}=0.426250$,
$y_{130}=0.371412$,	$y_{141}=0.334192$,	$y_{156}=0.290660$,
$y_{167}=0.263308$,	$y_{178}=0.239260$,	$y_{185}=0.225474$,
$y_{200}=0.199368$.		

10. The numerical and the exact solutions agree to at least six decimal places.

15. (a) $h \le 0.002$, (b) $h \le 200$, (c) $h \le 200$, (d) always unstable.

16. Stable: b,f,g; unstable a,c,d,e.

20. $y(0.8)=2.3167$.

22. 27.00022.

23. Exact solution is $x^2y+xe^y=1$.

25. Exact solutions are:

(a) $y=x$, $v=-x$, (b) $y=x$, $v=-x$, (c) $y=\sin(x)$, $v=\cos(x)$.

26. The exact solution is of period 2π.

CHAPTER 5

1. (a)   Exact: $y=(5x+x^3)/6$.

    Numerical: $y_0=0.000$, $y_1=0.168$, $y_2=0.344$,
                  $y_3=0.536$, $y_4=0.752$, $y_5=1.000$.

4. Numerical: $y_1=0.03130646$, $y_7=0.21748777$,
                 $y_{15}=0.45297718$, $y_{19}=0.56107487$,
                 $y_{24}=0.68367225$, $y_{32}=0.84379853$,
                 $y_{35}=0.89060645$, $y_{39}=0.94063185$,
                 $y_{44}=0.98218217$.

9. (a) -1,   (b) $1-x_i$.

10. (a)   1,   (b)   1,   (c)   1,   (d)   1.

13. Exact solution is $y=x^4$.

16. Exact solutions are:
    (a) $y=x^2$
    (b) $y=x^2$
    (c) $y=x^2$
    (e) $y = \frac{1}{2}(e^{-2x}+e^{-(1+\sqrt{3})x})$
    (f) $y=-x-1+2/(2-x)$
    (g) $y=1-e^{-100x}$
    (h) $y=1-e^{-10000x}$.

19. 3.5 .

CHAPTER 6

1. (a) elliptic (d) hyperbolic
    (b) hyperbolic (e) hyperbolic
    (c) parabolic (f) hyperbolic.

2. (a) Hyperbolic on the upper-half plane, elliptic on the lower-half plane, parabolic on the X axis.

    (b) Hyperbolic outside the unit circle $x^2+y^2=1$, elliptic inside, and parabolic on the circle.

    (c) Elliptic on all of E^2.

CHAPTER 7

1. (a) $u(1/4,1) \sim -16800$, $u(1/2,1) \sim 23759$, $u(3/4,1) \sim -16800$.

CHAPTER 8

2. (a) $u=1-y$

    (b) $u=x+\frac{1}{6}[(x+y)^3-(x-y)^3]$

    (c) $u=\frac{1}{2}[(x+y)^2+(x-y)^2 + \int_{x-y}^{x+y} e^{-r^2}dr]$.

4. $-3\leq x\leq 3$, $-2\leq x\leq 4$, $-6\leq x\leq 0$, $-1\leq x\leq 15$, $-8\leq x\leq -6$, $-9\leq x\leq 3$.

5. $y-3=\pm x$, $y-3=\pm(x-1)$, $y-3=\pm(x+3)$, $y-8=\pm(x-7)$,
    $y+1=\pm(x+7)$, $y-6=\pm(x+3)$.

9. All are hyperbolic.

Abramowitz, M. and I.A. Stegun, HANDBOOK OF MATHEMATICAL FUNCTIONS, National Bureau of Standards, Washington, D.C., 1965.

Ames, W.F., NONLINEAR PARTIAL DIFFERENTIAL EQUATIONS IN ENGINEERING, Academic Press, New York, 1965.

Amsden A.A., *The Particle-In-Cell Method for the Calculation of the Dynamics of Compressible Fluids*, Rpt.3406, Los Alamos Scientific Lab., L.A., N.M., 1966.

Anselone, P.M.(Ed.), NONLINEAR INTEGRAL EQUATIONS, University of Wisconsin Press, Madison, Wis., 1964.

Atkinson, K., ELEMENTARY NUMERICAL ANALYSIS, Wiley, New York, 1985.

Bernstein, D., EXISTENCE THEOREMS IN PARTIAL DIFFERENTIAL EQUATIONS, Princeton Univ. Press, Princeton, N.J., 1950.

Bers, L., *On Mildly Nonlinear Partial Differential Equations*, J.Res.Nat.Bur.Std., 51, 1953, pp.229-236.

Bulgarelli, U., V. Casulli and D. Greenspan, PRESSURE METHODS FOR THE NUMERICAL SOLUTION OF FREE SURFACE FLUID FLOWS, Pineridge Press, Swansea, U.K., 1984.

Casulli, V., *On the Stability of the Pressure Method for the Navier-Stokes and Euler Equations*, in Proc. of the 3rd Int. Conf. on BOUNDARY AND INTERIOR LAYERS, Boole Press, Dublin, 1984; *Eulerian-Lagrangian Methods for Hyperbolic and Convection Dominated Parabolic Problems*, in COMPUTATIONAL METHODS FOR NONLINEAR PROBLEMS, Pineridge Press, Swansea, U.K., 1987.

Casulli, V., and D. Greenspan, *Numerical Solution of Free Surface, Porous Flow Problems*, Int. Jour. For Numerical Methods in Fluids, 2, 1982, pp.115-122; *Pressure Method for the Numerical Solution of Transient, Compressible Fluid Flows*, Int. Jour. For Numerical Methods in Fluids, 4, 1984, pp.1001-1012.

Cheney, W. and D. Kincaid, NUMERICAL MATHEMATICS AND COMPUTING, 2nd Ed., Brooks Cole, Monterey, CA, 1985.

Churchill, R.V., FOURIER SERIES AND BOUNDARY VALUE PROBLEMS, McGraw-Hill, New York, 1941.

Clenshaw, C.W., *The Solution of van der Pol's Equation in Chebychev Series*, in NUMERICAL SOLUTIONS OF NONLINEAR DIFFERENTIAL EQUATIONS, Wiley, New York, 1966, pp.55-63.

Collatz, L., THE NUMERICAL TREATMENT OF DIFFERENTIAL EQUATIONS, Springer-Verlag, Berlin, 1966.

Corliss, G., and Y.F. Chang, *Solving Ordinary Differential Equations Using Taylor Series*, ACM Trans. in Math. Soft., 8, 1982, pp.114-144.

Courant, R. and D. Hilbert, METHODS IN MATHEMATICAL PHYSICS, Vol.II, Wiley, New York, 1962.

Courant, R., E. Isaacson and M. Rees, *On the Solution of Nonlinear Hyperbolic Differential Equations by Finite Differences*, Comm. Pure Appl. Math., 5, 1952, pp. 243-255.

Dahlquist, G. and A. Bjorck, NUMERICAL METHODS, Prentice-Hall, Englewood Cliffs, 1974.

Davis, P.J., INTERPOLATION AND APPROXIMATION, Blaisdell, Waltham, 1963.

Davis, P.J. and P. Rabinowitz, NUMERICAL INTEGRATION, Blaisdell, Waltham, 1967.

de Boor, C., A PRACTICAL GUIDE TO SPLINES, Springer-Verlag, Berlin, 1978.

Dennis, J.E., Jr., NUMERICAL METHODS FOR UNCONSTRAINED OPTIMIZATION AND NONLINEAR EQUATIONS, Prentice-Hall, Englewood Cliffs, 1983.

Fehlberg, E., *Neue Genauere Runge-Kutta-Formeln fur Differential-gleichungen n-ter Ordnung*, ZAMM, 40, 1960, pp.449-455.

Ferziger, J.H., NUMERICAL METHODS FOR ENGINEERING APPLICATION, Wiley, New York, 1981.

Feynman, R.P., R.B. Leighton and M. Sands, THE FEYNMAN LECTURES ON PHYSICS, Addison-Wesley, Reading, 1963.

Forsythe, G.E. and C. Moler, COMPUTER SOLUTION OF LINEAR ALGEBRAIC SYSTEMS, Prentice-Hall, Englewood Cliffs, 1967.

Forsythe, G.E. and W.R. Wasow, FINITE-DIFFERENCE METHODS FOR PARTIAL DIFFERENTIAL EQUATIONS, Wiley, New York, 1960.

Fox, L., THE NUMERICAL SOLUTION OF TWO-POINT BOUNDARY VALUE PROBLEMS IN ORDINARY DIFFERENTIAL EQUATIONS, Oxford University Press, New York, 1957; NUMERICAL SOLUTION OF ORDINARY AND PARTIAL DIFFERENTIAL EQUATIONS, Addison-Wesley, Reading, 1962; AN INTRODUCTION TO NUMERICAL LINEAR ALGEBRA, Oxford University Press, New York, 1965.

Franklin, J.N., *Difference Methods for Stochastic Ordinary Differential Equations*, Math. Comp., 19, 1965, pp.552-561.

Froberg, C.E., NUMERICAL MATHEMATICS - THEORY AND COMPUTER APPLICATIONS, Benjamin Cummings Pub., Menlo Park, CA., 1985.

Gear, C.W., NUMERICAL INITIAL VALUE PROBLEMS IN ORDINARY DIFFERENTIAL EQUATIONS, Prentice-Hall, Englewood Cliffs, 1971.

Golomb, M. and H.F. Weinberger, *Optimal Approximation and Error Bounds*, in ON NUMERICAL APPROXIMATION, University of Wisconsin Press, Madison, 1959, pp.117-191.

Golub, G.H. and C.F. van Loan, MATRIX COMPUTATIONS, Johns Hopkins Univ. Press, Baltimore, 1985.

Greenspan, D., INTRODUCTION TO PARTIAL DIFFERENTIAL EQUATIONS, McGraw-Hill, New York, 1961; LECTURES ON THE NUMERICAL SOLUTION OF LINEAR, SINGULAR, AND NONLINEAR DIFFERENTIAL EQUATIONS, Prentice-Hall, Englewood Cliffs, 1969; DISCRETE MODELS, Addison-Wesley, Reading, 1973; DISCRETE NUMERICAL METHODS IN PHYSICS AND ENGINEERING, Academic Press, New York, 1974; ARITHMETIC APPLIED MATHEMATICS, Pergamon Press, Oxford, 1980.

Greville, T.N.E., THEORY AND APPLICATIONS OF SPLINE FUNCTIONS, Academic Press, New York, 1969.

Hamming, R.W., NUMERICAL METHODS FOR SCIENTISTS AND ENGINEERS, McGraw-Hill, New York, 1962.

Hemmerle, W.J., STATISTICAL COMPUTATIONS ON A DIGITAL COMPUTER, Blaisdell, Waltham, 1967.

Henrici, P., DISCRETE VARIABLE METHODS IN ORDINARY DIFFERENTIAL EQUATIONS, Wiley, New York, 1962; ESSENTIALS OF NUMERICAL ANALYSIS WITH POCKET CALCULATOR DEMONSTRATIONS, Wiley, New York, 1982.

Heun, K., *Neu Methode zur Approximativen Integration der Differentialgleichungen einer unabhangigen Variable*, ZAMP, 45, 1900, pp.23-38.

Householder, A.S., THE THEORY OF MATRICES IN NUMERICAL ANALYSIS, Blaisdell, Waltham, 1964.

Isaacson, E. and H.B. Keller, ANALYSIS OF NUMERICAL METHODS, Wiley, New York, 1964.

Jain, M.K., NUMERICAL SOLUTION OF DIFFERENTIAL EQUATIONS, 2nd Ed., Wiley, New York, 1984.

Janenko, N.N., DIFFERENCE METHODS FOR SOLUTIONS OF PROBLEMS IN MATHEMATICAL PHYSICS, Amer. Math. Soc., Providence, 1967.

Kamke, E., DIFFERENTIALGLEICHUNGEN, Akademic-Verlag, Leipzig, 1959.

Keller, H.B., NUMERICAL METHODS FOR TWO-POINT BOUNDARY VALUE PROBLEMS, Blaisdell, Waltham, 1968.

Kelly, L.G., HANDBOOK OF NUMERICAL METHODS AND APPLICATIONS, Addison-Wesley, Reading, 1967.

Kutta, W., *Beitrag zur naherungsweisen Integration totaler Differentialgleichungen*, ZAMP, 46, 1901, pp.435-453.

Lax, P.D. and B. Wendroff, *Systems of Conservation Laws*, Comm. Pure Appl. Math., 13, 1960, pp.217-237.

Lieberstein, H.M., *Overrelaxation for Nonlinear Elliptic Partial Differential Equations*, Rpt.80, Math. Res. Ctr., Univ. Wisconsin, Madison, 1959.

Marchuk, G.I., METHODS OF NUMERICAL MATHEMATICS, Springer-Verlag, Berlin, 1975; DIFFERENCE METHODS AND THEIR EXTRAPOLATIONS, Springer-Verlag, New York, 1983.

Mitchell, A.R. and D.F. Griffiths, THE FINITE DIFFERENCE METHOD IN PARTIAL DIFFERENTIAL EQUATIONS, Wiley, New York, 1980.

Moore, R.E., INTERVAL ANALYSIS, Prentice-Hall, Englewood Cliffs, 1966; METHODS AND APPLICATIONS OF INTERVAL ANALYSIS, SIAM, Philadelphia, 1979.

Morris, J.Ll., COMPUTATIONAL METHODS IN ELEMENTARY NUMERICAL ANALYSIS, Wiley, New York, 1983.

Nakamura, S., COMPUTATIONAL METHODS IN ENGINEERING SCIENCE, Wiley, New York, 1977.

Noble, B., NUMERICAL METHODS, Oliver and Boyd, London, 1964.

Ortega, J.M., NUMERICAL ANALYSIS, Academic Press, New York, 1972.

Ortega, J.M. and W.C. Rheinboldt, ITERATIVE SOLUTIONS OF NONLINEAR EQUATIONS IN SEVERAL VARIABLES, Academic Press, New York, 1970.

Ostrowski, A.M., SOLUTION OF EQUATIONS AND SYSTEMS OF EQUATIONS, 2nd Ed., Academic Press, New York, 1966.

Parlett, B.N., THE SYMMETRIC EIGENVALUE PROBLEM, Prentice-Hall, Englewood Cliffs, 1980.

Potter, D., COMPUTATIONAL PHYSICS, Wiley, New York, 1973.

Protter, M.H. and H.F. Weinberger, MAXIMUM PRINCIPLES IN DIFFERENTIAL EQUATIONS, Springer-Verlag, New York, 1984.

Rall, L.B., COMPUTATIONAL SOLUTION OF NONLINEAR OPERATOR EQUATIONS, Wiley, New York, 1969.

Ralston, A. and P. Rabinowitz, A FIRST COURSE IN NUMERICAL ANALYSIS, 2nd Ed., McGraw-Hill, New York, 1978.

Richtmyer, R.D. and K.W. Morton, DIFFERENCE METHODS FOR INITIAL-VALUE PROBLEMS, 2nd Ed., Wiley, New York, 1967.

Roberts, C.E., ORDINARY DIFFERENTIAL EQUATIONS, Prentice-Hall, Englewood Cliffs, 1979.

Samarskii, A.A., AN INTRODUCTION TO NUMERICAL METHODS, (in Russian) Nauka, Moscow, 1982.

Sarafyan, D., *Continuous Approximate Solution of Ordinary Differential Equations and their Systems*, Comp. & Maths. with Appls.,10, 1984, pp.139-159.

Schechter, S., *Iteration Methods for Nonlinear Problems*, Trans. AMS, 104, 1962, pp. 179-189.

Schultz, M.H., SPLINE ANALYSIS, Prentice-Hall, Englewood Cliffs, 1966.

Shampine, L.F., and R.C. Allen, NUMERICAL COMPUTING: AN INTRODUCTION, Saunders, Philadelphia, 1973.

Stetter, H.J., ANALYSIS OF DISCRETIZATION METHODS FOR ORDINARY DIFFERENTIAL EQUATIONS, Springer-Verlag, New York, 1973.

Strang, G. and G.J. Fix, AN ANALYSIS OF THE FINITE ELEMENT METHOD, Prentice-Hall, Englewood Cliffs, 1973.

Stroud, R.H., and D. Secrest, GAUSSIAN QUADRATURE FORMULAS, Prentice-Hall, Englewood Cliffs, 1966.

Traub, J.F., ITERATIVE METHODS FOR THE SOLUTION OF EQUATIONS, Prentice-Hall, Englewood Cliffs, 1964.

Varga, R.S., MATRIX ITERATIVE ANALYSIS, Prentice-Hall, Englewood Cliffs, 1962.

Vemuri, V., and W.J. Karplus, DIGITAL COMPUTER TREATMENT OF PARTIAL DIFFERENTIAL EQUATIONS, Prentice-Hall, Englewood Cliffs, NJ, 1981.

Vichnevetsky, R. and J.B. Bowles, FOURIER ANALYSIS OF NUMERICAL APPROXIMATIONS OF HYPERBOLIC EQUATIONS, SIAM, Philadelphia, 1982.

Wendroff, B., THEORETICAL NUMERICAL ANALYSIS, Academic Press, New York, 1966.

Wilkinson, J.H., ROUNDING ERRORS IN ALGEBRAIC PROCESSES, Prentice-Hall, Englewood Cliffs, 1963.

Young, D.M. and R.T. Gregory, A SURVEY OF NUMERICAL METHODS, Addison-Wesley, Reading, 1973.

Zienkiewicz, O.C. and K. Morgan, FINITE ELEMENTS AND APPROXIMATION, Wiley, New York, 1983.